Alfred Henry Huth

The Marriage of near Kin

Considered with Respect to the Laws of Nations, the Results of Experience, and the

Teachings of Biology

Alfred Henry Huth

The Marriage of near Kin
Considered with Respect to the Laws of Nations, the Results of Experience, and the Teachings of Biology

ISBN/EAN: 9783337216337

Printed in Europe, USA, Canada, Australia, Japan

Cover: Foto ©berggeist007 / pixelio.de

More available books at **www.hansebooks.com**

THE MARRIAGE OF NEAR KIN

THE MARRIAGE OF NEAR KIN

CONSIDERED WITH RESPECT TO THE

LAWS OF NATIONS, THE RESULTS OF EXPERIENCE,
AND THE TEACHINGS OF BIOLOGY

BY

ALFRED HENRY HUTH

——" FRAGT NICHT DEN WIEDERHALL EURER KREUZGÄNGE, NICHT EUER VERMODERTES PERGAMENT, NICHT EURE VERSCHRÄNKTEN GRILLEN UND VERORDNUNGEN! FRAGT DIE NATUR UND EUER HERZ, SIE WIRD EUCH LEHREN, VOR WAS IHR ZU SCHAUDERN HABT, SIE WIRD EUCH MIT DEM STRENGSTEN FINGER ZEIGEN, WORÜBER SIE EWIG UND UNWIDERRUFLICH IHREN FLUCH AUSSPRICHT "

LONDON
J. & A. CHURCHILL, NEW BURLINGTON STREET
1875

PREFACE.

THE object of this work is to put in a collected form all that has been written on the Marriage of Near Kin; an object which the author believes to the full worthy of the trouble which the chaotic nature of his materials has given him in collection. Hitherto, while the subject has been exciting increased attention from all quarters, no work has been written on it from a modern point of view; and while the scientific journals have been full of articles taking up one part of the subject, and modern biological and medical works are continually touching on it, no one has yet come forward with a generalization of the whole of our knowledge, which, being dispersed in periodicals, most of them extremely difficult to get, partly because they are out of date, but more often owing to the absence of proper references, has as yet been unattainable by anybody who cannot devote a considerable time to the study.

In his treatment of the question the author has endeavoured to divest himself of all prejudice, whether of tradition or custom, which might tend to warp his judgment; and has, therefore, found himself often obliged to treat as debatable, assumptions which inveterate association has rightly made shameful to doubt, but which *undisturbed would make the discovery of truth impossible. The importance of the subject, the happiness of many people, the elicitation of truth, are sufficient to justify the author in this course of an inquiry which he would have been content to leave to the common sense of the present age, in which theories are no longer adored because our fathers have believed in them, had not all-prying Science lit upon the question, and after her fashion made men doubt truth to be a liar, in order that they might doubt those things also which are not truths.*

Of the many shortcomings of the work the author is fully aware; but he does not pretend to have attempted an exhaustive inquiry, for which, in the present state of our knowledge, a lifetime would be too short. He will be satisfied if he has only succeeded in showing how rotten is the foundation of what many people accept as demonstrated truths, leaving to others the task of building a worthier edifice. For every statement the author has given a reference to his authority, which with the index to Authors quoted is amply sufficient for verification; he hopes, therefore, that if his work does not help to decide the question as to the advisability of marriage between near kin, it will at least serve as a hand-book to the literature of the subject, and prevent certain authors from assuming as demonstrated, a question whose investigation is as yet hardly begun.

ALFRED H. HUTH.

17, Kensington Square, May, 1875.

LIST OF CONTENTS.

INTRODUCTION.

PAGE

The importance of the marriage laws—They should be secure and easy—Fallacy of " Free-love "—The prohibitions on marriage—Alleged evil results of consanguineous marriage—Importance of distinct premises—Plan of this work . . . . 1—8

CHAPTER I.

THE DEGREES WITHIN WHICH MARRIAGE WAS PROHIBITED AMONG ANCIENT CIVILIZATIONS WHICH HAVE INFLUENCED OUR LAW; AND THE PROHIBITED DEGREES RESULTING FROM THEM IN THE VARIOUS MODERN CIVILIZATIONS.

Marriage in Egypt—Incest of the Ptolemies—Persian marriage—Tribes of lesser importance—Marriage in ancient Greece—Marriage among the Jews—The curse of sterility—Disregard of the Jewish law by the Herods—The Roman family—The history of the Roman prohibited degrees—Influence of Christianity on marriage—The prohibited degrees under the Popes—The Greek Church—Russian marriage prohibitions—The Swedish—The Danish—The ancient Icelanders—The history of the marriage prohibitions in England—Marriage with a deceased wife's sister valid by English law—The ancient Germans — The prohibited degrees in Prussia — The Rhenish Provinces—In Wurtemburg—In Hanover—In Holland—In France—In Italy—In Spain and in Portugal—In the United States of America . . . 9—59

CHAPTER II.

THE INFLUENCE OF ASCETICISM ON THE PROHIBITIONS OF MARRIAGE.

The reaction of asceticism on immorality, and immorality on asceticism—Egyptian ascetics—Jewish ascetics—Moral extremes outside the pale of the Catholic Church—Moral extremes in the orthodox Church—The prohibitions made stricter afterwards for the sake of money and power—Mischievous consequences of the wideness of the prohibitions—Spiritual relationship . . . . . 60—84

CHAPTER III.

THERE IS NO INNATE HORROR OF MARRIAGE BETWEEN NEAR KIN IMPLANTED IN MANKIND; THE ORIGIN OF THE PROHIBITED DEGREES AMONG SAVAGES; AND THE ONLY NATURAL PROHIBITED DEGREES.

To discover a natural prohibited degree it is necessary to investigate the prohibited degrees of all races—The Arabs and Mahommedans generally—The Druses—The Circassians—The Armenians—The Georgians—The Ossetes—The Mingrelese—The Parsees—The Affghans—The Beloochese—The Hindoos—The Khonds—The Sodah—The Twana—The Ho—The Koch—The Bodo—The Dhumal—The Garrow—The Munipurees—The Warali—The Magar—The Moondah—The Oraon—The Toda—The Yerkala—The Doingnak—The Singhalese—The Siamese—The Chinese—Chinese Turkestan—The Thibetians—The Mongol—The Mantchu—The Tunguz—The Jakut—The Ostyak—The Samoyed—The Kalmuck—The Kirghiz—The Nogai—The Lapps—The Malays—The Benkulen—The Palembang—The Lampong—The Batta—The Kalang—The People of Bali—The Dyaks—The Formosans—The People of Wajo—The Caroline Islanders—The Ladronese

—The Polynesians—The Marshall Islanders—The Sandwich Islanders—The New Zealanders—The Australians—The Tasmanians—The Papuans—The Eskimo—The North American Indians—The Ancient and Modern Mexicans—The People of Yucatan—The Darien Indians—The Indians of New Granada—The Macusi—The Warrou—The Arrowak—The Yamanos—The Ancient and Modern Peruvians—The Mandrucu—The Coroados—The Tupi—The Abipones—The Guarani—The Charruas—The Yuracares—The Apachalites—The Caraïbs—The Canary Islanders—The Ashantees—The Fantis—The People of Dahomey—The Papels—The Bambarra—The People of Cape Palmas—The Jolof—The People of Aquapim—The Kaffirs—The Zulu—The Hottentots—The Akombwi—The Bogo—The Somali—The People of Uganda—The Galla—The Helebi, Ghagar, and Nuri Gypsies—The Ghawazee—The Moors—The Tuarik—The Hovas—Incompleteness of works on travel—History of marriage—Exogamy caused by female infanticide—Endogamy the result of pride of race—Exogamy never really prohibits marriage between near kin—Mr. Tylor's theory of the origin of the prohibited degrees—There is no innate horror of incest—Deductions from the habits of animals—Deductions from fact and fiction—History of the reasons given for the prohibited degrees—The only natural prohibition . . 85—157

CHAPTER IV.

OBSERVATIONS ON SOME ISOLATED COMMUNITIES WHO HAVE CONTINUALLY INTERMARRIED AMONG THEMSELVES.

The mutineers of the "Bounty"—A community in Java—The family of Souza—Isolated communities in England—In Scotland—St. Kilda—In Iceland—Westmannoë—The Faroë Islanders—In Ireland—The Scillys—In France—The Pyrenees—The Cagots—The Vaquéros—The Chuetas—The Azore Islanders—The Petits Blancs of Réunion—Flinder's Isle—The Samaritans—The Jews . . . 158—205

CHAPTER V.

The Value of Statistics hitherto Collected, concerning Marriage between Near Kin, Examined.

There are no means of estimating the proportion of consanguineous to non-consanguineous marriages—Goitre and crétinism—Idiocy, epilepsy, insanity, and chorea—Deaf-mutism — Defective sight—Barrenness and low viability—Congenital malformations, rickets, hydrocephalus, spina bifida, convulsions, phthisis, tabes mesenterica, tubercular meningitis—Leprosy, ichthyosis, and albinoïsm . . . . 206—259

CHAPTER VI.

The Results of In-and-in Breeding on the Lower Animals.

Observations on the breeding in-and-in of animals are particularly useful for determining whether consanguineous marriages can originate disease or sterility without inheritance; and these observations can be applied with safety to the solution of the question in man—The in-and-in breeding of sheep—Of cattle—Of elands—Of guanaco—Of pigs—Of horses—Of donkeys—Of goats—Of dogs—Of deer—Of rabbits, and experiments in the production of albinoïsm—Of fowls—Of ducks—Of pigeons—Conclusion . . . 260—306

CHAPTER VII.

The Alleged Benefits of Change of Blood in Mankind.

If a cross is beneficial as a cross, the more distantly related the two animals crossed are, the more beneficial will that cross be—Anglo-Indians—Topas—Dutch-Singhalese—

CONTENTS. xi

PAGE

Dutch-Malays—Portuguese-Malays—Lipplapen—Spanish-Malays—Euro-Polynesians—Anglo-New Zealanders—Tasmanian half-breeds—Australian half-breeds—Portuguese-Chinese—Griquas—Mulattoes—Mestisos—Paulistas—Gauchos—Zamboes—Cafusos—Negro-Caraïbs—Negro-Mulattoes—Indian-Mestisos—Maroons—Negro-Arabs—Persian-Arabs—Abyssinian-Arabs—Kuruglis—Turks—Chinese-Cambojias—Chinese-Tagals—European-Eskimo—The sterility of half-breeds is not incompatible with the unity-of-man theory—A cross is not good *as a cross*—Debased state of half-breeds—Their sterility—The more similar two races are, the safer is a cross . . 307—332

CHAPTER VIII.

WHY ARE THERE TWO SEXES?

The views of Mr. Herbert Spencer and Mr. Darwin on the use of dual sex—The theory of Pangenesis not sufficient to account for the division of the sexes—Objections to Mr. Herbert Spencer's theory—The identity of the sexes—Hermaphroditism—Parthenogenesis—Economy of the dual sex—Why some hermaphrodites are incapable of self-fertilization—Why are there no more than two sexes?—Cause of the sterility of hybrids—The only value of a cross is to remove an hereditary tendency to disease 333–352

CONCLUSION.

Recapitulation—The value of the census—The fear that an inquiry on the subject of consanguineous marriage would be "inquisitorial"—Which degrees should be prohibited and which permitted—General results of this inquiry . 353—359

APPENDIX.

Cases of consanguineous marriage, and statistical inquiries . i.—xlii.

ERRATA.

Page 26, note 5, *for* "Stern," *read* "Sterne."
,, 59, line 10 from top, *for* "Vaquiros," *read* "Vaquéros."
,, 110, line 2 from top, *for* "Gomora," *read* "Santa Cruz."
,, 213, note 2; p. 214, note 1; p. 215, note 1; p. 216, note 1; p. 217, note ; p. 219, note 3; *for* "crétinism," *read* "crétinisme."
,, 220, line 3 from top, *for* "Salpétrière," *read* "Salpêtrière."
,, 243, line 5 from bottom, *for* "16," *read* "17;" and line 4 from bottom, *for* "5·3," *read* "5·7."
,, 281, note 3, *for* "250," *read* "205."

THE MARRIAGE OF NEAR KIN.

INTRODUCTION.

LIKE the rest of our legislation, the law of marriage is a fortuitous agglomeration of Acts and Judgments, about which we trouble ourselves extremely little. And yet perhaps it is the most important of all our laws. On it depends the law of inheritance, by which all property is governed, and over which many nations have been divided among themselves; and what is more important still, on it depends their morality, and with their morality their strength as a nation. For if a man is not bound to cleave to one wife he certainly will not trouble himself to educate his children, or for their welfare pinch and save; wherefore, instead of servants they would grow up enemies to the State, a miserable, ignorant, and incapable set of beings, who could conduct their country nowhere but to its ruin. Indeed, without marriage there would be no State, and no people to govern. It is a fallacy, as every sensible man knows, to think that it is sufficient that infants should be born, and that it is the duty of the State to see that they are properly brought up. The experience of foundling-hospitals amply shows that this is not an economical method; compulsory attendance at Government schools would also, as a monopoly, fail, as monopolies do; and, lastly, but by no means of least importance, is the fact that such a system is contrary to human nature, and therefore could not succeed. But granting for the moment that we could do without the marriage tie, and the State could take care of the infants, it is more than doubtful whether the population would not actually dwindle to nothing. People would not care to have children, and they certainly would have but few children whether they cared to or not; and yet it is a vital necessity not only that the population should be as much as the State can support, but that it should be more. For in a population which remains stationary every one may find employment; while in a population which increases in excess of its means of support a certain proportion must either emigrate or die. In the first case, therefore, the clever and stupid are equally employed, the weak as well as the strong; in the second case, the State selects the strongest and cleverest, while the weak and the stupid die out, or emigrate. There can be no doubt that of two countries, one of which had a stationary population, and the other an increasing population, which would ultimately overcome the other, or which would exist the longest, enjoy the greatest amount of happiness, and progress the fastest.

Since, then, the marriage tie is of such supreme importance, all things tending to make it insecure or difficult should be got rid of at once. Marriage should not be taxed as it now is, the contract should be clear and easily proved, and no impediment should be thrown in the way of a marriage which is not likely to produce unhealthy offspring; since every impediment directly tends to produce immorality, and immorality is directly and fatally injurious to the community.

The first condition is easily secured, because it only requires a little mechanical arrangement; but how are we to know whether a given couple should be allowed to marry or not? Were either insane we should not allow it, because we know that the chances are the progeny would also be insane. If a parent of either had become insane after marriage then we should not prohibit it, because, though there would be a good chance that insanity was become hereditary in the family, yet there would also be a good chance that the children might escape any hereditary taint, and the risk of prohibition would therefore be too great, we could not, for the sake of possible good, drive to a positive evil. For the same reason, or possibly from motives of humanity, we do not dissolve a sterile marriage as so many nations have done, and still do. The benefit in a great and fertile population would be very doubtful; the evil most assured. And yet we prohibit marriages between near kin. Is then the immorality caused by certain of these prohibitions nothing; and is the gain so very great?

The sole reason why we now prohibit these marriages, is that our fathers did so, and their fathers did so before them. Yet our fathers permitted marriage between crétins, lepers, and between many other persons, in cases where we should now prohibit it.

Why, then, should they have prohibited marriage between near kin? Did they observe that the progeny were diseased in any way, as we observe that the children of the insane were insane? This they certainly did not; for we ourselves cannot show it to be the case, even with the aid of what statistics we have got; nor do they give that as a reason in their writings. It is not easy to find out the real cause. All we know is, that these prohibitions have constantly been questioned, that they have been made more severe by law-givers, and less severe, again, by powerful individuals, or by the steady weight of public opinion. But public opinion means public practice; and public practice, when it is against the law, tends to bring both that law and other, even beneficial, laws into contempt. When the Popes prohibited marriage within very wide degrees, the people only made use of these prohibitions, either to live immorally, or to get rid of their wives or husbands, as the case might be. Nothing is more evident than that it is useless to prohibit marriage beyond certain degrees. Marriages may be prevented in some measure, but immorality, the consequence, cannot be; the remedy is worse than the evil. It is true that marriages between near kin have been accused in modern times of producing all those diseases of the nervous system and malformations, which we in our ignorance cannot as yet satisfactorily account for. But they are not prohibited in wider degrees, simply because none of these assertions can be proved. We should, in our endeavours to prevent idiocy and deaf-mutism, madness, and malformations, greatly increase immorality and contempt of the law, and probably increase those very diseases; because immorality does produce them indirectly, as we shall see. At the same time we should be doing nothing but a gigantic experiment; and even if we succeeded in stopping these marriages we should not get rid of the diseases; because, even supposing for the moment that these marriages are a cause at all, no one supposes them to produce more than a fraction of the insane or malformed.

Since the question of the alleged harmfulness of marriages between near kin is the object of this work, it will be useless to argue it any further at this point. It is essential, however, that in a question of this kind, as indeed in any, the premises should be perfectly clear, or the argument will be worthless; and this question has had its full share of indistinct statement and irrelevant answer. The consequence is, that there is a widely spread uncomfortable feeling that consanguineous marriages are to be avoided if possible; but if Angelina insists on having Edwin, and Edwin is fortunate enough to make Angelina persist in her intention—why then, let them marry, and may they be happy!

We will therefore put the following questions before ourselves:—

I. Whether consanguineous marriages are themselves, by the mere fact of consanguinity, and irrespective of any inheritance, injurious to the offspring. Whether, in a marriage between two relatives who are both perfectly healthy, live under healthy conditions, and whose families are perfectly healthy, the children born will probably be unhealthy?

II. Whether consanguineous marriages give a

greater proportion of unhealthy children than non-consanguineous marriages; or, in other words, whether it is a fact that consanguineous marriages through intensification of a previously dormant hereditary family taint, give a greater proportion of unhealthy children?

The first is a question of creation. The second a question of inheritance. Hence we may safely apply experiment on the organic world to the former, and the result, when no harm is done, will be favourable to the harmlessness of these marriages; but, if there are any evil results, we cannot say they were created by the consanguinity, since we cannot know but what inheritance has come into play. Hence, for the first question, the only method of investigation is a negative one. For the second question, the only method is a statistical inquiry, so contrived as to include the whole population of a large country; that is, so contrived as to overpower, as far as possible, disturbing causes.

This last method is so obvious that we cannot but feel surprised that it has never been tried yet. These marriages have been accused of causing sterility, malformations, the diseases of the mind and of the senses; of rickets, albinoism, phthisis, and crétinism; and even of the production of hydatis![1] Surely this was enough to attract the attention of the most callous Government—surely such a question as this would enter into the census!

[1] This last extraordinary statement was made by M. Aubé in 1857, and therefore ninety years after the researches of Pallas, seventy-five years after their confirmation by Goeze, and thirty-six after Bremser! It is nevertheless gravely quoted by MM. Devay (*Du Danger des Mar. Cons.*, p. 52) and Chipault (*Etude sur les Mar. Cons.*, p. 103) without a word of comment.

The fact is that the common sense of the world rejected these theories. They had married cousins for generations time out of memory. They had rebelled when these marriages were forbidden, and only acquiesced in the law when forced on them by the whole weight of a powerful priestocracy, under condition that it should exist but as a tax. In these times, however, this inheritance of the Middle Ages has been revived, a spectre clothed in the garb of modern philosophy. It is a matter for surprise, therefore, that this question should not have entered into the census, despite the carelessness of the general public; but our surprise is changed to absolute amazement, when we find that in 1871 this method, when proposed, was rejected by Parliament. Well may Mr. Darwin say in the conclusion of his *Descent of Man*, "When the principles of breeding and of inheritance are better understood, we shall not hear ignorant members of our legislature rejecting with scorn a plan for ascertaining by an easy method whether or not consanguineous marriages are injurious to man."

We have no means, therefore, of ascertaining whether consanguineous marriages are so injurious that they ought to be prohibited, and yet it has been most positively asserted that they are. What are the grounds, what the proofs, that these authors have for their statements? Certainly none that can be relied upon; and yet they have contrived to excite a sort of sensation as fatal to our case, as the nurse's bogy stories are to that of a child.

They argue deductively that, because some prohibitions are found throughout the world, from the earliest to the present times, these marriages must therefore be harmful; and that because there are two sexes, therefore crosses must be necessary. They argue inductively that, because in a certain number of cases they have selected of consanguineous marriage, the results as regards the progeny have been unfortunate, therefore a far greater proportion of children born from consanguineous marriages are afflicted with disease than those born from non-consanguineous marriages; and that because they find that a certain number of deaf-mutes, idiots, crétins, or what not, are derived from consanguineous marriages, therefore these diseases were caused by the consanguinity and by nothing else.

I will attempt an answer to these allegations in the following order: I hope to show that the reason many nations have prohibited these marriages is not because they have observed any evil result; I hope to show that many communities have lived without crosses, and without any excess of disease; that the statistical evidence we possess as yet is worse than worthless; that as far as experiment in the animal kingdom goes, the evidence tends to confirm the harmlessness of these marriages; that crosses are seldom beneficial, and often harmful; and to show that there may be other reasons for the existence of two sexes besides any supposititious benefit from crosses.

CHAPTER I.

THE DEGREES WITHIN WHICH MARRIAGE WAS PROHIBITED AMONG ANCIENT CIVILIZATIONS WHICH HAVE INFLUENCED OUR LAW; AND THE PROHIBITED DEGREES RESULTING FROM THEM IN THE VARIOUS MODERN CIVILIZATIONS.

THE Egyptian, the Greek, the Jewish, the Roman, and perhaps in a smaller degree the Persian, were probably the only ancient civilizations which have affected our marriage law. Among these, the Jews of course stand pre-eminent from their connection, through the Old Testament, with Christianity. But the Romans, from their dominant position and talent for jurisprudence; the Greeks, with their nimble minds and subtle theology; the Egyptians, with their constant foresight as to a future life, and especially with the ponderous weight of their ancient civilization,— have all contributed in a marked degree to bring about the present curious medley of sentiment, custom, and law, which interferes with and confirms our present marriage contracts. Regarding the question, as to what degrees were permitted to intermarry, we find that the earliest civilizations were unusually free from all trammels. The Egyptians were accustomed to marry their sisters from the earliest times of which we have any record.[1] In the time of the Ptolemies we find this custom placed beyond any historical doubt; and even did not their mythological story of the marriage of Osiris and Isis confirm its occurrence before that period, yet so politic and conciliating a king as Ptolemy Soter the reputed son of Lagus, the first of a new dynasty of stranger kings, with a by no means undisputed title, and in a nation peculiarly intolerant of foreigners and of foreign customs, would hardly have dared to marry his daughter Arsinoë to her uterine brother Philadelphus; nor would Philadelphus have been able to deprive his elder brothers of the kingdom, or have been so powerful and popular, if his marriage had in any way shocked the ideas of his Egyptian subjects as to what was right and proper. What makes it more certain that this was an old Egyptian custom, is the fact that the Greeks did not at that time marry their uterine sisters, and that the Egyptians were an endogamous race, they were divided into castes, on much the same model as the Indian divisions, and they habitually married in these castes;[2] from which it appears extremely probable that, as among the ancient Peruvians, at least the kings and nobles customarily married their sisters, so as not to defile their line with ignoble blood.

It is impossible to give an idea of the extent to which the Ptolemies practised incest without a genea-

[1] Diodorus, i. 27, cited by Wilkinson, *Ancient Egyptians*, First Series, vol. ii. p. 63; and Seneca in his *Apocolocynth*, cited by Adam, *Fortnightly Review*, 1865, p. 714. Also Philo, see *Ibid*.
[2] Wilkinson, *Ancient Egyptians*, First Series, vol. i. pp. 239, 245.

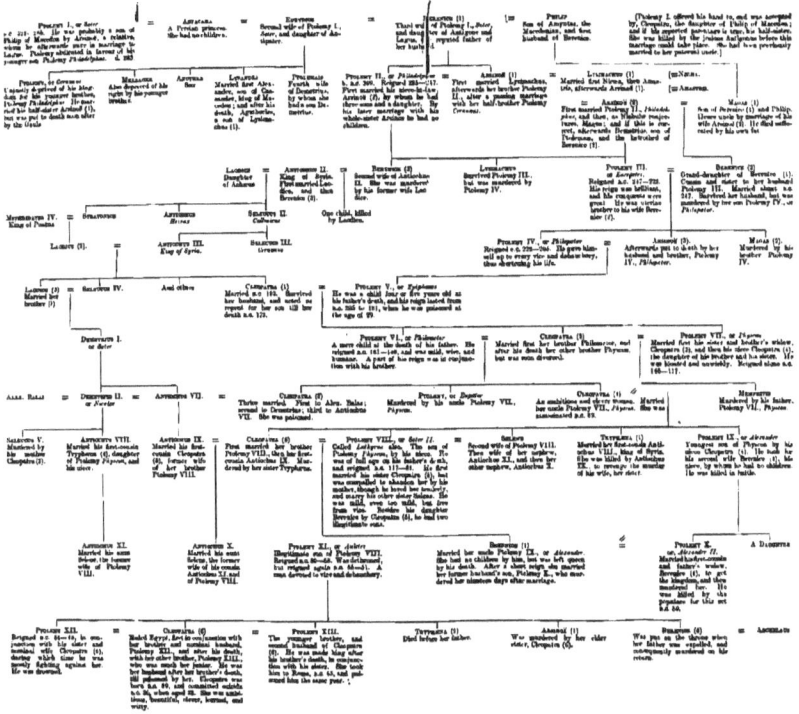

he was drowned.   her hu
                                    till pe
                                    born
                                    n.c. 30
                                    tious,
                                    witty.

[*To face page* 11.]

logical table, for the more unions of that kind there are, the more difficult is it to follow such complicated relationships. In the annexed table it will be seen how they habitually married their sisters, their nieces, their cousins; indeed in all degrees excepting those of the direct ascending and descending line. And with what result? Niebuhr assures us that a visible curse rested on the dynasty of the Ptolemies,[1] a conclusion he seems to have come to from that preconceived notion which so many people have; for he tells us that several physiologists assert that such marriages lead to scrofula, idiocy, etc., and that their evil results are clearly seen in those villages where the peasantry intermarry among themselves.[2] This assumption I have considered below; but that the Ptolemies were under a manifest curse is most certainly an unwarrantable assertion. In their genealogical table it may be seen that they were neither sterile[3] nor particularly short-lived. That they were not more subject to disease than would be expected from their luxurious habits, or stupider than the generality of people, we gather from history: nay, some of them were singularly sharp-witted; Cleopatra, the daughter of a brother and sister, great-grand-daughter of another brother and sister, and a great-great-granddaughter of Berenice, who was both cousin and sister to her husband, might with advantage compare in astuteness to Catherine de Medicis of France. While as for that hardness of heart and utter careless-

---

[1] Niebuhr, *Vorträge über alte Gesch.*, vol. iii. p. 570.
[2] *Ibid.*
[3] Mr. Galton, in his *Hereditary Genius*, p. 152, says, "The result of Ptolemaic experience was distinctly to show that intermarriages are followed by sterility." He seems to base this assumption on the fact

ness of human life, on which Niebuhr lays so much stress, as a clear manifestation of the curse under which they laboured, it was not peculiar to the Ptolemies, nor even to other barbarous civilizations of that time, but may be seen in the history of any despotism that has that the last queen, Cleopatra, was only an illegitimate descendant of Ptolemy I. The following table will show the facts more clearly:—

Name.	Relationship to		No. of Children by First Wife.	No. of Children by Second.
	First Wife.	Second Wife.		
Ptolemy II.	No blood relative.	Sister.	4	0
,, III.	Cousin and sister.	—	3	—
,, IV.	Sister.	—	1	—
,, V.	Scarcely connected.	—	3	—
,, VI.	Sister.	—	3	—
,, VII.	Sister.	Niece, the daughter of his sister *and* his brother.	1	5
,, VIII.	Sister.	Sister.	1	0
,, IX.	?	Niece, the daughter of a brother and sister.	2	0
,, X.	First-cousin.	—	0	—

By this table it will be seen that the Ptolemies had as many children when they married their nearest relatives, as when they married strangers. It is true that apparently they had few children, but we must bear in mind that their fertility cannot be compared to the present fertility of Europe; they had hareems like other Eastern nations, and we should compare their fertility therefore to that of other great Eastern potentates. There were besides many other causes for their apparent sterility; thus the sister of Ptolemy II. had only one child by her former husband, who was no relation to her, and she was besides much her brother's elder; the wife of Ptolemy IV. was murdered; Ptolemy VII. divorced his first wife soon after marriage; Ptolemy VIII. also soon divorced his first wife, and he was forced to abandon his second; Ptolemy IX. was killed in battle; while Ptolemy X. murdered his wife nineteen days after marriage. It is hardly to be expected, when we consider all these little circumstances, that they should have had many children.

ever been written. No one will deny that the Egyptians decayed in wealth and power almost from the accession of the Ptolemies; but the theory is untenable that this was due to their consanguineous marriages. When we consider that they did not introduce this custom, but followed it, and that it must have been customary for centuries before, we see at once how absurd such a theory must be.[1] The Egyptians began to decline when they ceased to be that exclusive people they formerly had been;[2] not probably from the admixture of fresh blood, of which there could be but little, but rather from the increase of vice and the general decay which always accompanies disregard of the marriage tie. For when Plato visited them not long before their conquest by Alexander, although they had been open to Greek intercourse for centuries, he was greatly struck by their strong conservatism,[3] a conservatism which was lost in its most important part, under the vicious example of their foreign conquerors, spoiled as they were by Asiatic luxury.

Sharpe's supposition that Cambyses was the introducer of the Egyptian custom of marriage into Persia,[4] is probably only half true. For apart from the fact that we find Semiramis, according to the legend, practised incest at a supposed date of 2,000 B.C., the custom was so widely spread, the influence of one mad king could have been so very slight, and the reasons that gave rise to the custom of marrying near kin, operated and operate so generally and widely,

[1] Wilkinson, *Ancient Egyptians*, First Series, vol. i. p. 239.
[2] Sharpe, *Hist. of Egypt*, vol. i. p. 202.
[3] *Ibid.*, vol. i. p. 207.
[4] *Ibid.*, vol. i. p. 130. See also Herodotus, iii. 31.

that we cannot suppose him to have been the first to introduce that practice in his kingdom. Agathias charges the Persians of his time with incest, and accuses Zoroastre of having introduced it.[1] Zoroastre recommends, above all other alliances, those between first-cousins, as marriages deserving the reward of Heaven;[2] but he says nothing about marriage with nearer relatives; and though this permission might have been in one of the lost books, we should certainly have expected to have found it in the Vendidad, if anywhere. Yet Theodoretus made the accusation before him,[3] and Al Beidâwi after him says that Mahommed means by "those that follow their lust," whom he holds up among others to condemnation, the Magians, whom their prophet Zerdusht (Zoroastre) permitted to marry their mothers and sisters.[4] That the Persians really did practise incest, however introduced, is beyond a doubt. Ctesias, according to Tertullian, says they married their mothers. Clemens Alexandrinus confirms this, and adds sisters; and in the *Recognitions*, erroneously ascribed to him, also daughters. Again, Cyril tells us that the Emperor Julian, arguing against the Christians, said, "There is a greater difference in the laws and manners of men than in their speech. For what Greek will say that it is proper to cohabit with a sister, a daughter, or a mother? This, however, is judged to be good by the Persians."[5] Diogenes Laërtius speaks of marriage with

[1] Lib. ii. c. 24, cited by Adam, *Fortnightly Review*, 1865, p. 716.
[2] Du Perron, *Zend Avesta*, vol. ii. pp. 556, 612.
[3] Cited by Freinsch in his edition of Q. Curtius, vol. ii. p. 578, note.
[4] Sale's *Koran*, p. 59, note o.
[5] By *sister* the Emperor must have meant *uterine* sister, because the Greeks were accustomed to marry their half-sisters.

a daughter only; but Antisthenes, according to Athenæus, accused Alcibiades of incest with his mother, daughter, and sister, "like the Persians."[1] Eudoxos of Cnidos, says, "Among us, it is forbidden to contract marriages with our mother; in Persia, on the contrary, it is considered an honourable action." Jerome, too, accuses the Medes and Persians of marriage with their mothers, their daughters, and their nieces. Minutius Felix says they permitted marriages with their mothers.[2] Plutarch claims for Alexander the Great the credit of having induced the Persians to give up these marriages.[3] The cause was probably, as in Egypt and in modern cases, that a man might form an union worthy of his rank. Thus the title of Berenice, the daughter of Ptolemy Soter II., was Queen and *Sister* to her husband, although she was really only his niece.[4] Mr. Adam considers that the Magians especially formed unions of this kind, that it was altogether an aristocratic custom, but that it also led to similar unions among the lower classes. "It seems probable," he says, "to a degree approaching to certainty, that marriage with mothers was long an established aristocratic institution; sometimes, as in the age of Alexander and under Hellenic influence, discountenanced, and at last partially abandoned; at other times renewed and invested with full practical force, and perhaps, with religious sanctions; and at all times conducing, with or without the authority of law, to other corrupt conjugal

[1] *Loc. cit.* Adam, *Fortnightly Review*, 1865, pp. 715-718.
[2] *Loc. cit.* Boudin, *Mém. de la Soc. d'Anthrop. de Paris*, vol. i. 1863, p. 550; also Adam, *ut sup.* p. 716.
[3] Adam, *ut sup.*
[4] Sharpe, *Hist. of Egypt*, vol. ii. p. 11.

unions, such as those with a daughter or a sister, and to a general depraved state of social life."[1] Thus he points out that Xanthus, in his lost work entitled *Magica*, from which Clemens Alexandrinus quotes, is more cautious than Clemens himself; for Xanthus only stated that the Magians were allowed to marry their mothers and daughters; while Clemens adds that it was also lawful to marry sisters. Again, Diogenes Laërtius, speaking of the Magians only, and quoting from Sotion, says they thought it right to marry a mother and a daughter;[2] and Catullus says a Magian must have sprung from such a marriage.[3] Strabo only mentions marriage with mothers.[4] According to Sextus, those especially were accustomed to marry their mothers who were considered the wisest of their race, namely the Magians.[5] Tatian says, "The Greeks regard cohabitation with the mother as a thing to be avoided, while such practice is held in the highest repute by the Persian Magians." Philo holds essentially the same view: "The magistrates," he says, "of the Persians marry their own mothers, and consider the offspring of such unions most noble and worthy of the highest sovereign authority."[6] We have many instances of its occurrence among the nobility. Thus Herodotus gives the well-known story of the marriage of Cambyses. It appears that he first married his half-sister and then his whole-sister. One of these, Atossa, afterwards became the wife of Darius, to whom she was related in the sixth

[1] Adam, *Fortnightly Review*, 1865, pp. 718-719.
[2] Adam, *ut sup.*, pp. 716, 717.   [3] Reich, *Ehe*, etc., p. 36.
[4] Adam, *ut sup.*, p. 717.
[5] Boudin, *Mém. de la Soc. d'Anthrop. de Paris*, vol. i. 1863, p. 550.
[6] Adam, *ut sup.*

degree; and Darius afterwards married his niece, Phratagune, the daughter of his brother Artanes. Plutarch asserts that Artaxerxes married one of his own daughters; and on the authority of Heraclides the Cumean, and others, he adds that the same monarch afterwards married another daughter. The former marriage, says Plutarch, he was induced to contract by the persuasion of his mother Parysatis; but, according to Agathias, he refused an offer of marriage from his mother, as altogether contrary to law and custom.[1] The Bactrian Satrap Sysimithres, who lived in the time of Alexander the Great, married his mother and had two children. The Carian Satrap Mausolus married his sister Artemesia.[2]

All the tribes round about seem to have done the same thing, if we are to believe the general statements of ancient authors. Thus Euripides in *Andromache*, makes the jealous Hermione cry: "Such is the whole race of barbarians; a father marries his daughter, a son his mother, and a maid her brother * * * * and the law prevents none of these things."[3]

Ptolemy says that most of the inhabitants of India, Ariana, Gedrosia, Parthia, Media, Persia, Babylonia, Mesopotamia, and Assyria, might marry their own mothers; while the inhabitants of Northern Africa,

---

[1] Adam, *Fortnightly Review*, 1865, p. 718.
[2] Adam, *ut sup.*, and Boudin, *Mém. de la Soc. d'Anthrop. de Paris*, vol. i. 1863, p. 551.
[3] τοιοῦτο πᾶν τὸ βάρβαρον γένος·
πατήρ τε θυγατρὶ παῖς τε μητρὶ μίγνυται
κόρη τ'ἀδελφῷ, διὰ φόνου δ'οἱ φίλτατοι
χωροῦσι, καὶ τῶν δ'οὐδὲν ἐξείργει νόμος.
Androm., v. 173.

namely, those of Cyrenaica, Marmarica, Ægyptus, Thebaïs, Oasis, Troglodytica, Arabia, Azanæa, and Æthiopia Media, might intermarry with their sisters. St. Jerome says the Medes, Indians, Persians, and Ethiopians marry their mothers and grandmothers, their daughters and grand-daughters.[1] Tiraquellus and Alexander assure us that the ancient Arabs, Persians, Assyrians, Parthians, Medes, Egyptians, Phrygians, Galatians, Ethiopians, Indians, Scotch, Irish, and the nomadic tribes of Naura, were accustomed to live in incest. Herodotus says the Machlyses and Auses knew no marriage or relationship. Plato, Lactantius, Aristotle, and Solinus, say the same of the Garamantes.[2] Strabo says the Scythians knew no marriage prohibitions, that they married their daughters and sisters.[3] The Canaanites seem also to have married their near relations; for the Jews are commanded, "After the doings of the land of Egypt, wherein ye dwelt, shall ye not do: and after the doings of the land of Canaan, whither I bring you, shall ye not do: neither shall ye walk in their ordinances." And then further, after enumerating the forbidden degrees: "Therefore shall ye keep mine ordinance, that ye commit not any one of these abominable customs, which were committed before you."[4] Justin informs us that the Phœnicians were allowed to marry their sisters.[5] A law of Constans, given at Antioch, and addressed to the Phœnicians A.D. 339, prohibits marriage with a niece,

---

[1] Adam, *Fortnightly Review*, 1865, p. 713.
[2] *Loc. cit.* Freinsch's edition of Q. Curtius, vol. ii. p. 578, note.
[3] Priscus, cited by St. Lager, *Du Crétin*, etc., p. 114; and Perier, *Mém. de la Soc. d'Anthrop.*, 1863, vol. i. p. 197.
[4] Leviticus xviii. 3, 30; xx. 23.
[5] Boudin, *Mém. de la Soc. d'Anthrop. de Paris*, 1863, vol. i. p. 551.

whether a brother's or sister's daughter, under pain of death.[1] The Greeks were essentially an endogamous race, as were most ancient civilizations surrounded by barbarians.[2] The Greek colonies even owe their existence to the law that the children of a mixed marriage had no right of citizenship; these were forced to emigrate, for they were no true sons of Hellas. Athenian law prohibited, under very severe penalties, the marriage of a citizen with a foreigner.[3] "No Spartan," says Grote, "could go abroad without leave, nor were strangers permitted to stay at Sparta; they came thither, it seems, by a sort of sufferance, but the uncourteous process called xenêlasy was always available to remove them, nor could there arise in Sparta that class of resident metics or aliens, who constituted a large part of the population of Athens, and seem to have been found in most other Grecian towns."[4] The old law which forbad a Heracleid to marry a foreigner was revived by Lysander to turn out Leonidas II., who had married an Asiatic wife. In the same way Themistocles was not a citizen because his mother was a foreign woman; and it was a charge of Æschines in his attack on Demosthenes, the orator, that his was no true Greek descent.[5]

[1] Devay, *Du Danger des Mariages entre Cons.*, p. 70.
[2] In Persia a marriage was not considered legitimate, unless both of the contracting parties were Persians. Hence the surname *Nothus*, which some of the Persian kings bore. Niebuhr, *Alte Geschichte*, vol. ii. p. 219.
[3] Art. *Matrimonium*, in Smith's *Dict. of Antiquities*, etc.
[4] Grote, *Hist. of Greece*, vol. ii. p. 153.
[5] Arts. *Leonidas II.* and *Themistocles*, in Smith's *Dict. of Greek and Roman Biography*; and Perier, *Mém. de la Soc. d'Anthrop. de Paris*, 1870, vol. iii. p. 215.

Pericles himself was no real citizen, because his parents were not both Greeks; but the son of Pericles, by his Greek wife, was a true Athenian citizen. The numerous class of aliens which thus arose in nearly every Grecian town, formed a real danger; for they could not be expected to remain content without that franchise which the lowest man of the populace possessed. They had, therefore, every now and then, to be driven forth, with directions to found a colony, by which expedient they remained loyal to the mother country, while they extended her power.[1]

Despite their antipathy, however, to foreign marriages, the Greeks were by no means so lax in their prohibited degrees as the Persians or Egyptians. We should have expected indeed to find at least the same laxity in Greece as among these latter, for according to Sir G. Wilkinson, the Athenians were originally an Egyptian colony,[2] while the Persians regarded no prohibitions, and the surrounding nations apparently respected them no more. But whatever was the reason, whether the parent Egyptian colony was purer at the time of its separation, or whether their law was formed by other considerations, they always prohibited marriage in the direct ascending line, as did the Egyptians, and besides marriage with a whole-sister.[3] According to Philo, the Spartans were allowed to marry a sister-uterine, but not a sister-german. But this, observes Mr. M'Lennan, is probably untrue, and is discordant with the habits of the Lacedæmonians.[4]

[1] Niebuhr, *Alte Geschichte*, vol. i. pp. 305, 306.
[2] Wilkinson, *Ancient Egyptians*, First Series, vol. ii. p. 63.
[3] Art. *Matrimonium*, in Smith's *Dict. of Antiquities*.
[4] M'Lennan, *Prim. Mar.*, pp. 221, 222.

A Spartan husband was forbidden the luxury of jealousy; husband and wife were brought together utterly regardless of likes or dislikes; and those were married together who were the finest physically.[1] Xenophon asserts that mother and daughter were sometimes married to the same individual;[2] but Grote says that though some women were the recognized mistresses of two hearths, yet this sort of bigamy was strictly forbidden to the men, and never, perhaps, permitted, except in the extraordinary case of King Anaxandrides, when the royal Herakleidan line was in danger of becoming extinct. Polybius also states that three or four Spartan brothers sometimes had only one wife between them.[3]

To this state of the marriage law, Grote partially ascribes the fall of Sparta, on the general ground that "it has been observed often as a statistical fact, that a close corporation of citizens, or any small number of families, intermarrying habitually among one another, and not reinforced from without, have usually a tendency to diminish."[4] But I have protested, and must protest against an assumption on such vague grounds, when there are so many other facts to account for the same thing. We find Xenophon pointing with "pride to the tall and vigorous breed of citizens which the Lykurgic institutions had produced. The beauty of the Lacedæmonian women was notorious throughout Greece, and Lampitô, the Lacedæmonian woman introduced in the Lysistrata of Aristophanes, is

---

[1] Grote, *Hist. of Greece*, vol. ii. p. 150.
[2] M'Lennan, *Prim. Mar.*, p. 298.
[3] Grote, *ut sup.*, vol. ii. pp. 150, 160, 161.
[4] Grote, *ut sup.*, vol. ii. p. 161.

made to receive from the Athenian women the loudest compliments upon her fine shape and masculine vigour."[1] Now this "beauty and masculine vigour" is just what the offspring of consanguineous marriages are said not to have; moreover, the old prohibition against the naturalization of strangers was no longer in force when Sparta began to decline.[2] Hence we find that at a time when marriages between near kin were frequent, and human beings were treated as a modern breeder would treat his cattle, the people were handsome and healthy, and there was no sign of degeneration. But at the same time we cannot be astonished that an institution like that of Lykurgus, the foundation of which was so utterly contrary to all human nature, should ultimately sink under the waves of a reform more gracious to human sympathies and more tender to its passions. Nor should we wonder that that reform should involve disturbance, and even anarchy.

Marriage was permitted at Athens between half-brothers and sisters. Thus, Mnesiptolema, the daughter of Themistocles' second wife, was married to Archeptolis, her half-brother. Cimon is said to have married his half-sister Elpinice, because she was too poor to obtain a husband suitable to her birth. Afterwards Callias, a rich Athenian, fell in love with and married her, while Cimon married Isodice.[3] Nepos says such marriages were both customary and lawful.[4] The marriage of an uncle and niece was especially com-

---

[1] Grote, *ut sup.*, vol. ii. p. 152. [2] *Ibid.*, p. 161.
[3] Smith's *Dict. of Greek and Roman Biography*, Art. *Elpinice;* and Langhorne's *Plutarch*, vol. i. p. 358; vol. iii. p. 299.
[4] Nepos, *Præf.* and *Cimon*, cap. i.

mendable, by reason of the near relationship, as Lysias instances of two brothers, one of whom proposed to the other that he should marry his daughter, and the brother was favourable to the proposal, since she was so nearly related.[1] Grote also says of Pericles that he married a woman "very nearly related to him," influenced by family considerations, that were considered almost obligatory at Athens.[2]

In the case of a man dying intestate, and without male issue, his heiress had no choice in marriage, but was compelled by law to marry her nearest kinsman, who must not, however, be related to her in the direct ascending or descending line. If the heiress were too poor to obtain a husband of her own rank, her nearest unmarried kinsman either married her, or portioned her suitably to her rank. When there were several co-heiresses, they were also married to their kinsmen, the nearest relative having first choice. In short, the heiress, together with her inheritance, seems to have belonged to her nearest kinsman; so that in the earlier times a father could not marry his daughter, if she were an heiress, to anyone else without her kinsman's consent.[3] But this was not the case according to the later Athenian law; for fathers were then allowed to dispose of their daughters just as they liked, whether they were heiresses or not; and after his death they might be married according to the provisions of his will, just as widows were disposed of. The father of the orator Demosthenes left directions in his will that his sister's son should marry his widow, and the son of his brother

---

[1] Adam, *Fortnightly Review*, 1865, pp. 719-720.
[2] Grote, *Hist. of Greece*, vol. iv. p. 230.
[3] Art. *Matrimonium*, Smith's *Dict. of Antiquities*.

should marry his daughter.[1] And though Jeremy Taylor points out that Æschylus calls a marriage of first-cousins a "marriage which the law forbids," and adds that "the family is dishonoured by it," yet he shows that that also is a confirmation of the above law, because the maid was an heiress, and she was married without her kinsman's consent.[2] Terence hinges his play of *Phormio* on this law, and brings it in again in the *Adelphi*.[3]

Even when an heiress was already married, her husband was obliged to give her up to any kinsman with a better title; and men often put away their former wives in order that they might marry an heiress.[4]

The same law was common to Sparta. Thus Leonidas married Gorgo, the heiress of Cleomenes, his half-brother, as he was her nearest relative; and Anaxandrides, the father of Leonidas and Cleomines, married his own sister's daughter. Should a father

---

[1] Smith's *Dict. of Greek and Roman Biography*, Art. *Demosthenes*.
[2] Jeremy Taylor, *Duct. Dubit.*, book II. chap. ii. rule iii. sec. 53, p. 233.
[3]   "Lex est, ut orbæ qui sunt genere proxumi,
    Eis nubant: et illos ducere eadem hæc lex jubet.
    Ego te cognatum dicam, et tibi scribam dicam:
    Paternum amicum me assimulabo virginis:
    Ad judices veniemus: qui fuerit pater,
    Quæ mater; qui cognata tibi sit; omnia hæc
    Confingam: quod erit mihi bonum, atque commodum,
    Cum tu horum nihil refelles, vincam scilicet."
                            Phormio, Act i. sc. 2.
    "Hæc virgo orba est patre:
    Hic meus amicus illi genere est proxumus:
    Huic leges cogunt nubere hanc."
                            Adelphi, Act iv. sc. 5.
[4] Smith's *Dict. of Antiquities*, Art. *Epiclerus*.

fail to decide on whom he would bestow his daughter's hand, the king decided who among the kinsmen should be the lucky man.¹

In earlier ages it is probable the law concerning the forbidden degrees was not so clearly defined. Thus Homer says:

> "At length we reached Æolia's sea-girt shore,
> Where great Hippotades the sceptre bore,
> A floating isle! High raised by toil divine,
> Strong walls of brass the rocky coast confine.
> Six blooming youths, in private grandeur bred,
> And six fair daughters, graced the royal bed:
> These sons their sisters wed, and all remain
> Their parents' pride, and pleasure of their reign."²

Valerius Maximus says that after Homer's time marriages between brothers and sisters ceased, but marriages between aunts and their nephews, uncles and their nieces, and first-cousins, were always allowed.³ He must, however, mean that marriages between whole-brothers and sisters were no longer allowed, for, as we have seen above, marriage with a half-sister was very frequent. In the very early times, the ancient inhabitants of Macedonia are suspected of practising promiscuous intercourse;⁴ and Hippias says that parents sometimes married their children;⁵ possibly from the

¹ Smith's *Dict. of Antiquities*, Art. *Matrimonium*.
² Pope's *Homer's Odyssey*, book x. 1-8.

τοῦ καὶ δώδεκα παῖδες ἐνὶ μεγάροις γεγάασιν·
ἓξ μὲν θυγατέρες, ἓξ δ'υἱέες ἡβώοντες.
ἔνθ' ὅγε θυγατέρας πόρεν υἱάσιν εἶναι ἀκοίτας.

Ovid also notices it, in the *Metam.*, lib. ix. 506:—
"At non Æolidæ thalamos timuere sororem."

³ Reich, *Ehe*, etc., p. 22.
⁴ See Freinsch, Edition of Q. Curtius, p. 578, note.
⁵ Adam, *Fortnightly Review*, 1865, p. 719.

curious but widely-spread idea, mentioned by Æschylus, that a mother is not related to her child. For he makes Orestes plead before the gods that he is not of kin to his mother, and they decide that she who bears the child is merely its nurse.[1] The Egyptians also considered a mother "little more than a nurse;"[2] and Menu considers her merely the field bringing forth the plant sown on it.[3] The Tupinamba of Brazil seem to have the same idea,[4] and it does not appear unknown even to English law, for the Duke of Suffolk's wife was adjudged not of kin to her son.[5] The mythology of the Greeks makes Jupiter brother and husband of Juno, as the Egyptians made Osiris the brother of his wife Isis, from whom they probably derived the story.[6] Byblis falls in love with her brother Caunus; "the gods fare better, for they have their own sisters in marriage," are the words Ovid puts in her mouth.[7] Oceanus married Tethys, his sister, and if we are to believe that he had 3,000 rivers, as many Oceanides, and three or four daughters, the marriage could not be considered unfruitful.[8] Canace married her brother Macareus.[9] Diomedes and Iphidamus are supposed to have married their mother's

[1] Tylor, *Early Hist. of Mankind*, p. 300.
[2] Wilkinson, *Ancient Egyptians*, First Series, vol. ii. p. 65.
[3] Tylor, *ut sup.* But Sanc'ha and Lic'hita seem to consider that the influence of the mother is great, and refer to what they evidently think with Jacob, to be a fact. See Colebrooke's *Hindu Law*, book iv. chap. i. sec. 21; and Genesis xxx. 37-39.
[4] Tylor, *ut sup.*, p. 299.
[5] See Stern, *Tristram Shandy*, vol. i. p. 393.
[6] Wilkinson, *Ancient Egyptians*, Second Series, vol. i. p. 268.
[7] *Metam.*, book ix. l. 496.
[8] Thetis, one of his daughters, is also said to have been his wife. See Smith's *Dict. of Biography*, etc., Arts. *Tethys* and *Oceanus*.
[9] *Ibid.*, Art. *Canace*.

sisters; Alcinous married his brother's daughter Arete; and Andromeda was promised to her uncle Phineus.[1] Jocasta married Œdipus, ignorant that he was her son, and though Homer makes them suffer in hell for this offence, yet their punishment is as much for the parricide as anything else.[2] The fable makes the sons born of this marriage fight and slay each other, not that they were supposed to be insane, but because they quarrelled about the division of their father's kingdom.[3] Thyestes and his daughter Pelopeia, Clymenus and his daughter Harpalyce, Oenomaus and his daughter Hippodameia, Erictheus and his daughter Procris, Nyctimene and her father, and Menophrus and his daughter Cyllene, and his mother Bliade, are further instances.[4] Myrrha commits incest with her father Cinyras; and what is particularly observable, Adonis, the most beautiful of men, is the result of that union; " even Envy herself would have commended that face, for he was just as the ideal of a cupid."[5]

Before the time of Moses the Jews were accustomed to marry very near relatives indeed. Thus we find Abraham married to his half-sister Sarah.[6]

---

[1] Jeremy Taylor, *Duct. Dubit.*, book II. chap. ii. rule iii. sec. 33. p. 228.
[2] Homer, *Odyssey*, book xi. ll. 270-279.
[3] Smith, *Dict. of Greek and Roman Biography and Mythology*, Art. *Œdipus*.
[4] Smith, *Ibid.*, Arts. *Clymenus, Oenomaus*, and *Nyctimene*; Adam, *Fortnightly Review*, 1865, p. 719.
[5] Ovid, *Metam.*, lib. x. ll. 515, 516, 520, 521.
" Laudaret faciem Livor quoque, qualia namque
Corpora nudorum tabulâ pinguntur Amorum.
Talis erat. . . . . . ille sororo
Natus avoque suo."

[6] Josephus makes her his niece. See *Jewish Antiq.*, book I. chap. vii. sec. 1. But this is only the common tradition, taken for granted by

Nahor married his niece, Milcah the sister of Lot, and his brother's daughter.[1] The incest of Lot needs only a reference;[2] but Mr. Adam points out that though the story of Lot and his daughters may have been invented by the Jews out of spite towards their enemies, the Moabites and Ammonites, yet Ruth was a Moabitess, and from her was David descended. If national hatred, he says, would have dictated, national vanity would have suppressed the imputation; we must therefore be contented to take it as we find it.[3] Jacob married his first-cousins, Rachel and Leah, for Laban thought it better to give his daughters to a kinsman;[4] and Isaac said, "Thou shalt not take a wife of the daughters of Canaan. Arise, go to Padan-aram, to the house of Bethuel thy mother's father; and take thee a wife from thence of the daughters of Laban thy mother's brother."[5] Esau had married two[6] wives "of the daughters of Heth," greatly to the anger of his parents. To please them he added his cousin, "Mahalath the daughter of Ishmael Abraham's son."[7] Amram, the father of Moses, married his paternal aunt, Jochebed.[8] After the law was defined by Moses, it was forbidden to marry a mother, a step-mother, sister, half-sister (whether legitimate or not),

Josephus and by Jerome, who supposed Sarah to be identical with Iscah, the daughter of Haran, and sister of Lot, called the "brother" of Abraham (Genesis xiv. 14, 16. See Smith's *Dict. of the Bible*, Arts. *Sarah* and *Iscah*). Abraham himself speaks of her as the daughter of the same father, but of a different mother (Genesis xx. 12).

[1] Genesis xi. 26-29, and xx. 12.   [2] Genesis xix. 31-38.
[3] Adam, *Fortnightly Review*, 1865, pp. 82-83.
[4] Genesis xxix. 19.   [5] Genesis xxviii. 1, 2.
[6] The ancient Jews were allowed polygamy, though the Jewish doctors, according to Salo, only allowed four wives. See his *Koran*, Introduction, p. 102.
[7] Genesis xxviii. 9.   [8] Exodus vi. 20.

THE MARRIAGE OF NEAR KIN. 29

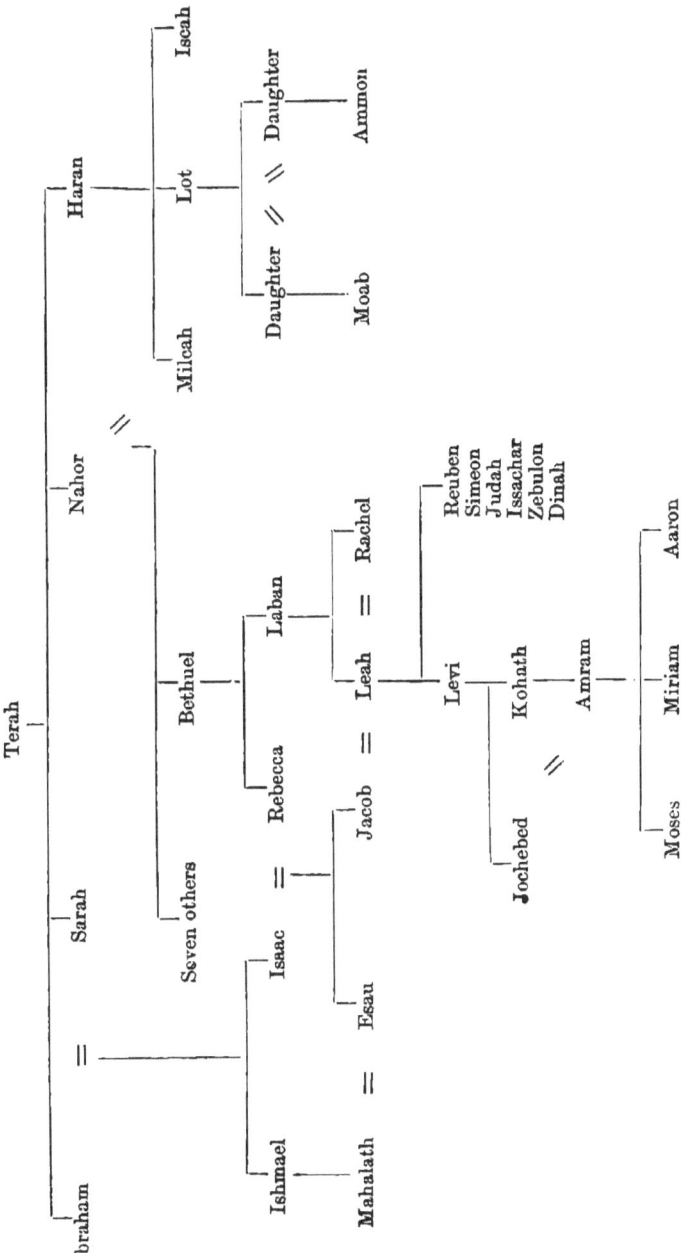

grand-daughter, aunt (whether by consanguinity on either side or marriage on the *paternal* side), daughter-in-law, brother's wife (except when she came under the provisions of the levirate law), step-daughter, wife's mother, step-grand-daughter, or wife's sister during the wife's life. To these degrees the Talmudists added grandmother, great-grandmother, great-grandchild etc.[1] Moses nowhere prohibits the marriage of an uncle and niece, though he prohibited such marriages as he himself was descended from; indeed, we have an instance of such a marriage in Othniel, the younger brother of Caleb the spy, who married Achsah, his niece, and the daughter of Caleb.[2]

Yet notwithstanding the law, the Jews seem to have continued to marry within the same degrees as before. Even if we pass over the complaints of the Prophet Ezekiel,[3] as the figurative language of an

[1] Smith's *Dict. of the Bible*, Art. *Marriage*.
[2] Joshua xv. 17. Moses, according to M. Chipault (*Etude sur les Mar. Cons.*, etc. p. 94), considered consanguineous marriages to be sterile, and he gives as his authority Leviticus xx. 20, 21: "And if a man shall lie with his uncle's wife, he hath uncovered his uncle's nakedness: they shall bear their sin; they shall die childless." "And if a man shall take his brother's wife, it is an unclean thing: he hath uncovered his brother's nakedness; they shall be childless." He does not see that the second verse completely refutes the meaning he might possibly force on the first. A sister-in-law is no blood-relation; moreover Moses absolutely enjoins a brother on certain occasions to marry his sister-in-law (Deuteronomy xxv. 5). The fact is, that Moses pronounced the sentence of sterility as a curse; and what could be a more appropriate curse on a forbidden marriage? No severer could have been uttered in that state of society which produced the levirate law; where a man is honoured according to the number of his children, and where sterility is almost universally counted as a reproach. It is perhaps on the same authority that Pope Gregory I. said of consanguineous marriage, "Ex tali conjugio sobolem non posse succrescere." See Chipault, *ut. sup.*
[3] Ezek. xxii. 11: "And one hath committed abomination with

oriental poet, yet it appears that marriage with a half-sister was still permitted: for Tamar says to Amnon, "Speak unto the king; for he will not withhold me from thee," although perhaps it was not generally allowed; because she says before, "no such thing ought to be done in Israel."[1] The commentators explain this by Deuteronomy xxi. 10–14, where directions are given as to foreign women taken in war,[2] by which their children before naturalization are considered no relations to any children they may have afterwards. Whether Tamar was thus no relation to Amnon in the eye of the law is very doubtful; at all events she seems to have considered herself related to her brothers; and Absalom, since he took upon himself to avenge her, seems to have thought so too.

Another instance of the neglect of the prohibitions of Moses, is the conduct of Absalom when he deposed his father and took his hareem; no more than was customary perhaps in oriental nations, but distinctly

his neighbour's wife; and another hath lewdly defiled his daughter-in-law; and another in thee hath humbled his sister, his father's daughter."

[1] 2 Samuel xiii. 12, 13.

[2] The Jews were forbidden to marry a foreign wife. See Exodus xxxiv. 15, 16; Deuteronomy vii. 3; xxv. 5; Ezra x. 2; Nehemiah xiii. 27. Only the tribe of Benjamin was permitted, on an exceptional occasion, when they had been nearly exterminated (See Judges xxi). Soldiers, however, might at once prostitute a woman taken in war, after which she was entitled to a month's mourning, and then he might marry her after naturalization (Deuteronomy xxi. 14). Now, say the commentators, Tamar's mother, Maachar, was a prisoner of war, Tamar was the offspring of the first intercourse, before her mother was naturalized or had become a wife of David; hence she was nobody's child, no relation to Amnon, and even no relation to her uterine brother Absalom. See Jurieu, *Critical Hist. of the Doct., etc., of the Church:* Trans. 1705, vol. i. pp. 208, 209.

contrary to the law which forbids marriage with a step-mother. Moses curses with the curse of sterility those marriages which were against his law;[1] a terrible curse in a nation who considered he was blessed most whose marriage was the most fruitful.[2] Hence he provided that a marriage which proved childless should not result in the extinction of the family, but the nearest relative should marry a childless widow, and her children should be considered to belong to the first marriage.[3] For the same reason Sarah gives Hagar to Abraham,[4] and Leah and Rachel are full of jealousy of each other.[5] The marriage of two sisters at the same time to the same man, was forbidden, to prevent jealousy,[6] but with this exception, the greatest care

[1] Leviticus xx. 20, 21.
[2] See preceding note, p. 30 of this work; and Genesis xxx. 23 ; Deuteronomy vii. 14 ; 1 Samuel i. 6 ; 2 Samuel vi. 20, 23 ; Psalm cxxvii. 3, 5 ; Hosea ix. 14.
[3] See Deuteronomy xxv. 5. It seems that since a man is forbidden to take his brother's wife (Leviticus xviii. 16 ; xx. 21), some persons doubt whether the Jews were allowed to marry their deceased wife's sister ; though such a marriage is nowhere expressly forbidden, excepting during the lifetime of the first wife (See Leviticus xviii. 18). All the Rabbim, however, from the Mishna to the Shulcan Aruch, Eben Ezer ; and all the Jewish commentators from Philo down to Zunz, agree that though marriage with a woman after the divorce of her sister is prohibited, marriage with a deceased wife's sister is certainly permitted. Such marriages frequently take place among the modern Jews, and are considered most proper and laudable (*Evidence of Dr. Adler, before the Royal Commission on Marriage Law of* 1848, Appendix, Art. No. 35, p. 152). The Kairites, a far inferior sect to the Talmudists, do, however, consider these marriages unlawful (See *Ibid., Evidence of Dr. Pusey,* Questions 440, 441). How they reconcile this idea with the levirate law, I am at a loss to see ; at any rate it does not behove an English judge to take this latter view because it suits his prejudices, in the face of such overwhelming evidence on the other side (See *Ibid.,* Appendix, Art. No. 7, p. 129).
[4] Genesis xvi. 1, 2.   [5] Genesis xxx. 8, 9.
[6] Leviticus xviii. 18.

was taken that no family should die out. Thus Tamar should have had Shelah, as her only brother-in-law left; but Judah either did not wish the marriage to take place, since she had already been married to two of his sons, or was neglectful of the law. Impatient of this denial of her rights she took measures for making Judah himself the father of a son to her family; yet Judah confessed she was more righteous than he.[1]

An heiress, as in Greece, was obliged to marry a relation on the father's side. This law, it is said,[2] was enacted on the death of Zelophehad. He had five daughters and no sons, and they, on his death, claimed their inheritance in the tribe of Manasseh, which was allowed, on condition that they married "their father's brother's sons."[3] So Ruth claimed Boaz as her kinsman, and therefore as bound to marry her; but he told her there was another more nearly related, who had the priority.[4] Again, we read that the sons of Mahli were Eleazar and Kish, "and Eleazar died, and had no sons, but daughters: and their brethren the sons of Kish took them."[5] St. Joseph and the Virgin Mary are reputed to have been first-cousins; thus:—

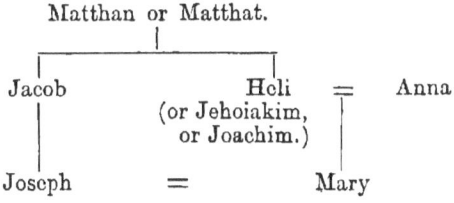

[1] Genesis xxxviii. 26.
[2] Smith's *Dict. of the Bible*, Art. *Zelophehad.*
[3] Numbers xxvii. 1-11; xxxvi. 1-12.
[4] Ruth iii. 12, 13.   [5] 1 Chronicles xxiii. 21, 22.

It is extremely doubtful whether Mary was the daughter of Heli or of Jacob. If the genealogy given by St. Luke refers to Mary, her father was Heli; if again Joseph was the son of the younger brother, and married his cousin, the daughter of his elder brother, she was the daughter of Jacob.[1]

The Herods, Greeks in their sympathies, and Jews only by policy, seem to have regarded no Jewish prohibitions. In the annexed genealogical table it will be seen that they took their brother's wives in the life time of the brother, and even though there were children of the first marriage. The table is otherwise instructive; for they frequently married their nieces, and still more frequently their first-cousins, while they laboured under no " visible curse," as Niebuhr says the Ptolemies did, for doing the same thing.[2]

Passing from the East to that great Western power, to which all those nations ultimately bowed their necks, we find the law concerning the marriage of near kin varied much at different periods. It will be necessary, however, to consider the character of relationship among the Romans, and how it was affected by the marriage tie, before we can go on to the prohibited degrees.

The *paterfamilias* was the grand centre around which all relationship was grouped. His power over his children was as the power of a master over his slaves; so long as he was alive, and did not choose

---

[1] She was connected by marriage with the tribe of Levi; hence it has been argued that she was not Joseph's cousin; but apparently on no sufficient grounds (See Smith's *Dict. of the Bible*, Art. *Mary the Virgin*).

[2] See above, p. 11, of this work.

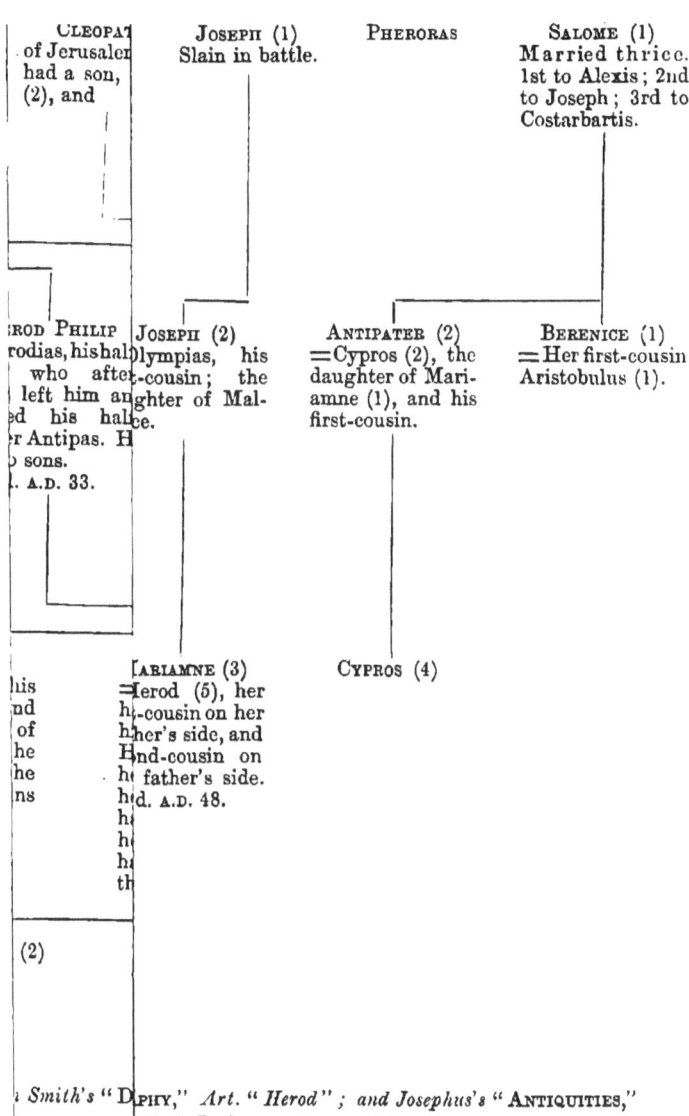

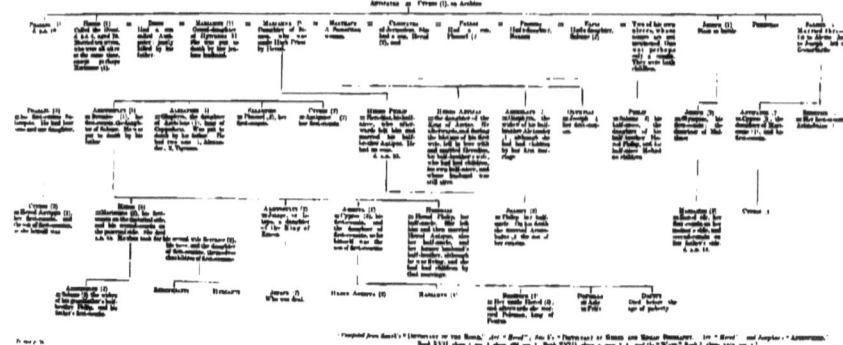

to manumit, or free them, so long were they absolutely subject to his commands. Marriage without manumission did not free any child from allegiance to his father; hence a husband was not even master over his own wife, unless part of the marriage ceremony had been manumission. As the character of this relationship was so essentially dependent on the father, an emancipated son or daughter was, until the time of Justinian, no longer any relation to him. The one had become the head of a new family; the other, by marriage after emancipation had entered into another family; and neither would be entitled to any inheritance from their father; while an unemancipated son, or a daughter married *sine conventione in manum*, since they still belonged to their father's family, would both be entitled to inherit after his death. Should there be no unemancipated children, or adopted child, the inheritance passed to the *consanguinei*, or brothers and sisters of the *paterfamilias*, to the exclusion of the female side; and should there be no *consanguinei* the inheritance went to the *agnati*, or male descendants of a common ancestor.

Marriage could be contracted in several ways, according to the position the wife was to occupy in her husband's house. A marriage *sine conventione in manum* did not free the wife from her father's control, or give her into her husband's power. She was still in the position of servitude to her father, still belonged to his family, and her property was not made over to her husband. On the contrary, she could even lend him money at interest, and send her slave to demand repayment, just as if he were a perfect stranger; while he, on his side, could bring an action against his wife for the recovery of any jewels he might have lent her. Indeed, so separate was the property of man and wife married *sine conventione*, that a law was passed which forbad either to give anything to the other, to the end that the richer of the two should be protected against the constant worry of the poorer.[1] A woman who contracted a marriage *cum conventione* went over to her husband's family, and was entirely in her husband's power. She still occupied the position she had formerly occupied in the eye of the law, but she had changed her allegiance. Though she was styled *materfamilias*, she occupied the same position as her own children, for no one could divide the authority with a *paterfamilias*. In both this and the former kind of marriage the husband was *paterfamilias* to, and hence absolute master over his own offspring; they only differed from each other in respect to the relation of the wife to her husband and her family.

Consent or cohabitation alone was sufficient to establish a marriage *cum conventione*, just as in Scotch marriages at the present time; but if the wife did not wish to enter into her husband's family, she could break the *usus* by absenting herself for three days in every year. This *usus*, which converted a marriage *sine conventione* into a marriage *cum conventione*, was, however, abandoned at a later period of their legal history.[2] There was, in addition to these forms, a sort of morganatic marriage, known as *concubinatus*, in which the parties were not legally married, and the children belonged to the mother. Neither mother nor children

---

[1] Laboulaye, *Cond. des Femmes*, pp. 17, 23, 26, 29, 33, 35, 37; and Smith, *Dict. of Greek and Roman Antiquities*, Art. *Matrimonium*.
[2] Smith, *ut sup*.

had any claim on the father, but in contracting such a marriage, a man was not allowed to transgress the prohibited degrees; nor, after he had contracted a marriage of this sort, could he enter into another of the same kind during the existence of the first. If he did so, the connection was called *stuprum*, and the offspring could not subsequently be made legitimate, as was allowed by Justinian in other cases.[1]

Like other proud races, the Romans were unfavourable to intermarriage with foreigners. In the earlier period there was no *connubium*, or permission to contract a lawful marriage, even between a patrician and a plebeian, while after the *Lex Canuleia*, which abolished this prohibition, there was yet no *connubium* between Romans and Latini and Peregrini, up to the time of Justinian; or between Romans and freedmen until the *Lex Julia et Papia Poppœa*, after which only senators were forbidden such marriages, while freeborn citizens were allowed to intermarry with the freedmen class, provided that the latter were not women of bad character, an exception which Constantine extended to women of the lowest classes.[2]

Concerning the prohibited degrees, " the elder the times were," says Jeremy Taylor, " the more liberty there was of marrying their Kindred."[3] The marriages of the Tarquins supply us with an early instance of the legality of marriages with nieces. Thus:

[1] Smith, *ut sup.*, Art. *Concubina*; and Sandar's *Inst. of Just.*, pp. 111, 113.
[2] Smith, *ut sup.*, Arts. *Matrimonium* and *Leges Juliæ*; and Sandar's *Inst. of Just.*, *ut sup.*
[3] Jeremy Taylor, *Duct. Dubit.*, book II. chap. ii. rule iii. sec. 59.

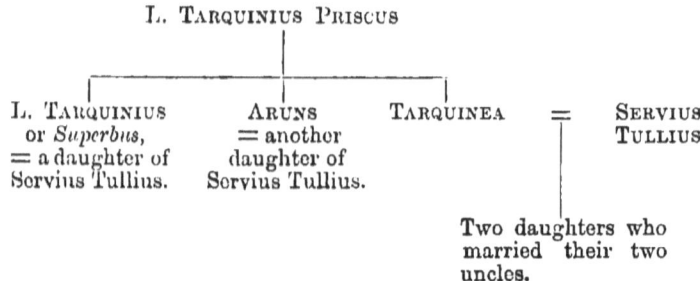

It is not clear whether the daughters of Servius Tullius were also the daughters of his wife Tarquinea, but he only appears to have been married once, and they therefore must have been. Livy is uncertain whether Tarquin and Aruns were sons or nephews of Priscus, and Dion Halicarnassus considers that they were nephews.[1] The marriages were therefore either between uncle and niece, or first-cousins, but most probably the former, or it would never have been mentioned as between the nearer relations. Plutarch says it was a practice before it became lawful for the Romans to marry their kindred.[2] This law was said to have been passed to enable a popular favourite to contract a very advantageous marriage,[3] but this is probably a fable. It seems rather, that as among the Greeks, an heiress was bound to marry her kinsman.[4]

[1] Jeremy Taylor, *Duct. Dubit.*, book II. chap. ii. rule iii. sec. 59; Smith, *Dict. of Greek and Roman Biography*, Arts. *L. Tarquinius Priscus* and *Servius Tullius.*

[2] It is possible that the people by whom they were surrounded paid no attention to prohibited degrees before they were subjected to Roman influence. John Taylor (*Elements of Civil Law*, p. 340) says the ancient Tuscans held wives in common.

[3] Jeremy Taylor, *ut sup.*; and Faber, *Viciss. Juris. Rom. de Incest. Nupt.*, p. 17.

[4] Jeremy Taylor, *Duct. Dubit.* book II. chap. ii. rule iii. sec. 53, says: "One *Cassia* was declar'd *inheritrix* upon condition, *Si Consobrino nupsisset.*" He refers to l. 2, *C. de instit. et subst.*; and Papinian,

It also follows from the constitution of the Roman family, for a woman must always be in a state of tutelage; hence, if she married out of her father's family, she would not be entitled after her father's death to any inheritance. John Taylor points out that there was an extra facility for a son to marry his first-cousin; for though his grandfather's consent without that of his own unemancipated father was not sufficient if he wished to marry out of his family, yet he needed not the consent of his father if he wished to marry his cousin on the paternal side.[1] There is no doubt that the marriage of first-cousins, at any rate, was always permitted by Roman law. In Livy's oration of Sp. Ligustinus there is this clause: "My father gave me to wife his own brother's daughter;"[2] and Cicero says that Cluentius' sister married her first-cousin, Aulus Aurius Melinus, a marriage which subsisted with all respectability and concord.[3] Quintilian, mourning for the death of his son, says he had intended to marry him to his brother's daughter.[4] M. J. Brutus married the daughter of his uncle Cato, a second marriage for both parties.[5] Augustus Cæsar gave his daughter Julia to M. C. Marcellus, the son of his sister Octavia,[6] a remarkable

"Conditionem illam, si consobrinam duxeris, hæreditatis institutioni utiliter adjici posse," l. 23 et 24, *D. de ritu nuptiarum.*
[1] John Taylor, *Elements of Civil Law*, p. 313.
[2] Livy, xlii. 34, cited by Jeremy Taylor, *ut. sup.*, sec 59. See also case No. 97, Appendix of this work.
[3] Cicero, *Pro Cluentio*, cap. v., cited by Jeremy Taylor, *ut sup.*
[4] Quint., *Annal*, l. 12, cited by Jeremy Taylor, *ut sup.*
[5] Smith, *Dict. of Greek and Roman Biography*, Arts. *Cato Uticenses* and *Porcia*.
[6] See Jeremy Taylor, *ut sup.* This marriage took place B.C. 25, and Marcellus died B.C. 23. Smith, *ut sup.*, Art. *Marcellus.*

marriage, because besides showing that marriages with first-cousins were permitted, Marcellus was adopted by Augustus also as a son. M. Aurelius married his first-cousin Faustina, the daughter of Antoninus Pius. Constantia, the daughter of Constantinus, was married twice, and each time to a first-cousin. His son also married a first-cousin, as may be seen from the accompanying genealogical table.

Claudius, the successor of Caligula, is the first who set the example of, and made it lawful to marry a niece, if we except the somewhat doubtful case of the marriage of the Tarquins. He married Agrippina, the daughter of his brother; and Tacitus makes him say, "It is a new thing among us to marry a brother's daughter, but it is customary among other peoples, nor is it forbidden by any law." The sentiment of the Romans was against such marriages, and it was not, therefore, carried further, or generally accepted at the time.[1] Suetonius notices only an obscure libertine who followed the example. Domitian forbad it; Nerva also tried to repeal the permission, but in vain. In the time of Gaius it was still unlawful to marry a niece, the daughter of a sister; but it was permitted by the Rules of Ulpian, and made lawful in the Civil Law.[2] Constantius and Constantine forbad the mar-

[1] "Nullo exemplo deductæ in domum patrui fratris filiæ .... Nova nobis in fratrum filias conjugia: sed aliis gentibus solennia, nec lege ulla prohibita" (Tacit., lib. xii. 5, 6). "Subornavit proximo senatu qui consererent, cogendum se ad ducendum eam uxorem .... dandamque ceteris veniam talium conjugiorum" (Sueton., lib. v. 26). Both are referred to by Jeremy Taylor, book II. chap. ii. rule iii. sec. 31.

[2] See Jeremy Taylor, ut sup., sec. 31; Smith, Dict. of Greek and Roman Antiquities, Art. Matrimonium; Royal Com. on the Laws of Marriage, 1848, p. 39, note. M. Chipault (Études sur le Danger des Mar. Cons., p. 91), and M. Mantegazza (Studj sui Matrim. Cons., p. 8), who

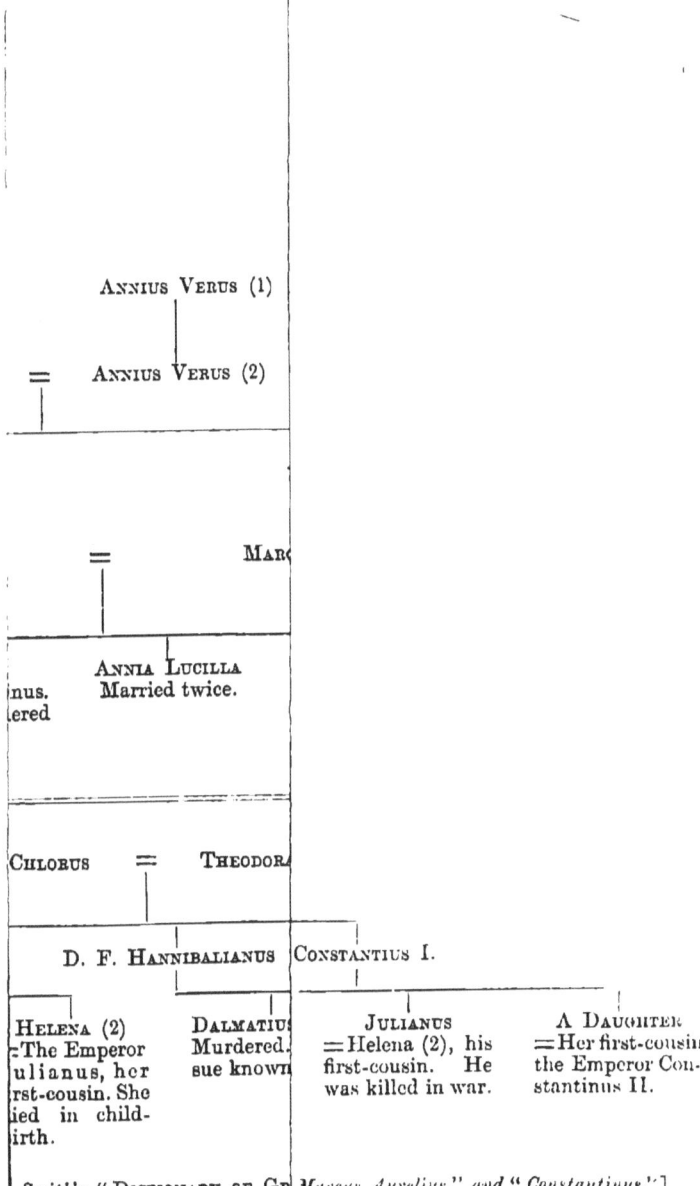

[Smith's "DICTIONARY OF GR*Marcus Aurelius*," and "*Constantinus*."]

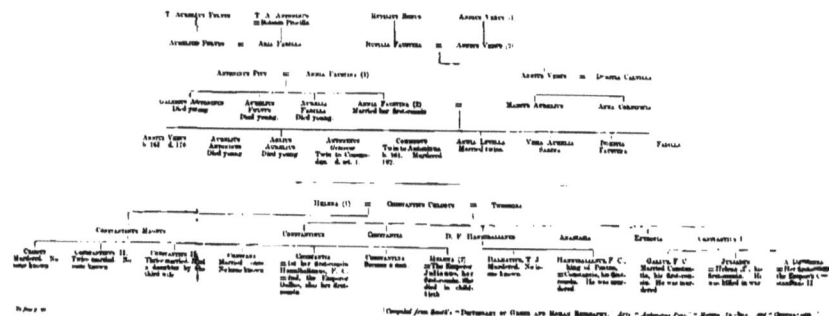

riage of uncle and niece under a penalty of death;[1] a prohibition which also appears in the *Institutes* of Justinian.[2] But Jeremy Taylor asserts that this prohibition was practically a dead letter until the time of Heraclius, when it was repealed, and marriage with a brother's daughter again made lawful.[3]

Marriage was never permitted between nearer relatives than uncle and niece.[4] There was no *connubium*, or faculty of marriage, in the direct ascending or descending line to the remotest degrees, whether the relationship was legal, natural, or adoptive, nor did the severance of the adoption make it possible. Brothers and sisters of whole or half-blood were not allowed to marry. Adoption, when it simulated this relationship, was also a bar to marriage so long as it existed or both parties were by adoption brought into the same *familia*. Hence an emancipated son could marry his sister by adoption; and it appears that a

copies him here, say that Antoninus Pius married his niece. I can find no confirmation of this: on the contrary, if the annexed genealogical table is correct, the relationship is impossible.

[1] Dr. Pusey, *ut sup.*, who refers to the *Cod. Theod.* l. iii. *tit.* 12.
[2] Sandar's *Inst. of Just.*, p. 107.
[3] Jeremy Taylor, *Duct. Dubit., ut sup.*, sec. 31.
[4] "The profane lawgivers of Rome," says Gibbon, "were never tempted by interest or superstition to multiply the forbidden degrees; but they inflexibly condemned the marriages of sisters and brothers, hesitated whether first-cousins should be touched by the same interdict; revered the parental character of aunts and uncles, and treated affinity and adoption as a just imitation of the ties of blood." (*Decl. and Fall of the Rom. Emp.*, vol. v. p. 399.) Gibbon has been apparently led away by his rhetoric in this passage. We have seen that the legality of the marriage of first-cousins was from time immemorial undisputed; that from the time of Claudius, so far from revering the parental character of uncles, marriage with a brother's daughter was lawful, and soon afterwards became so customary that prohibitions launched against the practice were laughed at. It was only with the spread of Christianity and asceticism that the prohibited degrees were again enlarged.

son-in-law might be adopted by his father-in-law, for Marcellus was thus adopted by Augustus, Claudius adopted his son-in-law Nero, and Marcus Aurelius married his adopted father's daughter Faustina. Marriage with a nurse, a wife's daughter, son's wife, wife's mother, father's wife, or the daughter of a divorced wife by a second husband, were all forbidden; but the son of a man by his former wife might marry the daughter of his father's second wife by a former husband.[1] Constantine forbad marriage with a sister-in-law, but it was permitted up to his time. The prohibition was renewed by Valentinian, Theodosius, and Arcadius; but Justinian does not seem to have noticed it.[2]

No slave could lawfully marry in the Roman Empire. All his unions, even those with a free person or freedman, were called *contubernium*, a word used for all unlawful unions, and hence for incest.[3] Before the *Lex Papia Poppæa* they were not allowed to marry a Roman citizen even if they had been freed; while after it they were still forbidden marriage with senators.[4]

The rise of Christianity had a lamentable effect in one way on marriage, while in another it was decidedly beneficial. On the one side it prohibited marriage within degrees ridiculously distant; on the other it doubtlessly raised the whole tone of morality, and with it the worthiness of the marriage state. At first the Church could certainly not interfere

[1] Smith, *Dict. of Antiquities*, Arts. *Matrimonium* and *Incestum*; Sandar's *Inst. of Just.*, pp. 106-110.
[2] See Sandar's *Inst. of Just.*, p. 109.
[3] John Taylor, *Elements of Civil Law*, p. 287; and Smith, *Dict. of Antiquities*, Art. *Concubina*.   [4] Sandar's *Inst. of Just.*, p. 110.

much, having enough to do to keep itself from extermination; but it soon directed its attention to marriage, and from its bias towards asceticism so bepraised the cœlibate state and threw such difficulties in the way of marriage, that it laid the foundation of its own overthrow.

The Emperor Theodosius I. (A.D. 379–395) was the first to bring in the iniquitous law that no man might marry his first-cousin, under penalty of death by burning, and confiscation of his property. St. Augustine seems to think that even before this law of Theodosius the marriage of cousins was discouraged by the Church, and that in consequence they were but of rare occurrence. St. Ambrose takes for granted that they were forbidden, though on what authority he grounded this theory no one knows. He reasons that since marriage with a first-cousin is forbidden, much more, therefore, is a marriage with a niece. The fact that some infatuated individual (probably at the instigation of the devil) actually dared to ask this redoubted saint, whether it was permitted to marry a niece, since such marriages were nowhere prohibited in the Scriptures, seems to indicate that before this enactment of Theodosius there was no settled law on the subject, but only a general feeling that it was unseemly to marry anyone near akin. It is supposed that Theodosius was acting under the advice of St. Ambrose, and in spite of the opinions of St Athanasius and St. Augustine, when he made this law; for these both declared that marriages between first-cousins were neither against the law of God nor of man.[1] Arcadius (A.D. 395–408),

[1] Πρόσταγμα τοῦ Κυρίου καὶ νόμον, νόμιμον εἶναι γάμον τὴν πρὸς ἀνεψιούς

and Honorius (A.D. 395–423), his sons who succeeded him, the former in the Eastern Empire, the latter in the West, confirmed their father's law, but mitigated the penalties. Arcadius soon afterwards permitted these marriages again, but his brother, however, still persisted in the prohibition, though he reserved to himself the power of giving dispensations. Without a dispensation the offspring were considered bastards, and the property of the parents was confiscated. Justinian (A.D. 527–565), confirmed the freedom granted to cousins by Arcadius to marry; and also, as we have seen, confirmed the previous decretals which prohibited marriage with a niece. But even this prohibition was repealed by Heraclius (A.D. 610–641).

In the West the Ostrogoth kings (A.D. 493–553) not only endured these prohibitions, but even added to them occasionally, until marriage was forbidden in the fifth and even sixth degrees.[1] From the

συζυγίαν. And St. Augustine, though he does not approve of these marriages for other reasons, says: "Nuptias has nec Divinam Legem et nondum humanam prohibuisse" (Cited by Samuel Dugard, *The Mar. of Cousin-Germans Vindicated*, pp. 87, 88).

[1] By civil law, second-cousins are related in the sixth degree; since every generation is counted between the two relatives, upwards from the one to the first common ancestor, and then downwards to the other. By canon law, only one line, and that the longest, is counted; from the first generation to the person in question. Thus: *K* is related in the fifth degree to *C*, *E*, *G*, *J*, and *L*, but in the sixth degree to *N*. The Romans counted relationships in the same way as we do now; that is according to the civil method. (See Sandar's *Inst. of Just.*, p. 107.) Pope Gregory I. (A.D. 590–604), according to M. Chipault, was the first to compute relationship by canon law. (See *Études sur les Mar. Cons.*, p. 93). By this means the prohibited degrees were made insidiously

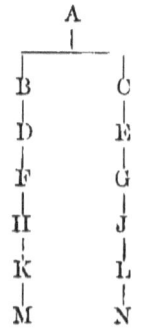

general isolation of the different parts of Europe the prohibitions were not severe everywhere alike. At a Council held at Epaune, A.D. 517, under Sigismund, King of Burgundy, marriage was prohibited in the fourth civil degree, and this canon was confirmed by the Councils of Clermont, A.D. 535; of Orleans, A.D. 538; and of Tours, A.D. 567. In the canons of these Councils, as now extant, the sixth degree is also forbidden; but Dr. Pusey observes that it has been omitted from the last edition of decretals, which has been, however, badly edited. At the Council of Auxerre, A.D. 578, marriage in the sixth civil degree, or between second-cousins, was forbidden.

A small Spanish Council which sat at Toledo, A.D. 531, went so far as to prohibit marriage wherever any relationship at all could be traced between the parties, whether by blood or affinity. But the Court of Rome does not seem to have adopted so impossible a regulation, until the reign of Pope Zacharia, A.D. 741. St. Gregory is said to have been the first to compute relationship by *generations* from a common parent, instead of by the rational method of counting the degrees between the parties in question, as was customary before him. He probably fixed the fourth canonical degree as the limit within which marriage was forbidden, or marriage between third-cousins was not permitted.[1] Pope Gregory III. (A.D. 731)

wider, just as a modern representative sought to increase his means by debasing his coin.

[1] St. Gregory has been accused of forbidding marriage within the seventh degree, an accusation founded on an epistle from, and answer to, St. Felix, Bishop of Syracuse. This is probably, as Dr. Pusey shows, a mistake, for these epistles are now considered spurious.

prohibited marriage within the seventh canonical degree, or between sixth-cousins, equivalent to the fourteenth civil degree; and Pope Zacharia (A.D. 741) informed Pepin of France, that marriage was prohibited wherever any relationship could be traced. Yet up to the ninth century, the prohibited degrees seem to have been still unsettled; for by the first Council of Mentz, A.D. 813, marriage was *thenceforth* prohibited within the sixth degree; in A.D. 847, the Bishop of Mentz recommends that marriage should be allowed in the fifth degree; but the Council held at Worms, A.D. 868, again confirmed the enactment of the Council of Toledo, that marriage should not be permitted wherever any relationship was traceable. Again, at a Council held at Douzy, A.D. 874, marriage was only prohibited within the seventh degree; and Pope Nicholas II. and a Roman Council, A.D. 1059, also prohibited marriage within the seventh degree, an enactment which was again confirmed by Pope Alexander II. (A.D. 1062–1073).

All these enactments make it evident that the people did not quietly acquiesce in these prohibitions. Indeed, even though dispensations might be bought, it was impossible to uphold so unnatural a law. At that time locomotion was not so easy as it is now, and even now a villager would find it difficult to contract a marriage outside the fourteenth civil degree. By the fourth Lateran Council,[1] marriage was again permitted outside the fourth canonical degree, that is, marriage was permitted beyond the degree of

(See *Royal Com. on Marriage Law*, 1848; *Evidence of Dr. Pusey*, p. 42, note.)

[1] Nov. 1215, the twelfth General Council.

third-cousins; and such is the nominal law at the present time wherever the canon law prevails.[1]

Though never so severe as the Roman, the prohibitions of the Greek Church were much more inflexible, and therefore not nearly so demoralizing. They permitted the marriage of second-cousins, according to Theodorus, who wrote A.D. 668—about the same degree that was permitted by the Roman Church of that time. Like the Roman Church, however, they got more severe under the influence of asceticism, for by the Council of Trullo, A.D. 694, they forbad the marriage of relations within the seventh degree, that is, third-cousins were allowed to marry, but no parties more nearly related; and, like the Roman Church, they held that relationship by affinity, adoption, or god-parentage, were equally bars to intermarriage with true relationship. At this time the Roman Church was beginning to launch out upon far wider

[1] The civil law permits marriage in the fourth degree, or between first-cousins; the canon law only permits marriage outside the fourth degree, that is, third-cousins are forbidden to marry. Hence, according to Taylor (*Elements of Civil Law*, p. 331), has arisen among the common people an idea that though first-cousins may marry, second-cousins may not; a belief which Samuel Dugard, writing in 1673, mentions as a common belief, and which appears even nowadays hardly to have died out (See Dugard's *Mar. of Cousin-Germans Vindicated*, pp. 91-94, and Mrs. Beeton's *Book of Household Management*, 1869, p. 1134, § 2856)—a curious instance of the extent to which our common sense may be misled by the influence of custom. For the history of these prohibitions I am chiefly indebted to the *Evidence of Dr. Pusey, before the Royal Com. on Marriage Law of* 1848, pp. 39-44; Sandar's *Inst. of Justinian*, p. 107; Buckle's *Common Place Book*, Art. No. 11; John Taylor's *Elements of Civil Law*, pp. 330-337; Chipault's *Etudes sur les Mar. Cons.*, pp. 92-94; Reich's *Geschichte, Natur- und Gesundheitslehre des Ehelichen Lebens*, pp. 72, 93, 94; Hallam's *Middle Ages*, vol. i. p. 512; the *Encycl. Méthodique*, Arts. *Empêchements du Mar.;* the *Algemeine Encycl. der Wissenschaften und Künste;* and Pierrot's *Univ. Lex.*, Art. *Ehe.*

prohibitions, but the Greeks remained fast to theirs, uninfluenced by the extravagance of the sister Church. The consequence was that, though the people occasionally disregarded them, as is shown by the excommunications of John the Good, A.D. 1079, on all those who had married their cousins; yet, since these prohibitions were neither so severe as the Roman, nor so often altered or permitted by dispensations, the people began to look upon them as Divine laws, and, as such, to wish for no change.[1]

The Russian practice at the present time is the same as that of the Greek Church. There was, however, and perhaps still is, a practice in some parts of Russia for a son's wife first to be married temporarily to his father.[2] M'Lennan says, that of the Muscovites and Livonians we have no certain information on their marriage law; but that it is implied by Magnus, that husband and wife invariably belonged to different kinships or village communities.[3] Quite lately a law has been brought in legalizing civil marriages between dissenters by birth, in Russia; but even here the Greek prohibitions are in force.[4]

In Sweden, marriages within the third civil degree are forbidden. A man may marry his sister-in-law by special permission of the king; and a few years

[1] See the *Evidence of Dr. Pusey*, Royal Com. on *Marriage Law*, 1848, p. 41; also the *Evidence of the Rev. W. Palmer*, pp. 54-57, *Ibid.*, and Reich's *Ehe*, etc., p. 98.

[2] Taylor's *Biblical Fragments*, appended to Calmet's *Dict. of the Bible*, vol. iii. p. 196: London, 1847. A custom analogous to that of the Indians of Darien. See below, and Reich's *Ehe*, etc., p. 408.

[3] M'Lennan, *Primitive Marriage*, p. 131.

[4] See the *Pall Mall Gazette* for Sept. 11th, 1873; and *The Times* for Oct. 5th and Nov. 6th, 1874.

ago, says Mayhew, writing about the year 1862, a law was abolished requiring the same formality for the marriage of first-cousins.[1]

The ancient Danes were probably in the habit of marrying their sisters.[2] At present, persons related in the direct ascending and descending line, or brother and sister in the collateral line, are forbidden to intermarry. Aunt and nephew may only marry by special permission from the king, but uncle and niece are always allowed to marry. There is, besides, no prohibitory relationship by affinity.[3]

The ancient Icelanders were no more particular than the ancient Danes; for in the story of the Volsungs and Niblungs, Signy, by a stratagem, makes her brother Sigmund father of her son Sinfjotli, and exults that he is both the son of Volsung's son and Volsung's daughter.[4]

The Anglo-Saxons, says Lingard, did not hesitate to marry a father's wife.[5] By the code of Theodore, Archbishop of Canterbury, A.D. 668–690, marriage is forbidden with a mother, daughter, sister, niece, great-grand-daughter, great-great-grand-daughter, step-mother, step-daughter, mother-in-law, sister-in-law, god-mother, nurse, the wife or mistress of any relation, two sisters in succession, a maid betrothed to another, or, indeed, any relative within the sixth canonical degree.[6] According, however, to the

---

[1] Mayhew's *London Labour and London Poor*, vol. ii. p. 175.
[2] Boudin, *Mém. de la Soc. d'Anthrop. de Paris*, vol. i. p. 552, 1863.
[3] Mayhew's *London Labour and London Poor*, vol. ii. p. 179.
[4] Translation by Magnússon and Morris, pp. 17, 18, 25.
[5] Lingard, *Antiquities of the Anglo-Saxon Church*, vol. ii. pp. 6, 7. See also Sharon Turner, *History of the Anglo-Saxons*, vol. i. p. 125.
[6] *Liber Pæn. de Incest.* xx., in Thorpe's *Early Inst.*, pp. 286-287.

Capitula et Fragmenta Theodori, marriage is only forbidden in the third degree.[1] This discrepancy is probably to be explained by the supposed permission of Pope Gregory I. to the English to marry in the fourth degree; and the canon of Ecgbert, Archbishop of York, which says, that though the canon law puts the limit at the sixth degree, for the very satisfactory reason that the Creation lasted six days, yet St. Gregory permitted the English, seeing that they were as yet but neophytes unhabituated to the curb, to marry in the fourth degree, so that they might be gradually habituated to greater severity; just as the apostle says: "I have fed you with milk and not with meat, for hitherto ye were not able to bear it, neither are ye yet able." In the same way Pope Innocent III. permitted the Livonians to continue the levirate law, as formerly practised before their conversion.[2] On this account Ecgbert allows marriage in the fifth degree; in the fourth the parties need not be separated; but marriage within the third degree is void, and the parties must be separated.[3] The English were, however, considered sufficiently stedfast in their faith at the beginning of the twelfth century; for in the year 1102, at a great council convened at Westminster by Anselm, marriage was prohibited within the seventh canonical degree;[4] and this prohibition remained in force till about the date of the Magna Charta, when

[1] Thorpe, *Early Inst.*, p. 317.
[2] *Evidence of Dr. Pusey before the Royal Com. on Marriage Law*, 1848, p. 45.
[3] *Pœn. Ecgberti*, lib. iv. 39; *Confess.*, p. 354; *Excerp.*, cxxxi. cxxxii. cxxxviii. cxl. cxlvi. in Thorpe's *Early Inst.*
[4] Buckle's *Common Place Book*, Art. 110.

marriage outside the fourth canonical degree was permitted.[1]

By the 25th Henry VIII. cap. 22, sec. 3, the prohibited degrees were declared to be the same as those mentioned in Leviticus, including a wife's sister; and by this Act it was declared that neither the Pope nor any one else could grant a dispensation to marry within any one of these degrees. Liberty was, however, left to the ecclesiastical courts to prohibit other marriages outside these degrees, or permit them as they chose. The 28th Henry VIII. cap. 7, repealed this Act, but the same prohibited degrees were re-enacted, with the modification that the religious ceremony does not constitute the validity of the marriage; only the consummation does; that therefore a man may marry his step-mother, uncle's wife, son's wife, or his brother's wife, provided that none of these marriages had been consummated. The same Act confirmed the prohibitory relationship by affinity, if a marriage had been consummated, even if there had been no religious celebration. The power of the ecclesiastical courts to enact further prohibited degrees was not touched by this Act; and the power of dispensation was left in the same state as enacted by the 25th Henry VIII. The liberty thus left in the hands of the Church to prohibit marriages outside the statutory degrees, or allow them, just as they thought fit, was abused, as it had always before been abused. In the preamble of the 32nd Henry VIII. it says: "Further also, by reason of other prohibitions than God's law admitteth, for their lucre by that court [the

[1] See p. 46 of this work.

Pope's] invented, the dispensations whereof they always reserved to themselves, as in kindred, or affinity between cousin-germans, and so to the fourth and fifth degrees, carnal knowledge of any of the same kin or affinity before in such outward degrees, which else were lawful, and be not prohibited by God's law. (4.) And all because they would get money by it, and keep a reputation of their usurped jurisdiction, whereby not only much discord between lawful married persons hath (contrary to God's ordinance) arisen, much debate and suit at law, with wrongful vexation, and great damage of the innocent party hath been procured, and many just marriages brought in doubt and danger of undoing, and also many times undone, and lawful heirs disherited, whereof they had never else, but for their vainglorious usurpation, been moved any such question, since freedom in them was given by God's law, which ought to be most sure and certain." The statute then declares[1] that "all and every such marriages as within this church of England shall be contracted between lawful persons (as by this act we declare all persons to be lawful that be not prohibited by God's law to marry). (7.) Such marriages being contract and solemnized in the face of the church * * * shall be, by the authority of this present parliament, aforesaid, deemed, judged, and taken to be lawful, good, just and indissolvable, * * * (8.) And that no reservation or prohibition, God's law except, shall trouble or impeach any marriage without the Levitical degrees. (9.) And that no person of what estate, degree or condition soever he or she be, shall,

[1] Cap 38.

after the said first day of the month of July aforesaid, be admitted in any of the spiritual courts within this the king's realm, or any of his grace's other lands, and dominions, to any process, plea, or allegation, contrary to this aforesaid act."

The law remained as defined by this Act for the remainder of Henry VIII.'s reign, and through that of King Edward VI., excepting that by the 2nd and 3rd Edward VI., cap. 23, the power of enforcing marriage upon sufficient proof of contract was restored to the spiritual courts, but "that in all other causes and other things therein mentioned, the said former Act of the 32nd year of the late king, do stand and remain in his full strength and power."

The first act of Mary, on her accession, was to legitimize herself, by declaring (1st Mary, Sess. 2, cap. 1), "the Queens Hygness to have been borne in a most just and lawful matrimonie," that the marriage of Henry VIII. with his deceased brother's wife was " according to the pleasure of Almighty God," that it was " a most lawful and Godly concorde," and "that the same marriage in every dede not being prohibited by the law of God, could not by any reason or equitie in this case be spotted, that it had its beginning of God, and by Him was continued, and therefore was ever and is to be taken for a most true, just, lawfull, and to all respects a sincere and parfitt marriage." The Act annuls the divorce of Katharine, the Act of 25th Henry VIII. cap. 22, "and every such clauses, articles, branches and matters contained and expressed in the foresayd Acte of Parliament made in the 28th year of the reign of the seyd late King, your father, or in any other Acte or Actes of Parliamente, as whereby your Hyguess is named or declared to be illegitimate." To this Mr. James Parker observes that since the marriage of a man with his brother's wife is expressly prohibited by Leviticus,[1] this Act was not itself consistent with "God's law," and that the marriage of Mary's parents alone was intended to be made lawful by this Act, because the marriage was under the Pope's dispensation, and the former marriage was said not to have been consummated. As for the first objection, this sort of marriage, as we have seen,[2] was not only allowed, but positively ordained when there were no children born, by the Old Testament, and permitted in other cases, provided two sisters were not married in succession, the latter in the lifetime of the former: an Act of Parliament would besides, even if this permission were contrary to the law as laid down in Leviticus, be amply sufficient to make such marriages legal. For the second objection, "it should be observed, that in this Act, not one word is said of a dispensation from the Holy See, nor of want of consummation of the marriage between Prince Arthur and Catherine."

The 1st and 2nd Philip and Mary, cap. 8, repealed so much of the 28th Henry VIII. cap. 7, as concerned the prohibition to marry within the degrees expressed in the said Act, as well as the whole of the 28th Henry VIII. cap. 16, and the 32nd Henry VIII., cap. 38, and others, which took away the ancient power of the Pope and spiritual courts over the law of marriage. At the end of Mary's reign, therefore, a man was prohibited marriage with a mother,

[1] Chap. xviii. 16. [2] See p. 32 of this work.

sister, grand-daughter, half-sister, or aunt; with a step-mother, daughter-in-law, or uncle's wife, if the marriage which constituted the relationship had been consummated. Marriage with a daughter or grandmother is not mentioned, but that they were unlawful is implied, as it is in the Old Testament,[1] whence these degrees are taken, and by the fact that marriages beyond these degrees are forbidden. For if a man may not marry his grand-daughter, neither may he marry his own daughter; nor if he is forbidden marriage in the second degree in the descending line, is it likely that he is permitted to marry in the same degree in the ascending line. Similarly, marriage with a niece was forbidden, because marriage with an aunt was.

These prohibited degrees were left as they stood by the 1st Elizabeth, cap. 1; for though the 28th Henry VIII. cap. 16, and 32nd Henry VIII. were revived, yet in neither of these Acts is there any definition of the prohibited degrees, nor were they interfered with even by the Act which declared Elizabeth's title to the crown.[2]

Archbishop Parker published a table, on his own authority, in the year 1563; which was of no legal value at the time, but received authority in the year 1603, from the 99th canon in ecclesiastical courts. This canon declared that " no person shall marry within the degrees prohibited by the laws of God, and expressed in a table set forth by authority in the year of our Lord 1563, and all marriages so made

---

[1] Leviticus xviii. 6-18; xx. 21; Deuteronomy xxv. 5, 6.
[2] Chief Justice Vaughan " evidently was not aware of the statute 1st Mary, Sess. 2, cap. 1 * * * as expressly repealing the 28th Henry VIII. cap. 7." Hence the authority relied upon by Mr. James Parker does not hold.

and conducted shall be adjudged incestuous and unlawful, and consequently shall be dissolved as void from the beginning, and the parties so married shall by course of law be separated. And the aforesaid table shall be in every church publicly set up and fixed at the charge of the parish." One of these tables, of the date of 1677-8, and signed by the Bishop of London, which I possess, states that on the authority of this canon, of the Roman law, and "by divers necessary consequences, from likeness, parity, or majority of reason," the marriage of a man with the wife of his brother—that marriage which is enjoined by the Bible, and most particularly permitted by the statutes of England—"to be incestuous; and all children begotten of such marriages to be illegitimate, or bastards, to all intents and purposes." Such does the canon law remain to this day; its only importance consisting in the astonishing amount of misery it has produced, and as an instance of the coolest and most unblushing effrontery.[1]

Heinneccius says it was customary among the ancient Germans to marry their sisters,[2] and perhaps mothers, or rather step-mothers, as appears from a little story cited by Gibbon: A princess of the Angles had been betrothed to Radiger, king of the Varni, a German tribe. The lover, however, from political motives, preferred his father's widow, the

[1] The reader should consult the *Report of the Royal Commission of* 1848, on the amount of misery and contempt of the law produced by this canon and the bad interpretation of the law on this subject. I am indebted for my information to the same Report, Appendix, Art. No. 7. Also to John Fry, *The Case of Marriage between near Kindred*, etc., 1773, pp. 121-146; Samuel Dugard, *Vindication of the Marriage of Cousin-Germans*, 1673, p. 90; and *The Case of Mr. Emmerton and Mrs. Bridget Hyde*, etc., 1682: London, printed for Richard Baldwin, pp. 31-43.

[2] Boudin, *Mém. de la Soc. d'Anthrop. de Paris*, vol. i. 1863, p. 552.

sister of Theodebert king of the Franks. The British heroine was not to be defrauded of her husband. She gathered a fleet and army, landed at the mouth of the Rhine, vanquished Radiger, dismissed her rival, and married her captive.[1] By the Prussian law marriage in the direct ascending or descending line, or between brothers and sisters, whether by half or by whole-blood, are forbidden, whether these relationships are legitimate or illegitimate. Step-parents, or step-brothers and sisters, are considered to be blood-relatives. Hence the marriage of uncle and niece is permitted, and there is no prohibitory relationship by affinity, except in the direct ascending and descending line. The Rhenish provinces are under the Code Napoleon, with the exception that a man is permitted to marry his sister-in-law, and that it is necessary to get a sort of formal permission from the Church to marry a niece. Such marriages, however, occur but rarely. They were freely allowed in Wurtemburg, but not in Hanover.[2]

The law of Holland adds to the Prussian prohibitions marriage between brothers and sisters-in-law, between an uncle or a grand-uncle, and a niece or grand-niece; and marriage between an aunt or grand-aunt, and nephew or grand-nephew; whether the parties are legitimately related or not. The king may, however, permit these marriages.[3]

In France marriages are prohibited in the direct ascending or descending line, as well as between

[1] Gibbon, *Decline and Fall of the Roman Empire*, vol. iv. p. 518.
[2] *Evidence of A. Bach, Esq., before the Royal Commission on Marriage Law of* 1848, pp. 84-86. By the new law of 1875, the prohibited degrees of all Germany are assimilated to the Prussian law.
[3] *Ibid.*, p. 86, note.

brothers and sisters, whether legitimate or illegitimate. Marriage between uncle and niece, aunt and nephew, or brother and deceased wife's sister, may be permitted by the Government. Adoption is no obstacle to marriage, but affinity by marriage is as prohibitory as the blood-relationship it simulates, and step-children are forbidden to marry a step-parent.[1]

Marriages are prohibited by the Italian Civil Code, in the direct ascending or descending line, and in the collateral line between brother and sister, aunt and nephew, or uncle and niece. Illegitimacy does not affect these prohibitions. Relationship by affinity or adoption in the direct ascending or descending line, or the first collateral degree, is considered to be true relationship, and marriage within those degrees is prohibited; but the king may permit marriage between relatives by affinity, or between aunt and nephew, or uncle and niece, if he chooses.[2]

By Spanish law, the old canonical prohibitions are still in force. All persons related in the ascending or descending line, and all persons of the collateral line related within the fourth canonical (or eighth civil) degree, are forbidden to intermarry. Relationship by affinity, by carnal knowledge, or by god-parentage, is considered to be the same as blood-relationship.[3]

[1] Adam, *Fortnightly Review*, 1865, p. 722; Devay, *Du Danger*, etc., pp. 84-87; *The Report of the Royal Commission on Marriage Law of* 1848, Appendix, No. 11, and Note, p. 135.
[2] *Codice Civile del Regno d'Italia*, lib. i. tit. v. 58, 59, 60, 68. Cardinal Wiseman asserted before the *Royal Commission on Marriage Law of* 1848, that marriages between cousins hardly ever occurred in Italy, since they are contrary to the canon law (see p. 107, Question 1194). This, however, does not seem to be the case, since Signor Mantogazza appears to have found no difficulty in collecting cases of these marriages (see his *Studj sui Matrim. Cons.*, pp. 20-26).
[3] Johnston's *Translation of the Spanish Civil Law*, pp. 50, 51, 128.

In Portugal the same canon law prevails, but in both countries dispensations are obtainable. Thus, in 1826, the two claimants of the Crown, Donna Maria da Gloria, and her uncle, Dom Miguel, proposed to settle their claims by marriage. A dispensation was obtained from the Pope, and it would have taken place, but for the alleged treachery of Dom Miguel.[1] The Vaquiros of the Asturian mountains habitually marry among themselves, and consequently have to buy dispensations for nearly every marriage.[2]

In the United States of America, the prohibited degrees vary somewhat in different States. In the State of New York there was formerly no statute defining these degrees. By the Revised Statutes of 1830, marriage in the direct line, or between brothers and sisters of half or whole-blood, is forbidden. In Louisiana, marriage is forbidden with an aunt or niece, and in Indiana and Kentucky the marriage of first-cousins even is said to be prohibited.[3] M. Devay asserts that these marriages are also forbidden in Ohio, but this has been shown by Mr. Darwin to be false;[4] there was merely an Act passed in the year 1855 or 1856, to ascertain the number of deaf-mutes, blind, insane, etc., and the relationship of their parents to each other if there was any.[5]

[1] Adam, *Fortnightly Review*, p. 722.
[2] Michel, *Hist. des Races Maudites*, vol. ii. p. 43.
[3] *The Lancet*, April 4th, 1868, p. 456; J. A. N. Perier, *Mém. de la Soc. d'Anthrop. de Paris*, vol. i. 1863, p. 230; Adam, *Fortnightly Review*, Nov. 1st, 1865, p. 723.
[4] Devay, *Du Danger*, etc., p. 141; Darwin, *Animals and Plants under Domestication*, vol. ii. p. 122, note.
[5] Bemiss, *Journal of Psych. Med.*, etc., p. 378, April, 1857.

CHAPTER II.

The Influence of Asceticism on the Laws of Marriage.

WE have seen in the preceding chapter that those difficulties which some people wish to see put again in the way of marriage, on the plea that immemorial experience has proved them to be useful, neither originated in the way they fancy, nor in the law of the Jews, but that they rose with the growing power of the Church, and fell with the fall of that power. The origin of the idea of the impropriety of marriages between near kin, as far as it concerns our law, is not therefore lost in antiquity. It was not the result of any observed evil effect of these marriages, but rather originated in the general asceticism which already thoroughly pervaded the whole Christian world in the second century, and continued to rise and fall in almost direct ratio to the amount of immorality to which it owed its origin. In all countries, and at all times, asceticism has been the direct result of dissolute manners, whenever these have become a hindrance and a canker on the progress of society. Hence, we do not meet with it in such primitive States as were the Sandwich Islands when first discovered; while in such States as Persia, India, or Ancient Egypt, it seems to have arisen spontaneously, as new life rises from corruption. In the former, there is no outrage on national life by this kind of immorality; in the latter, the immorality is so dangerous, that a reaction sets in, which however laughable and extravagant it may seem, is yet, perhaps, the sole means of moral reform.

The origin of Christian asceticism is lost in the earliest history of those nations from whom Christianity is derived. Even whether it originated spontaneously or separately among the ancient Egyptians and other Oriental nations, or these derived it from a common parent, we do not know, and can only guess that from the like influence of like causes, the former explication is the true one. The idea that our body is inherently evil, and that its desires and passions must be combated before we can attain goodness, can easily be explained on utilitarian principles, and might therefore occur to any man living in the society of other men; but that these principles should be generally recognized, and that the men practising them should be honoured, argues a state beyond the savage and below what we now call civilized. Hence ascetic ideas were generally known and acted upon before Christianity, by those nations who afterwards became the earliest Christians, and helped to make our law. There were Egyptians practising asceticism on the Island of Philæ, about B.C. 280. There were Jews practising conventual life, a considerable time before B.C. 110. There were Greeks imitating the Egyptian ascetics B.C. 164; and the Romans held many of them the doctrines of Zeno, and held in veneration the Vestal Virgins.[1]

The Therapeutæ, a sect of Essenes established in Egypt about A.D. 14, holding opinions which were a compound of Egyptian ideas as to resurrection with Jewish theism, may be looked upon as the prototype of Christian asceticism. They lived chiefly a hermit, instead of a conventual life like the Essenes, and this was the first form of Christian asceticism. Women were also held in greater honour than among the Essenes, for while the latter justified their tenets against marriage—"being aware of the lasciviousness of women, they are persuaded that none of them can keep true faith to one man"—the former allowed ancient virgins to join them in their retired life. Accordingly it was from Egypt that the first instance of Christian asceticism came, A.D. 181.[2]

In the Gospels cœlibacy is not recommended except any one feel impeded by marriage in the duty he owes to God. Christ puts that duty above all others, and next to it the duty to cleave unto one's wife.[3] St. Paul, however, distinctly puts the cœlibate state above the married, not because he considers marriage to be sinful, but because it brought trouble to the flesh, and was incompatible with entire devotion to God.[4] The author of the Revelation, moreover, saw a hundred and forty and four thousand saints round the Lamb, which were redeemed from earth

[1] Sharpe, *Hist. of Egypt*, vol. i. pp. 318, 400; Smith's *Dict. of the Bible*, Art. *Essenes*; Blunt's *Dict. of Sects and Heresies*, Art. *Essenes*.
[2] Smith and Blunt, *ut sup.*; and Sharpe, *Hist. of Egypt*, vol. ii. p. 204.
[3] Matt. xix. 1-12.
[4] 1 Cor. vii. 28, 29, 33, 34, 38; and 1 Tim. iv. 1-5.

because they had not defiled themselves with marriage.[1] Imbued as all reformers were at that time of general demoralization with ascetic ideas, it was difficult for Christianity to escape asceticism. Fortified with these unfortunate sayings, escape was impossible. Ascetic ideas gained ground among the Christians with wonderful rapidity, producing incalculable mischief, and working dreadful miseries for centuries.

The Gnostics had already set themselves against marriage about the time of St. Paul. Their tenets have much affinity with Buddhism, and were in all probability a graft of Christianity on Oriental theology. The Asiatic Gnostics thought that the body was an evil and should be tortured; while the Egyptian Gnostics held the comfortable doctrine that since the body was evil, everything it did must be evil, and hence have no concern with the better part.[2] The Nicolaitanes, another sect which rose about the same time, took advantage of Christianity to throw off the Jewish moral restraints, but were soon afterwards absorbed by other sects holding somewhat similar views. The Antinomians, on the other hand, were ascetic in their views of marriage, and to increase their virtue lived with spiritual sisters, in order that temptation might ever be present.

In the second century, asceticism and its complement libertinism spread into numberless unorthodox sects. Gnosticism was further developed in its ascetic form by Saturninus; by Valentinian, who considered that there were three types of men—1st, those re-

---

[1] Rev. xiv. 1-4.
[2] Mosheim, *Ecclesiastical Hist.*, vol. i. p. 38; and Blunt, *Dict. of Sects and Heresies*.

presented by Cain, or carnal nature; 2nd, those represented by Abel, or animal nature; and 3rd, those represented by Seth, or spiritual nature, who whatever they did must be saved, while the first kind must be damned necessarily, and the second kind would be saved or damned according to their behaviour. This idea also lay at the root of Antinomian profligacy as afterwards developed, for they held that they were incapable of sin, or rather that they must be saved in spite of any sin they had committed. Cerdo, and his disciple Marcion whose doctrines were widely spread in Italy, Egypt, Palestine, Arabia, Syria, Persia, and Cyprus, rejected all pleasures, contemned matrimony, and held virginity in honour. Montanus, also a Gnostic, entirely condemned second marriages as nothing better than fornication; even the authority of St. Paul could not vindicate them, since Montanus held that St. Paul had only a partial knowledge of what was right, and only an imperfect gift of prophecy. Apollonius says that Montanus even taught dissolution of marriage, and that the prophetesses Maximilla and Prisca abandoned their husbands in consequence. Marcus founded another sect who held views like those of Valentinus. Tatian, formerly a disciple of Justin Martyr, adopted the Gnostic heresy on Justin's death. Marriage, he taught, was a naughty devilish invention, and he was otherwise so ascetic and had so great a horror of incontinence that his disciples, the Encratites, were even forbidden the use of wine in the celebration of the Eucharist, or to eat meat, since it was necessarily engendered in sin. In these views he was succeeded by Severianus and his disciples.

Another Gnostic, Bardesanes, may also be numbered among the austerer sect-founders of this century. Basilides may be considered the chief of the Egyptian Gnostics. Himself probably austere to excess, his disciples hardly carried out his tenets as he had intended, and brought general discredit on them by their licence. Carpocrates, who adopted the views of his son Epiphanes, recommended to his disciples that they should lead the most vicious life, since our passions were implanted in us by our Maker, and everything He had made must be good. The Prodicians, an offshoot of the Carpocratians, were also a profligate sect holding Antinomian views without their austerity. These were afterwards called Adamites, because of their habit of meeting together as Adam and Eve are described before the Fall. The Cainites, who treated as saints Cain, Cora, Dathan, and the Sodomites; the Sethites, Ophites, Nicolaitanes, and Florinians, were other obscure Gnostic sects of the second century. The Abelonites of the same century were not exactly Gnostics, but though they married, they lived apart from their wives and adopted children. Another sect, called the Archontics, did not even allow marriage under these conditions; woman, they said, was a creation of the devil, and all who married fulfilled the works of the devil. The Elchasaites, on the other hand, compelled marriage, and despised virginity: their views were really praiseworthy, for they recommended early marriage that sin might be avoided.[1]

After the second century, the Gnostics are no longer heard of; but their place was soon taken by

---

[1] Mosheim, *Ecclesiastical Hist.*, vol. i. pp. 60-66; Reich, *Ehe*, etc. pp. 65, 68; and Blunt, *Dict. of Heresies*, etc.

the great and widely spreading heresy of the Manichæans. Manichæism rose in the latter half of the third century; its disciples were divided into the *Perfect* and the *Hearers;* of which the former practised the most rigid asceticism, were forbidden every sensual gratification, including the bath and marriage, and were allowed to do nothing but pray; while the latter were not bound to so ascetic a life, though moral conduct was insisted upon. The spread of this heresy under the influence of persecution was wonderful. From its cradle in Persia, it soon overran Asia Minor, Eastern Europe, and Northern Africa, and may be traced for centuries afterwards under the various names of Sakkophori, Solitaries, Encratites, Apotactics, Udroparastatæ, Paulicians, and a host of others. The Novatians, a heretic sect of this century, arose in Rome as reformers, who protested against the pardoning of sinners, and, among other things, held second marriages to be sinful. In Gaul and Spain rose a sect called the Abstinentes, who held, that though marriage was not absolutely wrong, yet that it was better to avoid it, an idea evidently founded on Matthew xix. 12. The Valesians, an obscure sect of eunuchs of this or of the next century, more nearly carried out this text.[1]

Of sects already in existence in the last century, the Manichæans were undoubtedly the most widely spread in the fourth century, generally hidden under a variety of names to preserve themselves from persecution. The Euchites, or Messalians, were perhaps the most marked of these. They gave themselves

---

[1] Mosheim, *Ecclesiastical Hist.*, vol. i. pp. 82, 84; and Blunt's *Dict. of Sects*, etc.

up wholly to an ascetic life, lived in communities, rejected marriage, and those who already had wives put them away. They never worked or gave alms, because, they said, they alone were the poor in spirit, and the only people to whom alms should be given. The Paterniani, another Manichæan sect, who were also called Venustians, were condemned by Damasus, in a council held at Rome, A.D. 367, for their immoral conduct. The Montanists cropped up again in Spain, under Priscillian, Bishop of Abila; the Novatians still flourished; and the licentiousness of the orthodox clergy induced Ardæus to found a reformed sect among the Goths. The Hieracites sought to restrict the Holy Communion to cœlibates; while Jovinian, backed up by Helvidius and Vigilantius, opposed the doctine of cœlibacy, and taught that virgins, widows, and married women have an equal degree of merit. We hear, also, in this century, of the Agapetæ or Dilectæ, a Spanish sect, which rejected marriage, but not for ascetic reasons. The name of the Agapetæ is, however, more generally applied to those monks and nuns who followed up the Antinomian doctrine, justifying their conduct by the text, "Have we not power to lead about a sister, etc.?"[1]

The Novatians and Manichæans were still numerous in the fifth century. In the seventh, we hear of a sect of the latter under the name of Agynians, who forbad marriage: but the chief Manichæan sect is the Paulician, which first appeared in Asia Minor, A.D. 660, and entered into a bitter dispute with the Greeks. They were persecuted almost continuously up to the

[1] 1 Cor. ix. 5. See Mosheim, *Ecclesiastical Hist.*, vol. i. pp. 82, 107, 114, 115; and Blunt's *Dict. of Sects*, etc.

middle of the ninth century, at the close of which they had converted the Bulgarian Church. In the tenth century they were still further reinforced, and became so strong that they gave rise to a sect of dissenters from their own body, the Bogomiles. At the close of the eleventh century, both they and the Bogomiles were systematically persecuted by the Emperor Alexius Comnenus; and from this time Paulicianism became insignificant, or, rather, it was diverted into several channels.[1] The Marcionites still existed in the East in the fifth century, and were not yet extinct in the eighth.[2] The Euchites appeared again in the tenth and twelfth centuries.[3] In the eleventh century Leutardus, a peasant of Virtus, near Chalons-sur-Marne, had a vision, telling him to introduce a reformed Christianity. He preached that the cross should be dishonoured, not worshipped, that no tithes should be paid, and that marriage was unlawful; while to put his precepts into practice, he divorced his wife. There soon appeared other sporadic examples of Manichæism in different parts of Europe —harbingers of the great heresies of the next few centuries to which Manichæism led.[4]

The Adamites appear again at the beginning of the twelfth century, but in so licentious a form that they were at once suppressed. The Henricians, founded by Henry of Lausanne, arose about the same time. They condemned cœlibacy, but only as a means for the prevention of sin, like the Elchasaites of the second

[1] Blunt's *Dict. of Sects*, etc.; and Mosheim, *Ecclesiastical Hist.* for these centuries.
[2] Mosheim, *Ecclesiastical Hist.*, vol. i. pp. 131, 187.
[3] Mosheim, *Ibid.*, p. 300; Blunt's *Dict. of Sects*, etc.
[4] Blunt's *Dict. of Sects*, Art. *Manichæans*.

century. At the same time, the immorality of the orthodox clergy called into existence the sect of Petrobrusians. Manichæism appeared in various places in its ascetic form, and spread greatly under the names of Albigenses, Cathari, Paterini, Albanenses, Bagnolenses and perhaps Waldenses. A sect called Apostolicals also appeared in the neighbourhood of Cologne, who held severe Antinomian views. They did not allow marriage, but each brother slept in the same room with a spiritual sister, to put their continence to a test.[1] The Beguins appeared first towards the end of this century; supposed to have arisen from the disproportion between the sexes caused by the Crusades. They professed extraordinary piety of the primitive Antinomian kind, and though always shameless, they were not at first unchaste. Captivated by their true austerity, which offered so marked a contrast to that of the orthodox Church, numbers of both rich and poor swelled their ranks to such a degree that in the thirteenth century they were spread nearly all over Europe, under the names of Bicorni, Beghards, Biguttes, Fraticelli, Turlupins, Picards, and the Brethren of the Free Spirit; and their ranks were further swelled by the Ortlibenses and Amalricians, kindred sects. Their undeniable asceticism greatly perplexed the Inquisition, who were driven to account for the coldness and insensibility of the Beghards to the promptings of sensual gratification, by the theory that it was given them by the devil that they might impose on others by a sanctimonious appearance. But this austerity soon

[1] Mosheim, *Ecclesiastical Hist.*, vol. i. pp. 301, 306; and Blunt, *Dict. of Sects*, etc.

gave way to a reaction, when under the profession of extreme godliness their practice became grossly immoral; for their worship was as the worship of the Adamites, and their private life as that of the Dilectæ.[1]

Besides looking after the Beghards, the Inquisition had plenty of other work on their hands in this century. The Cathari of Lombardy, called also Concorezences, Concordenses, Concoretii, or Concorenses, were estimated at 500,000 persons. Of the more ascetic Catharists, or *Perfecti*, there were only 4,000, according to a census taken by themselves; but the *Credentes* were "innumerable." The Stedingers, another sect of Manichæan origin, arose on the borders of Friesland and Saxony in this century. The Apostolicals of Cologne were still in existence; and another sect of Apostolicals, also known as Dolcinists from their leader, appeared in Lombardy and in some districts of the Tyrol with the same tenets as their contemporary namesakes, and, like them, each accompanied by a spiritual sister. Most of these sects were very persistent; while in the fifteenth century a new sect arose in the Netherlands—a branch of the Beghards—who called themselves Men of Understanding.[2] In the sixteenth century the Beghards, under the influence of the Reformation, had developed into the antisacerdotal sect of the Anabaptists; from these the Familists,

---

[1] Mosheim, *Ecclesiastical Hist.*, vol. i. pp. 343-347, 348 note 2; Blunt's *Dict. of Sects*, etc.; and see a quotation from Johannes Trithemius, in Reich's *Ehe*, etc., pp. 167, 168, giving a horrible account of their worship, but evidently exaggerated, like the account cited in Lecky's *Hist. of European Morals*, vol. i. p. 440, of the charge against the early Christians.

[2] Mosheim, *Ecclesiastical Hist.*, vol. i. p. 372; and Blunt's *Dict. of Sects*, etc.

or David Georgians, developed, who, under the guidance of Henry Nicolas, became extreme Antinomians, pretending " that they could, without evil, commit the same act which was sin in another to do;" just as the Valentinian Sethites. The Anabaptists were taken in hand by Simon Menno about the middle of this century, and consequently took the name of Mennonites. Everyone was to be married, divorce was not allowed, and marriage was entirely forbidden between near relations. The stricter sort even considered many harmless amusements to be sinful, and excommunicated those of their weaker brethren who indulged in them. Another sect called the Libertines, or Spirituals, arose in Flanders, and were patronized by Margaret of Navarre, whose principles like those of the Brethren of the Free Spirit, were that since God the Beneficent was the Creator of all things, none of our actions could be evil. Another sect arose in Spain, called the Illuminati, identical in their tenets with the Familists. They were soon suppressed by the Inquisition, but rose again, and had again to be suppressed at the beginning of the seventeenth century. A sect of the same name, and holding the same opinions, appeared about ten years afterwards in Picardy, whence they spread into Flanders, but were put down by Louis XIII.[1]

In England the Familists, or rather Adamites, were represented in the seventeenth century by the Ranters; in Russia by the Khlisti, who had a community of women. The Skoptzi, another Russian sect, still existing in considerable numbers, are great ascetics, a

[1] Reich, *Ehe*, etc., p. 157; and Blunt, *Dict. of Sects*, etc.

revival of the Valesians of the third century. About the middle of this century was founded the sect of the Labadists, a society of Dutch Protestants very like in character to the early Quakers, and professing great austerity, but they died out before the century was well over. A Swiss Mennonite founded the sect of the Amenites, a very strict sect, still existing in Switzerland as Hook Mennonites and Button Mennonites, according to the means employed for fastening their coats. The same sect exists in America as the Omish Church. The end of the seventeenth century witnessed the establishment of the Crispites in England, who held Antinomian views.[1]

The licentious sects were represented in the eighteenth century by the New Born, an American sect, and the Rosenfelders, a Prussian sect. The former were Antinomian, and existed only from the year 1720—1740; the latter were not even professedly Christians, and lived in the greatest profligacy between the years 1763—1782, when they were dissolved. The sects holding ascetic views are represented by the Tunkers, a sect of Mennonite origin, founded in America by Conrad Peysel in 1719, who live together at Ephrata much as monks and nuns. Marriage is not forbidden, but those who marry must leave Ephrata. Another was the Shakers, who lived chiefly at New Lebanon. Marriage was not allowed at all at first, unless it was platonic; but now the Shakers are permitted to marry, provided that they leave New Lebanon. The Angelic Brothers, founded by George Gichtel, were a Dutch sect which forbad marriage, as did also the Buchanites. These latter soon became

[1] Blunt, *Dict. of Sects*, etc.

extinct, as their faith led them to believe that the end of the world was at hand, and when the world did not end at the appointed time the sect was naturally too disappointed to fix another date. In the present century, besides the greater number of these sects, there exists the Perfectionists, Familists, or Free Lovers, the White Quakers, and the Mormons; all profligate sects.[1]

The mere enumeration of these sects is sufficient to demonstrate the immense influence of asceticism on the laws of marriage, and the constant reaction between libertinism and austerity. Nor was the orthodox Church free from either immorality or its concomitant asceticism. Southey says that nothing in ecclesiastical history is more certain than that no such obligation as cœlibacy was imposed on the clergy during the first three centuries;[2] but it was at all events considered a virtue long before. Justin Martyr retorted on those who argued that marriage could not be evil since God had created the sexes, that God made some women naturally barren. Clemens of Alexandria accused the Gnostic Basilides of not considering the cœlibate state far superior to the married.[3] Athenagoras and Tertullian both forbad second marriages;[4] the latter, especially, looking upon marriage as a state which we had far better avoid altogether. Origen was against marriage, and Lactantius considered that if men could not be prevented from entering into marriage, every effort should be made to

[1] Blunt, *Dict. of Sects*, etc.
[2] Buckle's *Common Place Book*, Art. No. 130.
[3] Reich, *Ehe*, etc., pp. 63, 65; Mosheim, *Ecclesiastical Hist.*, vol. i. p. 53; Buckle's *Common Place Book*, Art. No. 70.
[4] Buckle, *Ibid.*

render them as platonic as possible.[1] St. Ambrose, in the fourth century, though he did not positively condemn second marriages, yet condemned the tenets of Jovinian, that marriage and moderate living was as pleasing to God as asceticism, and considered cœlibacy absolutely necessary for priests and deacons. Jerome, too, raised his voice against the abominable doctrine of Jovinian, and the similar opinions of Vigilantius and Helvidius—"sacrilegious tenets," as he calls them, which he could not "hear with patience, or without the utmost grief."[2] Hieronymus only tolerates marriage at all as the least objectionable means for the production of monks and nuns. Gregorius of Nyssa, Cyrill of Jerusalem, Cyprian, and Epiphanius, all praise the cœlibate state; while Chrysostom, though he does not go so far as to stigmatize marriage as a hindrance of all virtue and happiness, yet strongly dissuades from it.[3] Augustine, on the other hand, allows second, and even third marriages. Eustathius prohibited marriage altogether; and Ignatius thought that no Christian ought to marry without the consent of his bishop, to the end that these unions should conform more to the spirit, and less to the flesh.[4] It became part of the canon law that second and third marriages should not be countenanced by the presence of the priest at the marriage feast. "What kind of presbyter would he be, who, for the sake of the dinner, countenanced such a union by his presence?" cries

[1] Reich, *Ehe*, etc., pp. 64, 66.
[2] Bucklo's *Common Place Book*, Arts. Nos. 70, 133; Mosheim, *Ecclesiastical Hist.*, vol. i. pp. 104, 129.
[3] Reich, *Ehe*, etc., pp. 66, 67.
[4] Reich, *Ibid.*, p. 67; Buckle, *ut sup.*, Art. 70; Mosheim, *ut sup.*, p. 103.

Theodore, Archbishop of Canterbury, A.D. 668–690, in an outburst of indignation at the idea.[1] Before the Reformation the honour of a garland was denied a widow who married again; and by the canon law, if a man married two virgins in succession, or if a man married a widow, such offender was deemed guilty of bigamy, and even by the law of the land incapable of benefit of clergy.[2] The light in which the Church looked upon marriage is shown more plainly in other restrictions; indeed, there is no doubt that if the clergy had had their way, they would have forbidden all mankind, as they forbad themselves, ever to enter into the bonds of matrimony![3]

The marriage of Leo the Philosopher, in the tenth

[1] Thorpe, *Early English Inst.*, etc., p. 283; Theodore, *Pœn.*, xvii. § 10.

[2] Buckle's *Common Place Book*, Art. No. 70. The original meaning of the word *bigamy* was marriage of this sort, not of two women at once.

[3] "Qui in matrimonio sunt, abstineant se in iii. xl^{mas.}, et in Dominica nocte, et in Sabbato, et feria iiii. et vi. quæ legitimæ sunt, et iii. noctes abstineant se antequam communicent, et i. postquam communicent, et in Pascha usque ad octabas * * * * In primo conjugio presbyter debet missam agere, et benedicere ambos, sicut in Libro Sacramentorum continetur, et postea abstineant se ab ecclesia xxx. diebus; quibus peractis, pœnteant xl. dies et vacent orationi, et postea communicent cum oblatione * * * * Si quis vir, aut si quæ mulier, Dominica die, vel in natale Sanctorum, panes Deo offerant, necnon et communicaverint, non debent sequenti nocte nubere * * * * Quod si propter ebrietatem acciderit, sine consuetudine, iiii. dies pœniteant." (*Theodori Arch. Cant. Liber Pœnitentialis*, xvii.) "Sancti libri docent quid cuique homini fideli faciendum sit, cum legitimam suam uxorem primum domum duxerit; id est, juxta librorum doctrinam, ut, per spatium trium dierum et noctium, castitatem suam servare, et tunc tertio die missæ suæ adesse, et ambo eucharistiam accipere debeant, et deinde conjugium suum tenere coram Deo, et coram mundo, uti ipsis necesse erit. Et conjuges omnes oportet castitatem suam servare xl. dies et noctes ante sanctum Pascha, et per totam hebdomadam paschalem, et semper nocte diei Dominici, et diei Mercurii, et diei Veneris." (*Pœnitentiali Ecgberti, Arch. Ebor.*, lib. ii. sec. 21.) "Si quis conjugem suam,

century, raised quite a schism in the Greek Church, for he actually wished to bring in a law to legalize his own and other fourth marriages. So great was the opposition, however, to such a horrible permission that though Leo was able to tide it over during his lifetime, yet his son, the very issue of that fourth marriage, was obliged to convene an assembly of the clergy, A.D. 920, which peremptorily forbad all fourth marriages, and only allowed third marriages under certain conditions.[1] The Eastern Christians very nearly imposed cœlibacy on their clergy in the fourth century.[2] Gregory the Great really did impose it, though a doubtful authority ascribes it to Pope Siricius, in the latter part of the fourth century. In the Council of Bourges, A.D. 1031, it was ordained that only the inferior ministers of the Church should be allowed to have wives or concubines. The Council of Rouen, A.D. 1072, forbad any minister who married to receive or dispose of any of the Church revenues; and the Council of Rome, A.D. 1074, ordered all those who were already married to divorce their wives. In England, however, this brutal enactment could not be

---

si fieri potest, non cupidine voluntatis, sed solummodo creandorum liberorum gratia utitur, iste profecto sive de ingressu ecclesiæ, seu de sumendo Dominici corporis sanguinisque mysterio, suo est relinquendus judicio; quia a nobis prohiberi non debet, cum ei juxta præfinitam sententiam, etiam ecclesiam licuerit entrare; veruntamen quia ipsa licita admixto conjugio sine voluntate carnis fieri non potest, ideo aliquando a sacri loci ingressu abstinendum est, quia voluntas ipsa esse sine culpa nullatenus potest." (*Excerp. Ecgberti Arch. Ebor.*, cxii.) See Thorpe's *Early English Inst.* It appears that at all events, at a later period, as in the case of marriages between near kin, a dispensation might be bought to cover these prohibitions. (See Buckle's *Common Place Book*, Art. 1091.)

[1] Mosheim, *Ecclesiastical Hist.*, vol. i. p. 228.
[2] *Ibid.*, p. 111.

enforced, for by a Council held at Winchester, A.D. 1076, the secular clergy who were married were formally allowed to retain their wives.¹ The injurious effects of this legislation soon became manifest. Even Gregory is said to have revoked his edict.² Laws had to be enacted again and again forbidding priests to have their mothers or sisters to keep house for them.³ A tax used to be systematically levied by rulers for several centuries, which was simply a licence to priests to keep concubines,⁴ and Henry III. of Castille, among other social enactments, ordered that the concubines of priests should wear a piece of scarlet cloth in their head-dress, in order that they might be distinguishable from honest women.⁵ In the beginning of the fifth century their concubines were legalized by the Council of Toledo; indeed, these women seem to have had a special name, "Focaria."⁶ A law of Charlemagne seems to imply that priests sometimes practised even polygamy;⁷ and, as we have seen, several of the heterodox sects were avowedly founded to reform the orthodox priesthood. Even the throne of St. Peter was reached by the prevailing immorality.⁸ "It was observed,"

¹ Buckle's *Common Place Book*, Arts. Nos. 75, 99, 130.
² *Ibid.*, Art. No. 130.
³ Lecky, *Hist. of European Morals*, vol. ii. p. 351.
⁴ *Ibid.*, p. 349. Compare also the following pages.
⁵ *Biog. Univ.*, Art. *Henri III*.
⁶ Buckle's *Common Place Book*, Art. No. 65.
⁷ "Si sacerdotes plures uxores habuerint, sacerdotio priventur; quia sæcularibus deteriores sunt" (*Capitul.*, A.D. 769). See Hallam's *Middle Ages*, vol. ii. p. 7, note.
⁸ Gibbon says of Pope John XXIII. of blessed memory: "The most scandalous charges were suppressed, the Vicar of Christ was only accused of piracy, murder, rape, sodomy, and incest" (*Decline and Fall*, etc., vol. viii. p. 423); and Disraeli, in his *Curiosities of*

says Mr. Lecky, "that when priests actually took wives, the knowledge that these connections were illegal was peculiarly fatal to their fidelity, and bigamy and extreme mobility of attachments were especially common among them."[1] The idea, so encouraged by the Church, that he who giveth not his virgin in marriage doeth better, led, as we have seen, on the one side to the wildest orgies of licence, and on the other to an equally immoral outburst of cœlibacy. St. Nilus, St. Ammon, St. Melania, St. Abraham, and St. Alexis were all married, and lived in continence. The Emperor Henry II., Edward the Confessor, and Alphonso II., of Spain, are all said to have been husbands but in name.[2]

What could be more natural, when the whole spirit of these ages tended to make marriage difficult, when the ascetic saints who had the social governance of civilized man in their hands cudgelled their poor brains to find riddles hidden in every word of the Bible, that the Levitical prohibitions against the marriage of near kin should be so magnified and distorted as almost to justify Beatrix's saying, "Adam's sons are my brethren, and truly I hold it a sin to match in my kindred!" Or what more natural that when a set of greedy priests found that people were ready to pay a price to be allowed to marry, they should have made marriage yet more difficult, that they should increase the number of meshes in their net, and with it the

Literature, vol. i. p. 307, gives this pasquinade on Pope Alexander VI., written on a certain lady's tomb:

"Hoc tumulo dormit Lucretia nomine, sed re
    Thais; Alexandri filia, sponsa, nurus."

[1] Lecky, *Hist. of European Morals*, vol. ii. pp. 350, 351.
[2] *Ibid.*, vol. ii. pp. 341, 342.

amount of their revenues? Luther and Henry VIII. both roundly accuse the Church of inventing prohibitions "for lucre's sake."[1] Nor was this, powerful as it was, the only incentive. Hallam points out that "they served a more important purpose by rendering it necessary for the princes of Europe, who seldom could marry into one another's houses without transgressing the canonical limits, to keep on good terms with the court of Rome, which, in several instances * * * fulminated its censures against sovereigns who lived without permission in what was considered an incestuous union.[2] Thus, King Charibert was excommunicated because he married two sisters.[3] Celestine III. threw the kingdom of Leon under an interdict, because Alfonso IX. married Garsenda, princess of Castille, his cousin Alfonso's daughter, and he was obliged to divorce her. Sancho IV. of Castille was excommunicated because he married his second-cousin Maria. The succession of his son Ferdinand IV. was consequently disputed until he was legitimized by a Papal Bull. Pedro the Cruel, of Portugal, was excommunicated for marrying three wives in succession. Eleanor, heiress of Guienne

---

[1] Luther says, in regard to marriages of the fourth degree: "Der Pabst aber hat sie verboten aus lauter Heucheley, und um Geldes willen dispensiret er, und lässts zu" (Cited by Reich, *Ehe*, etc., p. 135). In the preamble of Stat. 32, Henry VIII. cap. 38, it says that these prohibitions are not "God's law," and that they were "for their lucre by that court [the Pope's] invented, the dispensations whereof they always reserved to themselves * * * * and all because they would get money by it" (See J. Fry, *The Case of Mar. between near Kin.*, etc., pp. 135, 136; and also Samuel Dugard, *Vindication of the Mar. of Cousin-Germans*, p. 90). This last author also accuses the Pope of granting dispensations for marriages which were for "Holinesse sake" prohibited, for the sake of lucre (*Ibid.*, p. 89).
[2] Hallam, *Middle Ages*, vol. ii. p. 9.
[3] *Evidence of Dr. Pusey, Royal Commission on Marriage Laws*, 1848, p. 45.

and Poitou, who despised her husband as more of a monk than a man, got a divorce with her husband's joyful assent, on the plea of affinity, and then married the unfortunate Henry II. of England.[1] And yet much closer cases of intermarriage were quietly permitted by the Popes. In a short period of the royal Spanish line alone, we find that Ferdinand V. married his first-cousin; John his son, married his sister's husband's sister, Margaret; Maria, his daughter, married her deceased sister's husband, Emmanuel the Fortunate, of Portugal, who after his second wife's death married his niece-in-law. Catherine, daughter of Philip I., married John III., of Portugal, the brother of her brother's wife, son of her sister's husband, and her first-cousin by blood. Charles V. married his first-cousin, who was also his sister-in-law. His daughter Mary married her first-cousin, Maximilian II. His daughter Johanna married John, son of John III. of Portugal, her first-cousin on both sides. And his son Philip II. was not only married four times, but the first marriage was with his first-cousin and sister-in-law, Mary of Portugal; the second was with Mary of England, his father's first-cousin; the third was the grand-daughter of his aunt, though not by blood; but his fourth wife was his niece by blood, the daughter of Maximilian II. His daughter by his third wife Isabella married Albert, son of Maximilian II., who was her first-cousin by blood, uncle by marriage, and otherwise related by various channels, any one of which by strict canonical law should have been deemed sufficient to prevent that union.

In fact, the prohibited degrees were far too useful to

[1] *Biog. Univ.*

abolish. When Alfonso III. of Portugal divorced his wife for sterility and married Beatrix de Guzman, the Pope made no move; but when the same monarch attempted to reform some great abuses in the Church, his kingdom was promptly laid under an interdict on the score of that divorce.[1] We see even now, how a law, which, while nominally enforced is virtually broken with impunity,.such as Lord Lyndhurst's Act against marriage with a deceased wife's sister, or rather the interpretation of the law on this point, is really a putting of power into the hands of the wicked. Men go through the ceremony of marriage, desert their wives, and have the tradition of the law on their side.[2] It may readily be imagined to what abuses the canonical prohibitions led. "History," says Hallam, "is full of dissolutions of marriage, obtained by fickle passion or cold-hearted ambition, to which the Church has not scrupled to pander on some suggestion of relationship."[3] It was thus that Philip I. of France repudiated Bertha for Bertrade. It was on the plea of affinity that Philip Augustus divorced Ingeburga of Denmark for Agnes of Méran, though she was related to him only through eleven degrees, of which two were mere affinity, as may be seen by the accompanying genealogical table. It is true that in this case, as in that of Henry VIII. of England, the Pope did not lend himself to the project; but the evil was done through the edicts of the Church, nor could the Church soften the bitter rage Ingeburga had to endure for six long years, or prevent poor Agnes from dying of a

[1] *Biog. Univ.*
[2] See *Royal Commission on Marriage Law*, 1848, Questions 3a, 6a, 103b, 118-20, 148, 180, 787, 818, 819.
[3] Hallam, *Middle Ages*, vol. ii. p. 8.

broken heart. The legend of Chilperic's marriage with Frédégonde is founded on the same iniquitous prohibitions. Audouère, the first wife of Chilperic, king of Soissons, had in her service a lady named Frédégonde, who, by her talent and intrigues, first became the confidant of her mistress, then the mistress of her master, and finally aspired to be his wife. The divorce of Audouère was brought about by the following ingenious plot. Chilperic's queen gave birth to a child during his absence in a war against the Saxons. The ceremony of the christening was fully prepared, the company and the priest were waiting, but the lady who was to have stood godmother, having been gained over by Frédégonde, was absent. The queen could not conceal her vexation at this untoward event. "What prevents you holding the infant yourself," said the perfidious Frédégonde, "and to its love for you as a mother, it will add the love of a godmother, which otherwise it would bestow on a stranger?" Audouère fell into the trap, ignorant of the canon on which her rival had founded her plan, which decrees that godfathers and godmothers contract with the parent of the child an alliance which prohibits all others. Frédégonde hastened to inform the king, on his return, that he was free to marry whom he pleased; and Chilperic, as superstitious as he was licentious, compelled his wife to enter a convent, banished the bishop who had performed the ceremony, and crowned the plot of Frédégonde by making her his queen.[1]

This monstrous canon first appeared under the auspices of the first Nicene Council, A.D. 325, by which marriage was prohibited with a wife's god-daughter,

[1] *Biog. Univ.*

god-child, or bridesmaids; or with the mother, sister, or daughter of a god-child.¹ It appeared in the *Institutes* of Justinian, A.D. 529, and by the Council of Trullo, A.D. 692, marriage with a relation by god-parentage was forbidden within the seventh degree, but only in the right line descending. Hence, as the Greeks reckon by the old Roman, or civil method, supposing a to be god-parent to B; B, is considered a brother to a's daughter β; and only when they come to the generation D, δ, supposing no other alliance has been contracted by their relatives, would intermarriage between the two lines be allowed. The same prohibition was enacted in the West, by the Council held at Rome, A.D. 721. Pope Zachary, A.D. 745, and the Council of Mentz, A.D. 813, confirmed these prohibitions.² The result was exemplified in the divorce of Audouère, and in the divorce of a man in the ninth century for the same thing; but this last case was pardoned by Pope John VIII. Innocent III. confirmed these prohibitions again. The Council of Trent, A.D. 1545–1563, did the same; and further declared that the person baptized, his parents, god-parents, and the priest who baptized him, were as much interrelated as though they were relatives by blood to each other. Accordingly no tolerably near relative of the priest could marry the god-relations or relations of any child that priest might have

$$
\begin{array}{cc}
A & a \\
| & | \\
B & \beta \\
| & | \\
C & \gamma \\
| & | \\
D & \delta
\end{array}
$$

---

[1] *Evidence of the Rev. A. P. Percival, Royal Commission on Marriage Law*, 1848, Question 317.

[2] *Evidence of the Rev. W. Palmer, Royal Commission on Marriage Law*, 1848, p. 51; *Evidence of Dr. Pusey, Ibid.*, Question 447.

baptized.[1] Boccaccio touches on these prohibitions more than once, and ridicules the idea that marriages of this sort were sinful.[2] Before the Council of Trent, A.D. 1545-1563, fornication with any relative within the fourth degree of the bride would prevent the marriage; but this Council reduced it to the second degree.[3]

[1] See Pierrot's *Univ. Lexikon*, Art. *Ehe*; The *Encyc. Method.*, Art. *Affinité*; and also the *Algem. Encyc. der Wissenschaften und Künste*.

[2] In the one case, Madonna Agnesa replies to the advances of Frate Rinaldo, her *compadre*: "Egli sarebbe troppo gran male; e io ho molte volte udito che egli è troppo gran peccato." Her compadre, however, proves to the satisfaction of his god-daughter "che loica non sapeva," that this was not the case (*Novella III., Giornata Settima*). Again, Tingoccio was in love with his *comare*, but kept this love secret from his friend Muccio, "por la cattivatà che a lui medesimo pareva fare, d'amare la comare; e sarebbesi vergognato che alcun l'avesse saputo." His friend, however, who was also in love with her, finds it out. Tingoccio dies; and according to a promise comes as a ghost to tell his friend about our future life. He assures him that no account is taken in the next world of these sort of unions (*Ibid., Novella X.*) Voltaire aptly expresses the surprise of an *Ingénu* untutored in the subtleties of priestly legislation : "Morbleu, mon oncle, pourquoi serait-il defendu d'épouser sa marraine, quand elle est jeune et jolie?"

[3] *Evidence of the Right Rev. N. Wiseman, D.D., before the Royal Commission on Marriage Law*, 1848, Question 1203.

CHAPTER III.

THERE IS NO INNATE HORROR OF MARRIAGE BETWEEN NEAR KIN IMPLANTED IN MANKIND; THE ORIGIN OF THE PROHIBITED DEGREES AMONG SAVAGES; AND THE ONLY NATURAL PROHIBITED DEGREES.

HAVING traced the origin of our present laws concerning the prohibited degrees, it remains for us to discover, if possible, whether there is any natural law which prohibits certain marriages, and if there is such a law, what marriages are contrary to it, and what are not. The only way in which this can be done is to examine the marriage customs of primitive nations, whether there is any practically universal law restricting marriage in certain relationships, and if so, to investigate the causes of that law. Now, I think we shall find that the evidence is contrary to the theory that man has any innate horror of incest. We shall see, that though some tribes of men habitually marry into foreign tribes and never into their own, others habitually marry into their own, and never into foreign tribes. We shall further see that these contradictory customs do not arise either from any observed evil effect of consanguineous marriage, or directly from any observed evil results of marriages between distinct races; but their probable origin in the one case is the scarcity of women caused by the custom of female infanticide; and on the other hand pride of race, which disdains any alliance with a foreigner, and hence inferior. While finally, we shall see that if there is a natural prohibited degree, this is more dependent upon the relative ages of the parties than on their relationship.

The Arabs, as we have already seen, were numbered among those nations who habitually practised incest.[1] Valerius Maximus affirms that they married their mothers, and this practice was probably continued down to the era of Mohammed.[2] But Sale gives a much higher picture of their moral state. They not only did not marry their mothers, but daughters, and paternal and maternal aunts were included in their code of prohibited degrees. They considered it most scandalous to marry two sisters at once, or to take a deceased father's wife, though up to the time of Mohammed, public opinion was not strong enough to prevent the frequent occurrence of marriages of this sort.[3] Mohammed forbad them to marry a father's wife, their mother, daughter, sister, half-sister, or aunt, whether on the paternal or maternal side; the daughter of a brother or of a sister, daughter-in-law, or two sisters. He permitted marriage with a step-daughter, if the marriage with her mother had not been consummated. By the Hanafee code, a man might not marry a woman from whose breast he had received a single drop of

[1] See p. 18 of this work.
[2] Maracus, *Refut. Alcorani.* Cited by Boudin, *Mém. de la Soc. d'Anthrop. de Paris,* vol. i. 1863, p. 552.
[3] Sale's *Koran,* Introduction, p. 105.

milk; but Esh-Sháfe'ee allows the marriage, provided he has not been suckled more than five times by her during the first two years of his life. He may not marry any of his foster-mother's relatives, related to him by milk in degrees in which he would be forbidden to marry them were they related by blood. A man might take four wives and any number of concubines,[1] but he might not even possess a slave related to him within the forbidden degrees, nor could he take as his concubines at the same time two persons so related that he could not have married them both.[2]

Mohammed himself took fifteen wives and eleven concubines.[3] Of these wives, Soudah was the former nurse of his daughter Fatimah; Hindah was his first-cousin on the maternal side; a third, Zeenab, also his cousin on the maternal side, was first married to an adopted son of his; but Mohammed, vanquished by her charms, made him divorce her.[4] His daughter Fatimah married Ali, first-cousin to her father,[5] and his daughter Rakiyah married his third-cousin Othman.[6] Indeed, though the prophet's marriage with Zeenab much scandalized his followers, yet marriage with cousins was always regarded with peculiar

[1] Sale says that four was the whole number of women he might possess, whether as wives or concubines (Introduction to the *Koran*, p. 102); but Lane denies this, and asserts that the number of concubine slaves was not limited (*Modern Egyptians*, vol. i. p. 123, note).
[2] Sale's *Koran*, p. 58; Lane's *Modern Egyptians*, vol. i. pp. 123, 127, 128.
[3] Chalmer says that this is the lowest computation, the highest is twenty-one wives (*Biographical Dict.*, Art. *Mahomet*).
[4] *Biog. Univ.*, Art. *Mahomet;* and Taylor's *Hist. of Mohammedanism*, p. 714.
[5] *Biog. Univ.*, Art. *Ali;* Chalmer, *Biog. Dict.*, Art. *Ali.*
[6] *Biog. Univ.*, Art. *Othman ibn Affan.*

favour. Their custom was almost identical with the Jewish and Greek law, which gave a nearest relative the first right to a girl's hand. But since by Mohammedan law marriage with a niece was not permitted,[1] a first-cousin had the first right to his cousin's hand, and no one else may marry her till he gives permission with the phrase: "She was my slipper, I have thrown it off."[2] This custom is common among all the Bedouins. A price is always demanded for a wife, but a cousin pays less than a stranger. Fathers do not like this, and sometimes trick their nephew out of his birthright by suddenly demanding the marriage price during his absence, and if it is not immediately forthcoming, they may marry their daughter to whom they please. A cautious man, therefore, who has a pretty cousin, always takes care to leave four of his relations as trustees in charge of the marriage price, whenever he may have occasion to go on a journey.[3] The stories in the *Thousand and One Nights* teem with instances of marriages between first-cousins.

The Bedouins are exceedingly proud of their race, and never marry out of their own tribe. They do not even intermarry with slaves, and hence they keep their blood far purer than the generality of Mohammedans.[4] The Arabs do not practise infanticide, but women are rated at a far lower value than men. Lane says, they "still show relics of that feeling which often induced their ancient ancestors to

[1] A restriction which tends to confirm our belief that these marriages were not permitted among the Jews.
[2] Compare Ruth iv. 7, 8.
[3] Burckhardt, cited by Reich, *Ehe*, etc., p. 287.
[4] *Ibid.*, p. 289; and M'Culloch's *Geographical Dict.*

destroy their female offspring." It is for this reason that the festivities at the birth of a boy are always greater than at the birth of a girl; that boys are often dressed as girls, when about to go in procession to be circumcised, and hold a handkerchief to the face, so as not to excite envy, and thus draw upon themselves the Evil Eye; that a male gets double the share of a female in an inheritance; and that to kill or maim a woman costs only half as much as to kill or maim a man.[1]

The Druses, according to De Sacy and Wolff, do not marry their mothers, sisters, or aunts, whether paternal or maternal.[2] Volney asserts that they took the religion of a mad Egyptian khalif, who, with an impostor, Mohammed-ben-Ismaël, acted entirely contrary to the Koran, forbidding pilgrimages or circumcision, and allowing the eating of pork, the drinking of wine, and marriage with sisters and daughters.[3] These assertions are, however, not of much value; for little or nothing is known of the religion and customs of this curious people.

The Circassians believe themselves all to have sprung from a common stock. Formerly they were divided into "fraternities," all equal in social standing. No one was allowed to marry within his own fraternity; and where, as sometimes happened, several fraternities joined, all the members included in these fraternities, even though they might number

[1] Lane, *Modern Egyptians*, vol. i. pp. 71, 129, 132, 134; vol. ii. p. 242.
[2] Reich, *Ehe*, etc., p. 57.
[3] I regret that I have forgotten whence I had my attention called to this passage. It occurs in Volney's *Voyage en Syrie*, etc., Paris, 1737, vol. ii. p. 35.

several thousands, were considered too nearly related to intermarry. Marriage between two members of the same fraternity used to be punished with death; but owing to the unnatural arrangement that all these individuals were considered relatives, and that therefore both sexes might visit each other without scandal, unions which were considered incestuous became too common to punish in this way, and the punishment of death was commuted into a fine of two hundred oxen, and the restitution of the girl to her parents. They imposed the same laws upon their serfs. Family pride is very great; hence a chief always marries the daughter of another chief; and marriages with inferiors are extremely rare. Many of the Circassians are now Mohammedans, and consequently marry according to the precepts of the Koran.[1]

Bodenstedt says that among the Georgians and Armenians it is considered a bad omen if the first-born child is a female, and the couple consider themselves the miserable sport of Fate, if several female births follow. A Georgian mother, who had only given birth to girls, would be ashamed to appear in public; but the birth of a boy is the occasion of great festivity for mother and child. They are accustomed to sell their daughters at the age of twelve or thirteen to any one who will pay their price.[2]

The Ossetes allow a father or brother to take the deceased husband's place; and a chief may succeed to his father's hareem.[3]

[1] Reich, *Ehe*, etc., pp. 261, 263; and M'Lennan, *Prim. Mar.*, pp. 101, 102.
[2] Reich, *Ehe*, etc., p. 271, note.
[3] Smith's *Dict. of the Bible*, Art. *Marriage*, note.

Chardin says of the Mingrelese, a tribe living on the east coast of the Black Sea, that they marry without scruple aunt and niece, or both at the same time. Marriage within nearer degrees is by no means rare; yet he says the Mingrelese are marvellously well made, have an admirably beautiful visage and form, a majestic air, and plenty of spirit and subtilty.[1]

The Parsees and Guebres keep their blood so pure that they still resemble in beauty the ancient Assyrian sculptures.[2] They never marry except among themselves; polygamy is forbidden, and divorce only allowed after seven years' barrenness. Of late years a party has arisen who wish to prohibit consanguineous marriages; probably influenced by Christian and Mohammedan ideas in India.[3] The modern Persians are mostly Mohammedans, who conform to the law of their prophet. A male child is weaned at the age of two years and two months, while a female is weaned at the age of two years.[4]

The Affghans are divided into a great number of tribes, excessively proud of their lineage. They will hardly acknowledge any one a member of their tribe who cannot prove his pedigree for six or seven generations. No male, with rare exceptions, marries out of his tribe; but no Affghan female will condescend to marry with a foreigner; and they practise the levirate law. Physically they are a hardy and robust race.[5]

[1] Cited by St. Lager, *Du Crétinism*, etc., p. 115.
[2] J. A. N. Perier, *Mém. de la Soc. d'Anthrop. de Paris*, vol. i. 1863, pp. 73, 80.
[3] Blunt, *Dict. of Sects*, etc.; M'Culloch, *Geographical Dict.*, Art. Bombay.
[4] Roich, *Ehe*, etc., p. 254.
[5] Roich, *Ibid.*, pp. 240, 241; J. A. N. Perier, *Mém. de la Soc.*

The people of Beloochistan have much the same customs as their Affghan neighbours. The Kamburani males may marry foreigners, but the females are never permitted thus to defile their blood. The Brahoos are even more severe, and oblige both sexes always to marry within the tribe. The Gypsies of Beloochistan, called Luri, have no marriage, and look upon their children as common property. Their fertility is poor; to compensate which they practise child stealing, and so bring new blood into their stock.[1]

The Hindoos of India are divided into about 133 castes, most of them with several subdivisions, and derived, according to tradition, from four original castes of the Brahmin, Kshutry, Wys, and Soodru. At first these four castes were allowed to intermarry, under certain conditions; and it is certain that the original castes did occasionally intermarry, sometimes lawfully and sometimes not; the offspring of which unions have greatly multiplied the number of the castes.[2] The higher castes reckon relationship first from the same ancient sage or Rishee; and, secondly, from the same Gotr, or family stock within seven generations. Nearly all the Indian castes are now divided into nations, which are not allowed to intermarry; the nations are divided into sects, some of which do not intermarry; and no caste may now intermarry with another. Persons who wish to intermarry may not, however, be of the same Gotr, that is, direct male descent from a common ancestor,

d'*Anthrop. de Paris*, vol. ii. 1865, p. 298; M'Culloch, *Geographical Dict.*
[1] Reich, *Ehe*, etc., pp. 240, 249.
[2] Steele, *Hindoo Castes of the Dekhun*, pp. xii. xiii.

or Rishee, within seven generations; and Brahmins of the same Rishee, or descendants of the Rishee's brothers and connections within three or four degrees, may not intermarry. Mayhew gives an instance of two Rajpoot families who were not allowed to intermarry, though their common ancestor dated from eight hundred years before. Marriage with a mother, sister, or daughter is absolutely prohibited, and also with a father's sister's daughter, a mother's brother's or sister's daughter, or a sister's daughter; but these latter prohibitions are sometimes broken through poverty. In the Wys caste all cousins may intermarry; and the Komtees even permit the marriage of nephews and nieces with aunts and uncles. The Brahmunjaee, and R. Josee marry a mother's brother's daughter, and the Kykaree marry the same, but they are not allowed to marry the daughter of a mother's sister. The wife belongs to her husband's stock, and hence marriage is not prohibited between two families who are already connected by marriage; indeed, marriages of this sort are preferred. An adopted son is forbidden to marry within the prohibited degrees of either his real or adopted family; and should he be re-adopted into his own family, he must not intermarry with the last, nor may his descendants for three generations.[1]

On the death of the husband, his nearest kinsman has authority over the widow if she have no son, for

[1] In some castes of Poona, intermarriage is forbidden between *all* cousins, and with a nephew or niece on *both* sides. Steele, *Hindoo Castes*, etc., pp. 26, 27, 47, 80, 163, 346, 347; M'Lennan, *Prim. Mar.*, pp. 105, 106, note; Mayhew, *London Labour*, etc., vol. ii. book iii. p. 119; Tylor, *Researches into the Early Hist. of Mankind*, etc., p. 282.

like the Romans they consider that a woman should be under perpetual tutelage. If she be not of high caste, and consequently obliged either to burn herself or lead a devotee's life, the levirate law comes into play. As soon, however, as a son is born to the family of the deceased, the kinsman must live with her as father and daughter-in-law.[1]

Females are considered much less desirable as children than males. They are by their nature evil, and must be constantly watched. They love their bed, their couch, their ornaments; have impure appetites; and are prone to wrath, weak flexibility, desire of mischief, and bad conduct; they are like leeches, ever exhausting their husbands—nay, wind, death, the infernal regions, the fury of the ocean, the edge of a razor, poison, venomous serpents, and devouring fire, all united, are no worse than women.[2] When this is the opinion of their law-givers we may readily anticipate the relative value of the sexes in the Hindoo mind. No female is entitled to investiture with the sacred thread, a ceremony analogous to our baptism; and women, together with children and idiots, are excluded from caste deliberations on caste offences. No woman can be security, nor may she give evidence, and a female only inherits one quarter of what is inherited by a male. A man is allowed to marry again if his wife either proves barren, or what is considered equivalent, if she only bears female children, and Menu expressly forbids marriage into a family where there are no male children. More-

[1] Steele, *Hindoo Castes*, etc. pp. 30, 31; Colebrooke, *Digest of Hindu Law*, Book IV. secs. xiii. cxlvi.-cl. clvi. clvii.
[2] Colebrooke, *ut sup.*, secs. xxiii.-xxix.

over, a woman becomes pure again twenty days after the birth of a boy, while after the birth of a daughter she is not pure for a month.[1]

The Gosawees, a devotee sect, do not marry, though females belong to the sect; but a degenerate branch of them, the Ghurbaree Gosawee, do. These have essentially the same laws of marriage as the other Hindoos, and though their castes permit disciples to join them, a Gosawee may only marry a woman belonging by birth to his caste, and female disciples, not by birth Gosawees, are forbidden to marry. The Gosawees are further divided into ten sects, one of which, the Geeree, are not allowed to marry among themselves.[2]

Of the various Indian tribes, Herodotus says, that some practise promiscuous intercourse like beasts,[3] and two Mussulmans who travelled over some part of India in the ninth century assert that the Indians never marry within their own family, but consider that marriage with strangers improves their offspring.[4] Both are probably right to a certain extent, for the aboriginal hill-tribes are but little removed from beasts; and, as we have seen, though the Hindoos do not marry out of their divisions, they do not marry strictly within their own family. The Khonds consider intermarriage between members of the same tribe, however large and scattered, as incestuous, and punishable by death. Marriage can only take place between members of different tribes, not even with strangers who have been

[1] Colebrooke, *ut sup.*, Book IV. secs. lxvi. lxx. lxviii. clxxxv. cxxviii.; and Steele, *ut sup.*, pp. 23, 57, 126, 275, 285, 30.
[2] Steele, *ut sup.*, pp. 444, 445.
[3] Herodotus, Book III. 101.
[4] Reich, *Ehe*, etc., p. 211.

long adopted into or domesticated with a tribe. They consider it degrading to give their daughters away into their own tribe, and deem it more manly to seek wives abroad.[1] The English have partially induced them to give up their ancient custom of female infanticide, due, partly to their religious belief, and partly to social causes. For women are regarded as very unpleasant relatives. A father is bound as a surety for his daughter's behaviour, and he, with his family and clan, are called upon to make good any sin she may commit to her husband, his family, or his clan. For this reason girls are generally murdered on the seventh day after birth, thousands yearly being thus sacrificed. Another reason for this female infanticide is that noble families deem it shameful to marry their daughters to any one of inferior rank, and equally shameful to have a nubile unmarried daughter; while many of them have neither the will nor the means of marrying their daughters in a manner considered proper to their rank. A very convenient belief here steps in: they consider a female child a most acceptable offering to the infernal deities; hence pride, avarice, and superstition form a cabal more than strong enough to conquer even a mother's love for her offspring.[2] The Sodha are also strictly exogamous. They cannot take a wife from the division to which they belong.[3] The Twana of the Punjaub commonly practise incest. They are a fine set of people, but are gradually dying out.[4] The Ho are forbidden marriage out of their

[1] M'Lennan, *Prim. Mar.*, pp. 95, 96.
[2] Reich, *Ehe*, etc., p. 210. See also Browne's *Indian Infanticide*.
[3] M'Lennan, *Prim. Mar.*, pp. 103, 147.
[4] Conversation with an Indian friend.

own tribe, but at the same time are forbidden to marry within their own family division. The Koch, Bodo, and Dhumal, are all forbidden marriage out of their own tribe.[1] The Garrows are divided into clans, and no one may marry into his own clan. They have the levirate law, and if there are no brothers left for the widow to marry, she marries her father-in-law.[2] The Munipuree and the Koupooe, Mow, Muram, and Murring tribes inhabiting the hills round Munipur, are each divided into four families. A member of any of these families may marry into any save his own.[3] The Warali tribes are divided into sections, and no one may marry in his own section. The Magar tribes are also divided into sections, all members of a section are supposed to belong to the same stock, and may not intermarry.[4] The Moondah and Oraon, again, are divided into clans, and marriage within the clan is forbidden.[5] The Todas are divided into five classes which never intermarry. The Yerkala of Southern India consider that the maternal uncle has a claim on the two first daughters of his sister as wives for his sons. His claim, like that of the Bedouins,[6] has a money value; he pays a lesser price for his nieces, and if he forgoes his claim on them, gets part of their price. The Doingnaks abandoned their parent stem, the Chukmas, rather than intermarry with the tribe in general as their chief, Jaunbux Khan, wished them to do.[7]

In Ceylon, marriage from a higher into a lower

[1] M'Lennan, *Prim. Mar.*, p. 147.
[2] Sir J. Lubbock, *Origin of Civilization*, p. 96; Eliot, *Asiatic Res.*, vol. iii. p. 28.
[3] M'Lennan, *ut sup.*, p. 109.
[4] *Ibid.*, p. 104.
[5] Lubbock, *ut sup.*, pp. 95, 96.
[6] See page 88 of this work.
[7] Lubbock, *ut sup.*, pp. 96, 102, 103.

clan was peremptorily forbidden. The people see no harm in incest, and it is largely practised, though it is said that under the native kings no one was allowed to marry any one related to him more nearly than a second-cousin. Polyandry was practised on the score of economy for the poor, and expediency for the rich; since in the first case money was saved in the price of the bride, and in the second division of property was avoided. This custom is now confined to the province of Kandy, as the English have altogether much modified the law of marriage. Although women are regarded more as companions than is usual in Asia, they are sold by their parents to whomsoever will pay their price, and as might be imagined, infanticide is consequently not practised.[1] From a recent census report it appears that 1 in every 723 is insane; 1 in 860, deaf; and 1 in 357, blind.[2]

Bowring asserts that marriage is not permitted within the seventh degree (nearer than third-cousins) of blood-relationship among the commoners of Siam, but that the king may marry his sister or his daughter.[3] Loubère narrates the facts more clearly; he says that in his time the king of Siam married his sister, and then his daughter, the issue of that marriage.[4] Yet Dr. Campbell tells us that the Siamese have no prejudice against consanguineous marriages, nor do they believe that the offspring suffer from them.

[1] Mayhew says that female infanticide was common, but M'Culloch, on the authority of Davy, does not believe in the practice. (Mayhew's *London Labour*, etc., vol. ii. book iii. pp. 126, 127; Reich, *Ehe*, etc., p. 212; J. A. N. Perier, *Mém. de la Soc. d'Anthrop. de Paris*, vol. i. 1863, p. 218; M'Culloch, *Geographical Dict.*)
[2] *The Times*, Nov. 24th, 1873.
[3] Tylor, *Researches into the Early Hist. of Mankind*, p. 283.
[4] J. A. N. Perier, *ut sup.*, vol. i. 1863, p. 218.

"B. dying, a few of his widows might be asked to join the hareem of A.; and on A.'s death, his heir might retain a few of the widows of A. and B. A son in this way may espouse his step-mother or step-aunt, or his niece."[1]

Martinus Martinius tells us that up to the time of Fohi, B.C. 2952,[2] the Chinese had no idea of incest, and kept their mothers among the rest of their hareem, or rather, practised promiscuosity. Fohi first separated the sexes, instituted marriage and marriage laws, and prohibited marriage between persons bearing the same family name. The number of their family names, according to Davis, is not more than 100, but others say 300, or even 1,000. Fohi himself is said to have divided them into 100.[3] Marriages between persons of the same family name are not only null and void, but punishable by fine and blows. Even a marriage between persons who are only related by affinity is considered incestuous, if within the fourth degree. Thus, a man may not marry his father's or mother's sister-in-law, his son or daughter-in-law's sister, his father's or his mother's aunt's daughter, or his mother's brother's or sister's daughter. Death by strangulation is inflicted on him who marries a brother's widow; while marriage with a father's or grandfather's wife is considered worse still, and the culprit is beheaded.[4] Yet a man may marry the aunt of his wife; for the late Emperor of China did so,[5] and it will readily be seen that

[1] Dr. J. Campbell, in the *Journal of Anthrop.*, London, Oct. 1870, p. 196.
[2] Reich, *Ehe*, etc., p. 189. Mr. Tylor gives this date as B.C. 2207 (*Researches into the Early Hist. of Mankind*, p. 282.)
[3] Tylor, *ut sup.*, pp. 282, 283.
[4] Mayhew's *London Labour*, etc., vol. ii. book iii. p. 131.
[5] One of his wives was the sister of another wife's father, Saishanga. See *The Times* for May 16th, 1872.

like so many other communities, the Chinese really take account of only one side of relationship, and hence a man may marry not only his first-cousin, but even his niece.[1]

From the evidence of the same two Mussulmans who travelled over India in the ninth century,[2] it appears that the Chinese were then polygamists. Davis, however, asserts that now at least there is no polygamy there. A man may at once take a concubine, if his wife is childless, or bears him only female children; but if she has a male child he is obliged first to get his wife's consent. It often happens that Chinese midwives are bribed with large sums to substitute a male changeling for a new-born female babe —a process they somewhat ungallantly describe as changing a dragon for a phœnix. They also have a saying that for a female infant a common tile is good enough as a toy, but to a male a gem should be given; and a Chinese proverb says that ten daughters are not equal to one son.[3] Under the circumstances, therefore, it is not astonishing that female infanticide is

[1] MM. Devay and Chipault, who have not recognized the real state of the Chinese law on this subject, hold up their prohibited degrees as models to European States (Devay, *Du Danger*, etc., p. 133; and Chipault, *Etudes*, etc., p. 85). Luckily for themselves, European States do not seem eager to follow the admonitions of these gentlemen. Setting aside for the moment the grave doubt whether marriages between near kin are really harmful, what possible gain would there be were marriages between persons only related by affinity to be forbidden in a further degree than they already are? Have we not already seen how every prohibition beyond a certain point is defied by the public? Do we not daily see marriages of this kind contracted in spite of the law and with very general approbation?

[2] See p. 95 of this work.

[3] Mayhow's *London Labour*, etc., vol. ii. book iii. pp. 129, 133, *et seq.*; Reich, *Ehe*, etc., pp. 190, 191, 211; and Waitz, *Anthropologie der Naturvölcker*, vol. i. p. 380.

largely practised. From recent accounts some 80 per cent. of the female children are drowned. Indeed, the evil is so great that it has more than once forced itself on the notice even of the Chinese Government.[1]

In Chinese Turkestan the law is very different, since every marriage is permitted save those between parent and child.[2]

Polyandry is customary in the neighbouring country of Thibet; the wife belongs to all the brothers of the family into which she marries.[3]

Among the Mongols, according to the Russian monk Hyacinth, marriage with a woman related on the paternal side is considered incestuous; but there is no relationship through the female side. A man may marry three sisters at once, and two families may intermarry for centuries.[4]

The Mantchu Tartars have just the opposite regulation as the Chinese. Marriage is forbidden between persons of different family name.[5]

The Tunguz, according to Mr. Tylor, prohibit marriage between second-cousins; but he says no more about them.[6]

The Jakuts are not allowed to take a wife from their own clan;[7] and the Ostyaks are forbidden marriage either in the ascending or descending line of their wife's relations, or even with a brother's widow; but it is considered an honourable thing to marry several

[1] See Mayhew, *ut sup.*; and the *Spectator* for Aug. 23rd, 1873.
[2] St. Lager, *Etudes sur les Causes du Crétin.*, etc., p. 115.
[3] Reich, *Ehe*, etc., p. 239; M'Lennan, *Prim. Mar.*, pp. 193, 194; Sir J. Lubbock, *Origin of Civilization*, p. 101.
[4] Reich, *Ehe*, etc., p. 245.
[5] M'Lennan, *Prim. Mar.*, p. 146.
[6] Tylor, *Early Hist.*, etc., pp. 283-284.
[7] Sir John Lubbock, *ut sup.*, p. 97.

sisters at once, and, as among other peoples, a woman is sold to her brother-in-law at a cheaper rate. The wife takes her husband's name; and no one may marry a woman of the same family name as himself. Hence a man may marry his sister's daughter; and every union is lawful, provided the father or deceased husband of the woman had a different name to the person who wishes to marry her. Infanticide is only practised on deformed children.[1]

The whole of the Samoyed nation is divided into three tribes, and intermarriage in the same tribe is forbidden. Since these tribes are generally in parts of the country far remote from each other, young men have to go great distances in search of a wife.[2] This is just the state of society in which we should expect to find female infanticide practised, and the following passage from Castren makes it more than probable:—
"If we ask a Samoyed bard how it is that he makes despicable woman the subject of an heroic poem, he answers at once, that since from ancient times it has always been customary to take a wife from a foreign stock, and never from their own tribe, and these stocks were usually hostile to each other, it was by no means easy to get a wife by fair means, or without paying a price which, owing to the practice of polygamy, was far beyond the means of any but rich men. We were therefore constrained to use force and gain a wife by individual valour."[3] This passage

[1] Sir J. Lubbock, *Origin of Civilization*, pp. 96, 97; Mayhew, *London Labour*, etc., vol. ii. book iii. pp. 168, 170.

[2] M'Lennan, *Prim. Mar.*, pp. 102-103; Tylor, *Researches into the Early Hist. of Mankind*, p. 284; Sir J. Lubbock, *Origin of Civilization*, etc., p. 96.

[3] Cited by Roich, *Ehe*, etc., pp. 279-280.

is the more valuable, as it gives a key to the origin of exogamy. An artificial dearth of women was produced by infanticide or polygamy, and hence young and poor, but fighting men, took them by force from their neighbours, and from this it became so customary to take wives by individual valour, that whoever did not do so was looked upon as a milksop. The Samoyeds are said to be now nearly extinct.[1]

The Kalmucks have a great abhorrence of marriages between near kin. Their wives must always be three or four degrees removed; and nobles must marry the daughters of nobles of a different stock; and though a proverb of theirs says that "Great folks and dogs know no relationship," it is only in consequence of the occasional marriage of a noble with his sister-in-law.[2] The Kalmuck population is now diminishing, the number of women especially is very disproportionate to the men, there being only 51,000 women to 68,000 men. That they diminish is attributable to the great mortality among the children; for though very few Kalmucks remain unmarried, and every mother has on the average four children, only two as a rule reach maturity; partly because in some places the people feed on fish, and have neither pastures nor herds; and partly because of female infanticide.[3]

Like most of the Tartar races, the Kirghiz and Nogais may not marry in their own clan.[4]

---

[1] *Pall Mall Gazette*, May 27th, 1871.
[2] M'Lennan, *Prim. Mar.*, pp. 98, 99.
[3] *Pall Mall Gazette*, May 29th, 1874.
[4] M'Lennan, *Prim. Mar.*, p. 103. Du Perron, talking of the marriage of cousins being considered by the Parsees as "L'alliance la plus recommandée," adds in a note, "Le même usage a lieu chez les Tartares;"

The Lapps never marry a girl descended from the same common parent as themselves, however distant the relationship may be. They are a polygamous, but not a fertile race.[1]

The Malays have no relationship excepting through females. When a man marries, his wife follows him, but still belongs to her own family, and so do all her children. The children acknowledge some relationship on the father's side, but not beyond the third degree, and that only in the direct ascending and descending line. No one is allowed to marry into his own family;[2] hence a man may not marry any relative on his mother's side, but he is allowed to marry his half-sister, his niece, his aunt, or more distant relatives on the paternal side.

The people of Benkulen, the Palembang, and Lampong have much the same marriage customs as the Malays proper; but the Batta have relationship through males. Hence, though they may not marry into their father's family, they may marry into their mother's family in the same close degrees. If a wife does not bear male children, her husband may send her back and demand her sister instead; yet infanticide is not practised.[3]

The Kalang of Java never marry a daughter to any one who cannot prove his descent from her particular stock.[4]

In the Island of Bali, the Hindoo religion exists

but he does not say which. See his *Trans. of the Zendavesta*, vol. ii. p. 556, note 3.

[1] Tylor, *Early Hist. of Mankind*, p. 284; Reich, *Ehe*, etc., p. 281.
[2] Waitz and Gerland, *Anthropologie*, vol. v. part i. pp. 141, 142.
[3] *Ibid.*, pp. 147-150, 186, 187, 190, 191.
[4] Sir J. Lubbock, *Origin of Civilization*, p. 103.

in great purity, with division into castes as in India.¹ From which I presume that they have the same laws of marriage as the Hindoos.

Among the Dyaks of Borneo, much immorality prevails, as is the case in nearly all the large islands by the sea; yet Dyak girls seldom intermarry with foreigners or even Malays. The Hill Dyaks are still more strict. Even the marriage of first-cousins is prohibited, and second-cousins may only marry on payment of a jar, a severe fine for them. Ruin and darkness had descended on the land, said they, when one of them married his grand-daughter; and it had remained so ever since. Yet they always marry in their own tribe, and are therefore all blood-relations.²

Near blood-relations may not intermarry on the Island of Formosa, and no man may marry unless he can produce the head of an enemy slain in battle.³ This latter, explains the former prohibition.

The State of Wajo is perhaps the most advanced on the Celebes, and here a high spirit of caste prevails. As usual in these cases, women alone do not dare to marry into a lower caste, while men may.⁴

Arrago asserts that on the Caroline Isles it is customary for brothers to marry their sisters. Gerland, however, is inclined to doubt this on the authority of Chamisso, who says that the marriage customs on the Carolines are the same as on Ponape,⁵

[1] Waitz and Gerland, *Anthropologie*, vol. v. part i. p. 40.
[2] Mayhew, *London Labour*, etc., vol. ii. book iii. p. 103; Tylor, *Early Hist. of Mankind*, p. 283.
[3] Reich, *Ehe*, etc., p. 226.
[4] Mayhew, *ut sup.*, vol. ii. book iii. p. 107.
[5] Ascension Island, or Seniavin. Some day, perhaps, these islands in the Pacific will have dropped a few of their superfluous names.

where first-cousins are forbidden to intermarry. It is customary for widowers to marry their deceased wife's sister, and widows to marry their deceased husband's brother. Rank was always inherited through the mother. The people of the Eastern Carolines were divided into classes, which were not allowed to intermarry; and as in other islands the chiefs were all members of that society known elsewhere as the Areoi, who alone were permitted to practise incest.[1]

On the Marianne (or Ladrone) Islands the people are divided into three classes, the Matuas, Atshaots, and the Mangatshangs; or nobles, gentlemen, and common people. Besides other disabilities, the lowest class never intermarry with the nobles, nor was there any intermixture of blood between them. The latter were a far superior race to the former, who were liars, faithless, and inhospitable. A wife here had extraordinary power. All her children, even those which were not her husband's, were considered legitimate; all relationship was on the female side; but the relationship to a female was nearest on the male side, or relatives would be near in the following order:—paternal-aunt, maternal-aunt, maternal-uncle, paternal-uncle. No marriage could take place between very near relatives on the female side, but here, as in all the Micronesian Isles, the chiefs belonged to the society of Areoi, called here Ulitaos, and could practise incest without hindrance.[2]

The inhabitants of the Radack and Ralick chains, forming the Marshall Islands, are divided into several

[1] Waitz and Gerland, *Anthropologie*, vol. v. part ii. pp. 106, 108, 121.
[2] *Ibid.*, vol. v. part ii. pp. 107, 111-113, 148.

clans. Here also relationship is through females only. Their customs are much the same as on the other groups.[1]

The Polynesians count relationship through the female line, the husband taking the wife's name. In most cases girls lived a free life before marriage, and polygamy was everywhere common. Throughout Polynesia existed a sort of spiritual relationship, a custom of exchanging names with a great friend who thus became another self, and who shared his wife with, and might not marry a blood-relation of his friend's. Wilson asserts that marriages with blood-relatives were everywhere avoided; but they do occur from political motives nevertheless, among the chiefs and nobles. Thus Tamehameha married a relation, while his son married a sister, and from love to his father, one of the latter's widows. The wives of chiefs throughout Polynesia were the equals of men, but all other women, except on certain occasions, were *noa*, or common to all. They had a curious society called the *Areoi*, which we have already had occasion to notice, who were a fraternity of nobles with branches in many of the islands, living like beasts in promiscuous intercourse. All the unmarried, widowed, or separated women belonging to the society could have as many lovers as they liked; and the offspring of these unions, if females, were killed at birth; if males, the firstborn alone were sacrificed, and the rest might be saved.[2] Besides in Tahiti, the Areoi

[1] Waitz and Gerland, *Anthropologie*, vol. v. part ii. pp. 105-106, 111, 121, 122.

[2] Gibbon doubts Cook's assertion as to the practices of this society (*Decline and Fall of the Roman Empire*, vol. viii. p. 191), and with some

were to be found in Raratonga, Nukuhiva, Hawaii, in the Carolines, and in the Mariannes. The mythology of the Polynesians makes the year bear months by her own father. An old legend, again, makes Taaroas take to wife his daughter, or by some accounts his sister Hina; their son married his mother or grandmother Hina, who became young again, and by her became the father of Ouru and Fana, from whom mankind are descended. Infanticide is everywhere very prevalent. Indeed in Hawaii, owing to the practice of female infanticide, there are far more males than females, but infanticide of both sexes is altogether largely practised.[1]

In Tahiti, as many as two-thirds of all children born were murdered, though chiefly females; and all children of mixed unions were killed, especially when the mother was of higher rank than the father; for the Polynesians consider that there are two races of men, one of heavenly, the other of earthly origin; hence mixed unions were considered a defilement of heavenly blood, and the children must not live. It is a rule, however, not to marry among kindred, from which I presume they do not marry among relatives of their mother's, for everywhere inheritance goes through the female line. The great chiefs belonging to the Areoi were only allowed to keep their eldest son; while the upper class Areoi were obliged to kill their eldest son, and

justice as regards *married* women of the higher class Areoi. (See Wilson, Forster, Ellis, and Morenhout compared in Waitz and Gerland, *Anthropologie*, vol. vi. p. 366.)

[1] Waitz and Gerland, *Anthropologie*, vol. v. part ii. p. 111; vol. vi. pp. 122, 222, 123, 127, 130, 131, 348, 366, 137, 138, 233, 234, 321, 322, 139, 140; Reich, *Ehe*, etc., p. 353; J. A. N. Perier, *Mém. de la Soc. d'Anthrop. de Paris*, vol. i. 1860, p. 218.

all their daughters; and the rest of the Areoi were obliged to kill all their children. In Tukopia, on the other hand, boys are frequently killed from a fear that the island may become overcrowded, while girls are sometimes saved because of the practice of polygamy.[1]

In the Markesas, as elsewhere in Polynesia, the succession is through females, and the princesses were allowed to practise polyandry. In Hawaii, Nukuhiva, and Tahiti, it is customary for the princes to marry a sister.[2]

In Tonga marriages between relations were usual. Infanticide was also largely practised, chiefly as a sacrifice during a parent's illness, those being preferred who were born from mixed marriages, or from parents of different rank. Inheritance went through females; and this was pushed to so ridiculous an extreme that the eldest sister and the aunts of the Tuitonga, together with their children, were considered more holy than the Tuitonga himself; while his daughter was so holy that though she virtually practised polyandry, she could not enter into wedlock; and her daughter was even more holy than she herself.[3]

In Samoa marriage between near relatives on the female side is not allowed.[4] In Vate, or Sandwich Island, they seem generally to marry into a different family, though these families are usually hostile to

---

[1] Waitz and Gerland, *Anthropologie*, vol. v. part ii. p. 191; vol. vi. pp. 113, 139, 219; Gerland, *Aussterben der Naturvölker*, p. 55.
[2] Gerland, *Aussterben der Nat.*, p. 46; Waitz and Gerland, *Anthropologie*, vol. vi. pp. 215, 216.
[3] Waitz and Gerland, *Anthropologie*, vol. vi. pp. 131, 138, 171, 177, 178.
[4] *Ibid.*, p. 127.

each other. Infanticide is here very common.[1] On the Nitendi (or Gomora) Group, the Salomon Isles (including Bougainville, Choiseul, Isabelle, and Malayta), Tanna, and the islands of Torres Straits, the custom of exchange of names is universal. The populace of most of these isles is divided into many families, who are hostile, and do not intermarry. In Fiji the chiefs have great power, for every man whose mother is a member of some chieftain's family, may travel about from isle to isle, and take any woman or property he likes, provided neither she nor it belongs to any other member of this privileged class.[2]

Great opposition is made in New Zealand to any one who wishes to take a wife from another tribe, unless it be for a political purpose. Marriages are not usual between brother and sister; but a man may marry several sisters at the same time, or he may marry first-cousins; and altogether marriages between near kin are far from uncommon. Mr. M'Lennan is however of opinion that at one time they must have been exogamous, since in the *Curse of Mania* there is an instance of a child fleeing from the tribe of its birth to that of its mother; and this seems the more likely since female infanticide is very common.[3]

The general state of health in all Polynesia and in New Zealand was very good at the time they were first brought under European notice. Dieffenbach

---

[1] Waitz and Gerland, *Anthropologie*, vol. vi. pp. 639, 657.
[2] *Ibid.*, pp. 622, 657, 663.
[3] *Ibid.*, pp. 131, 138; Gerland, *Ausst. der Nat.*, pp. 57, 58; M'Lennan, *Prim. Mar.*, p. 125; Sir J. Lubbock, *Origin of Civilization*, pp. 103-104; Mayhew, *London Labour*, etc., vol. ii. book iii. p. 72; Adam, *Fortnightly Review*, 1865, p. 720.

found in the inner part of New Zealand some cases of club-foot; Hasenscharte saw cases of hereditary polydactylism; and Beechey saw the same in Mangareva. Thompson even asserts that malformations of this kind are as common in New Zealand as in England; but he refers only to the coast, where the health of the people had suffered much. Malformations were most frequently found on the Sandwich Isles, where King saw crooked persons, a young man without feet or hands, many who squinted, and a man who declared his blindness was congenital; while they were rarest on the Markesas. Albinoes were seen on all the islands, and madness was not unknown; but the chief epidemic diseases were first brought by Europeans. Of all diseases that they were subject to before the arrival of Europeans, those affecting the skin were the most common—such as ichthyosis, elephantiasis, irruptions, and boils; but in New Zealand boils were not common, and most of the diseases were cured simply by salt. Scrofulous glands and sores were common in Tonga, and in both Tonga and Uevea the children suffered much from malignant postules—an affection which disappears of itself at nine years of age. The Tongese also had occasionally a mild inflammation of the eyes. Nevertheless the people are healthy, as a rule; disease is uncommon; and nature's power for healing is here exceptionally strong. There is one disease which is peculiar to the nobles who alone are allowed to drink an intoxicating liquor called *cava*. Cava drinkers suffer first from weakness and torpor; and if they continue the habit of drinking they grow thin, their eyes inflame, they get an irruption, which is marked at first by the appearance of a white scurf, then the skin tightens and cracks, and leaves a large cicatrix on healing, which is looked upon as a mark of honour, because it is a mark of nobility. Persons who have sufficient self-control to break themselves of this pernicious habit, speedily and entirely recover.[1]

The Polynesian women are remarkably fruitful; Cheevor saw a woman in Hawaii, who had twenty-five children; Forster gives many examples of great fertility from Tahiti; and Dieffenbach tells us that in New Zealand all marriages are fruitful, and twins are frequently born. It is the same in Tonga, Tukopia, and Samoa. Nor is the general viability lower than in Europe, while the old are free from that decay which so often accompanies old age in civilized countries. They are of medium height, well formed, but those who are badly fed and have to work hard are not so fine a race as the nobles, who are to the workers as giants to dwarfs.[2] Hale describes the former as of colossal build, stout, proud, brave, and impudent; the latter, he says, are thin, poorly, timid, and servile. In Niva, the usual way in which a chief disposes of his captives is to break their backs across his knee. Indeed everywhere the chiefs are both finer physically and better taught than the common herd, and in Fiji it is most strikingly shown that the poor state of the latter is due to the bad food and excessive

[1] Waitz and Gerland, *Anthropologie*, vol. vi. pp. 24, 25, 60.

[2] We may perhaps trace a similar state of things among the ancient Egyptians, who always distinguish their chiefs in their wall pictures by their superior size. Were the Polynesians to draw themselves, they would infallibly distinguish their nobles in the same way.

work; for the chiefs alone here are allowed to marry their relatives. For the same reason the women do not generally grow up handsome; but all writers agree in the splendid build of most Polynesians, a build often rivalling the finest statues of antiquity. It varies somewhat on the different islands; on the higher ones the inhabitants are bigger, stronger, whiter, and better developed; and the opposite is the case under reversed circumstances.[1] Now, indeed, all is changed; disease is common since intercourse with Europe has been established, and the character of the natives is much corrupted and debased.[2]

In East Australia, says Lang, through a large extent of the interior, and among tribes speaking different dialects, there are four names for men, and four for women, to wit:—Ippai and Ippata, Kubbi and Kapota, Kumbo and Buta, Murri and Mata. Now, an Ippai will marry a Kapota; a Kubbi will marry an Ippata, a Kumbo a Mata, and a Murri a Buta. The child of an Ippai and Kapota will not take either of these names, but considers itself a Murri or Mata, according to its sex. The child of a Kubbi and Ippata, again, will consider itself a Kumbo or Buta; the child of a Kumbo and Mata will be a Kubbi or Kapota; and the child of a Murri and Buta will be an Ippai or Ippata.[3] This

---

[1] Waitz and Gerland, *Anthropologie*, vol. vi. pp. 2, 6, 25, 26, 203, 542, 544; vol. v. part ii. pp. 47, 49, 55, 56, 113, 171, 173, 174, 193. M. Chipault, indeed, talks of disease in Isabello Island, Guam, Amboyna, Tahiti, and Barabora, as resulting from consanguineous marriage (*Etude sur les Mars. Cons.*, pp. 80-83). The authority against this supposition, based as it is on very slight authority, is overwhelming.

[2] Waitz and Gerland, *Anthropologie*, vols. v. and vi. generally.

[3] Tylor, *Researches into the Early Hist. of Mankind*, p. 285.

at first sight looks as if the tribes had made elaborate rules, to prevent consanguineous marriage; and so they have, but only in the same degrees as among ourselves, as may be seen by the accompanying table:

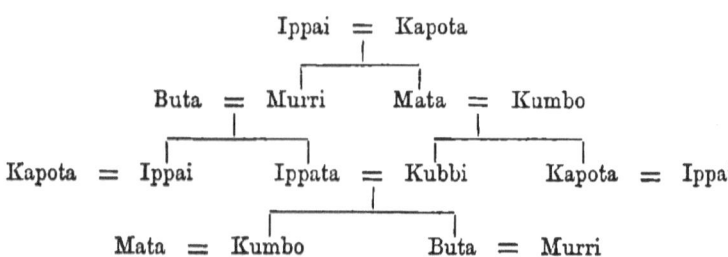

for first-cousins are allowed to marry, and very possibly marriage might be contrived within much nearer degrees, and still not break the rule, even were not marriage permitted also in the same tribe; for instance, an Ippai might marry an Ippata, and his children would be called Kumbo, or Buta.[1] This arrangement, therefore, though it might possibly indicate a former stage of civilization when it was customary always to practise exogamy, modified to a more peaceful stage when only the form of exogamy was kept up, certainly was not made to avoid marriage between near kin, for in that case it might be done in a much simpler and more effective way; indeed the succession through females countenances the first supposition, as much as it discountenances the second. In the North-west persons more nearly related than first-cousins are not allowed to marry. The people are polygamous, and since children take the mother's name, half-brothers and sisters may intermarry; marriage being only forbidden between persons of the

[1] Tylor, *Researches into the Early Hist. of Mankind*, p. 285.

same family name. Incest is punished with death.[1] Polygamy is universal throughout Australia, but it is rare at Port Essington. In the West girls are betrothed to married men while quite infants, when they are taken care of by his family, and the parents trouble no more about them. In the South the same custom is in force; but here they remain until full grown in their mother's care. Women are so scarce that the discarded wives of old men are taken by the young, who cannot get wives of a suitable age at all; indeed it is seldom that any man can marry before thirty. The reason is that female infanticide is largely practised; the third daughter being always killed, and the second generally. At Cape York, among the Muralugs, children of mixed unions, according to Dr. Gerland, are always killed; but here, if any other children are killed, it is the boys. In New South Wales every girl, the offspring of a mixed union, is killed. A man's widows, betrothed, and children, are all inherited by his uterine brother.[2]

The Australians often attain the age of seventy or eighty years; and Sturt frequently saw very old men with snow-white hair. Of illnesses, they have an eye disease, provoked by the desert state, dust, insects, and heat of the South, but as the causes, so the disease is not confined to that part. Skin diseases are common, such as lepra, and a widely spread disease like the itch, and boils, headaches, etc., are also known. On the whole, however, they are very healthy, and recover in a marvellous manner from wounds and disease.[3]

[1] M'Lennan, *Prim. Mar.*, pp. 117-120, 122; Mayhew, *London Labour*, etc., vol. ii. book iii. p. 69.
[2] Waitz and Gerland, *Anthropologie*, vol. vi. pp. 771-781.
[3] *Ibid.*, p. 718.

The Tasmanians were accustomed to seize their wives from other clans whose language was different; and though their wives were virtuous, promiscuosity was recognized. There were more men than women on the island, yet infanticide was very uncommon. Hombron stands alone in his assertion that the Tasmanians are ill-favoured; while on the other side Labillardière, Cook, Breton, Nixon, Jeffreys, Bibra, d'Urville, Peron, and Holman, combine in giving us an agreeable picture of them. According to Dr. Gerland they are probably of the same race as the Australians, but not further connected with them.[1]

In New Guinea (or Papua), a woman on marriage enters her husband's family; but relationship and inheritance go through the female line. A widow is obliged to marry her deceased husband's brother. Infanticide is common, and boys are much more thought of than are girls; indeed, the claim of the former to an inheritance is so much greater than that of the latter that if a man dies and leaves only daughters, his brother's sons get most of the property; while his daughters, or if he have none, his nieces, get only a very small portion. The inhabitants of Triton Bay were, according to Hombron, a mixed race, between Makassas, Balinese, Timorese, and Papuas; and he accounts for their beauty on this score. To this Dr. Gerland, however, well remarks that " such cross-breeds may possibly exist, but there must have been some aboriginal population to give birth to them, of the beauty of whom Hombron says nothing at all. It is besides impossible that these strangers could ever have been numerous enough for

[1] Waitz and Gerland, *Anthrop.*, vol. vi. pp. 719, 720, 813, 815, 816.

their offspring to form any important fraction of the population. The fact is, he has here followed his prejudices."[1]

The North-Western American Eskimo, the Kadjak, pay no attention to relationship in marriage; neither do the Aleuts;[2] but the Greenland Eskimo forbid marriage between persons related in the first three degrees of consanguinity, nor do they consider it reputable for persons, even though not related, to marry if they have been brought up in the same house. A man may marry two sisters or a mother and daughter at the same time; but such an action is looked on with disfavour.[3]

The North American Indians are divided into clans varying in the several nations from three to ten, the members of which are dispersed among the different nations. Every clan has its distinguishing emblem, such as a bird, beast, fish, etc., called a *totem*. The Algonquins never intermarry among themselves, for they have all the same emblem or family name. The Iroquese and the Delaware never marry into their own stock. The Canadian Indians do not marry in the first degrees of relationship, and greatly prefer sons to daughters, for the former add to their fighting power. The Kenai, Omaha, Atmah, and Kolosh, choose their wives from another stock; their offspring, as everywhere else among these Indians, belonging to the mother's family. Now, however, the Kenai no longer care to follow the old custom; marriages occur in the same tribe, and though consanguineous marriages are no commoner in conse-

---

[1] Waitz and Gerland, *Anthropologie*, vol. vi. pp. 536, 537, 634, 639, 661.
[2] Waitz, *Anthropologie*, vol. iii. pp. 313, 314,
[3] Mayhew, *London Labour*, etc., vol. ii. book iii. p. 173; Reich, *Ehe*, etc., p. 361.

quence than they were before, owing to the one-sided idea of relationship, yet the old people assert that the mortality has risen since this innovation was introduced. The Nutka have a curious trade which looks as if they, at least, did not pay much attention to forbidden degrees; for certain men make a trade of buying girls when six years old, and selling them again as brides when they are old enough. Among the Tinneh no man may marry a woman of the same family name as himself; if he does, he is said to have married his sister, even though she be really no relation. The Choctaw are divided into two great septs, each of which is subdivided into four clans, and no man may marry into any of the clans belonging to his sept. The restriction among the Creeks, Natches, and Cherokees, does not extend beyond the clan to which the man belongs. A Cherokee may marry a mother and daughter at once, but if he commits what they consider to be incest, he is burnt alive. The Ayowas consider that every husband has the first right to his wife's younger sisters, but he is allowed to hand them over to his friends if he likes. Uncles and aunts are deemed the same as fathers and mothers, and first-cousins consider themselves brothers and sisters. In the north-easternmost portion of the United States, in that part known formerly as New England, marriage was allowed between brother and sister in a chief's family, when there was no one else of equal birth to marry; and in other parts chiefs sometimes arrogated to themselves the right of marrying within degrees forbidden to the rest of the tribe. But these were the exceptions: the general law being everywhere the same, that persons of the same family name were forbidden to intermarry. Relationship was everywhere through the mother alone, who was considered to have a far greater share in the child than its father.[1] Half-brothers and sisters were therefore not related to each other; while maternal uncles and aunts were styled father and mother, and maternal cousins brother and sister.[2]

In Mexico, under the Chichemeks, and before the foundation of the Mexican Empire, it appears that a man might only marry one wife, who must not be a "near relation." This custom seems to have persisted in the youth of the Mexican Empire: marriages between persons related in the first degree or even between step-parents and their step-children, were forbidden under penalty of death. About the time of Montezuma I. the nobles first became a caste apart and privileged, and this class soon became better formed and stronger than the common people, who also were divided into castes since their professions were hereditary. It was the custom of the king and aristocracy to take their female relatives of the third degree as concubines, at least, in the time of Montezuma II., and in all probability also before him.[3] Among the Misteks there does not seem to have been any restriction on marriages between near kin; it was even customary for the ruler to marry a relative.[4]

[1] See p. 26 of this work.
[2] M'Lennan, *Prim. Mar.*, pp. 121, 122; Sir J. Lubbock, *Origin of Civilization*, etc., pp. 97, 98; Waitz, *Anthropologie*, vol. iii. pp. 106-109, 328, 329, 333; Mayhew, *London Labour*, etc., vol. ii. book iii. p. 86; Reich, *Ehe*, etc., p. 378.
[3] Waitz, *Anthropologie*, vol. iv. pp. 27, 130, 43, 71, 64, 81, 103, 146, 89; Reich, *Ehe*, etc., p. 405. [4] Waitz, *ut sup.*, p. 130.

The Comanche Indians[1] increase mostly by theft of children and women from their neighbours, and especially from the Creoles. The Apaches do the same. Hence these tribes are greatly mixed, and few persons belonging to them can boast of pure blood.[2]

The people of Yucatan never marry a wife of the same family name as themselves. But here the family seems to follow the male line, for marriage with maternal cousins is permitted.[3]

Relationship outside the first degree is no hindrance to marriage in Nicaragua, but rather a recommendation.[4]

The Indians of Darien marry a woman first for a week to her future father-in-law, and afterwards in succession to all her future husband's male relatives.[5]

The Panches of New Granada never marry a woman belonging to the same village as themselves. Female infanticide is practised, for it is customary to kill all infants born before a male; and therefore the marriage remains childless if no male child is born at all.[6]

The Indians of the Orinoko district may marry their sister's daughter, says Gili, but never a brother's. It is not at all uncommon for a man to inherit his step-mother on the death of his father, or his sister-in-law on the death of his brother.[7]

[1] Known also as Tetaus, Nauni, Niyuna, and Jamparika.
[2] Waitz, *Anthropologie*, vol. iv. pp. 213, 214.
[3] Tylor, *Researches into the Early Hist. of Mankind*, p. 287.
[4] Reich, *Ehe*, etc., p. 406.
[5] *Ibid.*, p. 408. Reich suggests that something like this may have been the origin of the European *Jus primæ noctis*.
[6] Waitz, *Anthropologie*, vol. iv. p. 376.
[7] Reich, *Ehe*, etc., pp. 418, 419.

THE MARRIAGE OF NEAR KIN. 121

The Macusi of Guiana have no relationship on the male side: a paternal uncle may not marry his niece, but he may marry his sister's daughter, his step-mother, or sister-in-law. They do not practise infanticide. A Warrou widow is inherited by her husband's nearest relation, but she can free herself by a refusal and consequent savage beating. They deduce relationship solely through the female line, and a child is therefore not considered to be any relation to its father.[1] The Arrowak are divided into several clans, and no one may marry in his own clan. But this, of course, does not prevent consanguineous marriage, because relationship is as usual one-sided, and in this particular case only through females.[2] The Yamanos of Ecuadore have the same prohibited degrees as the Macusi of New Guiana.[3]

The ancient Peruvians, according to Garcilasso de la Vega, before the dynasty of the Incas, used to marry indifferently whomsoever they chose; their mothers, their sisters, or even their daughters. The children of the first Inca, Manco-Capac (A.D. 949–1006) married among themselves, and ever afterwards the Inca's chief wife was his sister, so that the blood of the royal Peruvian line, the blood of the Sun, might always be kept pure; for they held that the Sun himself had married his sister the Moon, and from that union they were descended. According to Fernandez, the sister they married might not be an uterine sister. No female of the royal line was allowed to marry a vassal. The son

[1] Waitz, *Anthropologie*, vol. iii. pp. 391, 393; Reich, *Ehe*, etc., pp. 421, 422; Sir J. Lubbock, *Origin of Civilization*, pp. 98, 99.
[2] Waitz, *Anthropologie*, vol. iii. p. 392.
[3] Reich, *Ehe*, etc., p. 450.

of the Inca's sister was his heir, or failing a direct heir, probably the eldest male relative on the female side was chosen. Trades and offices were generally hereditary; no one could change his domicile; the members of every community were forced to marry within that community; nay, so paternal was the Government, that couples were paired by regular officials whose business it was. Gomara asserts that the soldiers customarily married their sisters after the royal example; and both he and Herrera say that the Orejones were permitted by Huayna-Capac (A.D. 1475–1525) to marry their female relatives on the paternal side. Hence, as polygamy was only permitted to the high nobles, by far the larger proportion of marriages in ancient Peru were between near kin, for it is impossible that a community always marrying among themselves and never allowed to change their domicile could have done otherwise. Far from being reduced to either physical or mental decay by these laws, there resulted a degree of skill in the arts apparently equal to any ancient civilization of that latitude; while after the first surprise was over, the people resisted with such bravery the attack of the Spaniards, that these were more than once reduced to very unpleasant straits.

Infanticide was very rare in Peru; indeed, it probably never occurred at all except for religious service, and then only on extraordinary occasions, such as on the beginning of a war, or the sickness of a parent. In these cases boys were generally sacrificed. In some of the provinces the first-born child was always killed; in others, one of twins, for the birth of twins was thought to be an omen of evil. The Peruvian Indians at present never marry within the fourth degree, probably a result of their Spanish teaching. Infanticide is hardly known, since formerly parents who wished to get rid of their offspring had only to place it at the door of some well-to-do person, whom custom compelled to take care of it; while they now have a large foundling hospital.[1]

The Brazilian races differ greatly from one another as to their marriage regulations. In some of the very scattered tribes who live in small communities far removed from each other, the nearest relatives intermarry. In more populous districts, on the contrary, the tribes are divided into families, and no one may marry into his own family.[2] The Mandrucus have the levirate law. The brother of a widow is obliged to marry his sister's daughter, if she is marriageable, and he cannot find a substitute. An uncle on the paternal side is not allowed to marry his niece. Lerius, who travelled towards the end of the sixth century in Brazil, says, "No one marries his mother, sister, or daughter; further consanguinity have they none. A paternal uncle may marry his niece, and all degrees outside this are open."[3]

[1] Waitz, *Anthropologie*, vol. i. p. 204; vol. iv. pp. 411, 415, 416, 412, 417, 461, 473-477, 478, *et seq.*; Reich, *Ehe*, etc., pp. 445, 446; J. A. N. Perier, *Mém. de la Soc. d'Anthrop. de Paris*, vol. i. 1860, p. 217. On the science of the ancient Peruvians, see also Godron, *De l'Espèce et des Races*, vol. ii. pp. 238, 239. There seems to be a very great resemblance between the ancient Egyptians and Peruvians, in every way; it has always struck me that much more might be done in the elucidation of history by means of the law of analogy than appears as yet to have been done.
[2] Sir J. Lubbock, *Origin of Civilization*, p. 99.
[3] Reich, *Ehe*, etc., pp. 434, 436.

Freycinet asserts that it is by no means rare that a Coroado is at once father, brother, and even son-in-law to the same person.[1] There is a sort of spiritual relationship among the Tupinamba like that in Polynesia: two men who adopt each other as brothers, are real brothers to all intents and purposes.[2] The Tupi only consider marriage to be forbidden in the first degree, and with a maternal aunt or an uterine sister; but an uncle has a prior right to his niece's hand.[3]

"Long experience," says Father Martin Dobrizhoffer, "has convinced me, that the respect to consanguinity, by which they are deterred from marrying into their own families, is implanted by nature in the minds of most of the people of Paraguay."[4] But Dobrizhoffer studied chiefly the Abipones, and therefore we must consider his remarks as applying only to them. He says that since the women are obliged to suckle their infants for three years, they often practise infanticide to prevent their husbands seeking other wives in the meanwhile. It is the boy who is generally sacrificed thus, for when a son grows up it is necessary to buy a wife for him; while a grown-up daughter will always command her price. Rengger tells us that many of the Indian races of Paraguay are too proud to intermarry with any race of a different colour, or even of a different stock; and in consequence these races are free from syphilis. The Guanias, according to the same author, practise

---

[1] J. A. N. Perier, *Mém. de la Soc. d'Anthrop. de Paris*, vol. i. 1860, p. 218.
[2] Tylor, *Researches into the Early Hist. of Mankind*, p. 290.
[3] Waitz, *Anthropologie*, vol. iii. p. 422.
[4] Cited by Tylor, *Researches into the Early Hist. of Mankind*, p. 287.

infanticide, generally on females. Azara says of another tribe, the Charruas, a brother is never allowed to marry a sister; from which I presume that marriages outside the second degree are permitted. The Yuracares do not seem to respect relationship in marriage, and practise infanticide.[1]

In the Argentine Republic, says Mantegazza, consanguineous marriages are frequent "with all their unhappy consequences."[2] Infanticide is very rare, for well-to-do people always adopt foundlings, of which probably there is a large proportion, since the looseness of morals is so great that it almost amounts to promiscuosity.[3]

The Patagonians marry in every degree save the "first."[4]

Du Tertre informs us that the ancient people of the Antilles had no idea of consanguinity; fathers married their daughters; mothers, their sons; or a man might marry two sisters or mother and daughter at once. The Apachalites avoided intermarriage with other stocks, and looked upon children born of exogamous unions as bastards; indeed, excepting in the first degree they married whom they would. The Caraïbs generally married their father's or mother's sister's daughter, who, according to Labat, were considered to be the wives appointed by nature for them; and often married several sisters at

---

[1] Reich, *Ehe*, etc., pp. 455-458; Waitz, *Anthropologie*, vol. iii. p. 534.
[2] Signor Mantegazza has written, besides his *Lettere Mediche Sulla America Meridionale*, a pamphlet entitled *Studj sui Matrimononj Consanguinei*, in which he does not mention anything about the unhappy consequences here. I have not seen the former work.
[3] Reich, *Ehe*, etc., p. 461.
[4] Waitz, *Anthropologie*, vol. iii. p. 505.

once. Davies asserts that, like the Bedouins and several other peoples, a girl belonged of right to her cousin, who on his part was obliged to marry her himself, or find her a husband. Yet Walker says, on the authority of Dr. Hancock, "the Caribes are the only American tribe who, without restraint take wives from the other tribes adjacent; and their superiority over all their neighbours is too well known to require a word in illustration." Probably Dr. Hancock refers to the Black Caraïbs, a mongrel stock. Relationship was through females alone; but the sons, says Herrera, generally inherited something from their father. The superior value of male children is shown in the custom that on the birth of the first boy the father had to take to his bed, and avoid certain kinds of food, a custom found in various and widely distant communities.[1]

Minutoli relates that the people of Lanzarota and Fuertaventura formerly practised polyandry, a wife usually having three husbands, who were but little more than her slaves. In the isles of Palma and Gomera the king had the *jus primæ noctis*, and the first-born were therefore considered nobles. Beauty, in these islands, is measured by the pound, yet the women were considered fruitful, though they had to suckle their children by means of goats. No woman was obliged to bring up more than two daughters, all the rest might be killed.[2]

Du Chaillu asserts that in Western Equatorial Africa there is no intermarriage in the same clan,

[1] Waitz, *Anthropologie*, vol. iii. pp. 383, 384; Reich, *Ehe*, etc., pp. 411, 412; Adam, *Fortnightly Review*, 1865, p. 720; Walker, *On Intermarriage*, p. 364.   [2] Reich, *Ehe*, etc., p. 318.

however far removed the relationship may be; but there is no objection to taking a brother's or a father's wife. He adds: " I could not but be struck with the healthful influence of such regulations against blood-marriages among them."[1] Monrad says the exact contrary. Blood-relationship, according to him, is no hindrance to marriage, except it be in the first or second degrees.[2] The fact is, that marriage is forbidden in the same clan, but that the relationship is one-sided, so that the nearest relatives on one side may marry, while intermarriage is forbidden between the most distant relatives on the female side. A Krooman inherits not only his father's wives, among his other property, but even his own mother.[3] The Fantis leave their wives, with the rest of their property, to their sister's son; and on other parts of the coast the son inherits all his father's wives, with the exception of his own and his father's mother. He must wait a year, however, before he may take possession. The Papels, the people about Cape Palmas, and the Bambarras, have all a similar custom.[4] These latter are divided into several castes, and no one may marry out of his own caste excepting the village chiefs, who are all of royal blood.[5] The Jolofs are also divided into castes, which probably do not intermarry. The castes of the Weavers and the Singers are considered so very low that nobody ever marries into them.[6] In Aquapim, two families who have the

---

[1] Sir J. Lubbock, *Origin of Civilization*, p. 95.
[2] Reich, *Ehe*, etc., p. 303.
[3] Mayhew, *London Labour*, etc., vol. ii. book iii. p. 63.
[4] Waitz, *Anthropologie*, vol. ii. p. 115.
[5] *Ibid.*, pp. 134, 135.
[6] *Ibid.*, p. 137.

same fetish do not intermarry;[1] the fetish doubtlessly being the same only for relatives on the female side.

Alberti informs us that the Southern Kaffirs consider it unseemly that blood-relations should be much together, and do not allow marriage between them. Kay says they intermarry much with neighbouring peoples, the chiefs of the Amakosa always marrying with Amatembu women, and the Amapondo though not always with neighbouring people, yet never marry in their own village. Affinity seems no hindrance to marriage, for a man may marry two sisters at once. The Kaffirs generally value girls more highly than boys, since a girl may be sold for from ten to seventy head of cattle, according to her rank. A father's widow belongs to his son, and if there is no male heir, she becomes the property of the community.[2]

The Zulu only permit marriage between the most distant relations. A widow belongs to her brother-in-law.[3]

The Hottentots only marry outside the third degree, punishing incest with death. They rejoice exceedingly when male twins are born to them or to their relations; but should female twins be born to them the weakest is put out to die.[4]

In a tribe of hippopotamus hunters settled on the little island of Nyamotobsi, on the river Zambesi, called Akombwi, or Mapodzo, the men rarely, and

[1] Tylor, *Researches into the Early Hist. of Mankind*, p. 284.
[2] Waitz, *Anthropologie*, vol. ii. pp. 355, 388, 390; Reich, *Ehe*, etc., pp. 322, 323.
[3] Waitz, *ut sup.*, p. 390; Mayhew, *London Labour*, etc., vol. ii. book iii. p. 64.
[4] Reich, *ut sup.*, pp. 323, 324.

the women never intermarry with other tribes. " They are rather a comely-looking race, with very black smooth skins."[1]

Of all tribes in Africa the Bogo seem to bear away the palm for prohibited degrees. Persons related to the seventh degree, says Munzinger, may not marry, whether the relationship be on the paternal or maternal side. Affinity is not held in such respect, for a widow falls with the rest of her husband's harcem, to his son or brother; but she never marries her own son. The Bogo midwives hail the birth of a boy with a quintuple shout of joy; a girl being received in ominous silence. The time during which a husband is forbidden to see his wife after her confinement is only three weeks in the case of a girl, and four when the infant is a boy.[2]

The Somali frequently marry into foreign stocks, forbid marriage between first-cousins, and in all nearer degrees of relationship excepting that of uncle and niece. A widow is generally married by her deceased husband's brother.[3]

In Uganda, Speke saw King Mtesa at a levée, attended by women who were at once his sisters and his wives.[4]

The king of the Assubo-Galla always marries a relative; and the Seers, who constitute a separate caste, always marry among themselves, considering themselves the only true Gallas. Women are considered very inferior to men, and may be murdered

[1] Livingstone, *Zambesi*, pp. 38, 39.
[2] Reich, *Ehe*, etc., pp. 333, 334.
[3] Waitz, *Anthropologie*, vol. ii. p. 522.
[4] Adam, *Fortnightly Review*, 1865, p. 721.

at half price. Widows are inherited by the brother-in-law.[1]

The Gypsies of Egypt marry exclusively among themselves. According to Newbold they are divided into three stocks, called respectively the Helebi, the Ghagar, and the Nuri, or Nawer. The women of the Helebi are modest, those of the Ghagar rather the reverse. The Helebi are most strictly endogamous, not even marrying into other Gypsy tribes; but the Nuri intermarry with the Egyptian Fellahin. Seetzen says that the Gypsies of Egypt always intermarry among themselves; nor do they allow their daughters to marry out of the tribe unless among the Ghawazee.[2] These latter are a distinct people, of a different caste of countenance to the Arabs, who fancy themselves directly descended from the Barmekees. They hardly ever marry out of their tribe, the women perhaps never; for they pride themselves on perpetuating their race unmingled. Lane considers them to be the direct descendants of the ancient Egyptian dancers, and says, "Upon the whole, I think they are the finest women in Egypt."[3]

The northern part of Africa is peopled chiefly by Mohammedans, who marry according to their Prophet's law, but often restrict intermarriage with other tribes. The Moors of Tunis are an instance of this; and their dislike of intermarriage with the Arabs is heartily reciprocated.[4] The Hal Ben Ali are also very proud of the purity of their blood, and therefore never contract marriage out of their tribe, except per-

[1] Waitz, *Anthropologie*, vol. ii. pp. 516, 518.
[2] Reich, *Ehe*, etc., pp. 248, 249.
[3] Lane, *Modern Egyptians*, vol. ii. pp. 86-91.
[4] Reich, *Ehe*, etc., p. 295.

haps with one of the fair Abd el Nour females. The same is the case with the Ouled Sidi Sheikh, renowned for their beauty, who never give their daughters in marriage out of the tribe, unless to some marabout of great family. The Tuarik will not ally themselves even with the Arabs, but always marry among themselves; and are said to be very handsome.[1]

The people of Madagascar, although they may neither marry a blood-relation, nor a relative by affinity up to the sixth generation, have really a much greater licence than appears at first sight, for the principal restrictions are on the female side, since collateral branches on the male side are allowed to marry in most cases on the observance of a slight but prescribed ceremony, which is supposed to remove any disability caused by relationship. The Hovas, the chief race, compel the king always to marry a near relative, when possible his sister's daughter, since otherwise his nearest relative's eldest son would inherit the throne instead of his own son. Thus, Radama's first wife was his sister, and another was his first-cousin Ranavalou. As a rule a man receives all the younger sisters of his wife into his harem. A female infant is not considered a pleasing gift by a Madagascar husband, boys being greatly preferred to girls.[2]

This little is all that I have been able to collect in the short time at my disposal, on the prohibited degrees.[3] It does not pretend to be more than a

[1] J. A. N. Perier, *Mém. de la Soc. d'Anthrop. de Paris*, vol. i. 1863, p. 191; Godron, *De l'Espèce et des Races*, vol. ii. pp. 166, 167.

[2] Waitz, *Anthropologie*, vol. ii. pp. 432, 433, 438, 446; Tylor, *Researches into the Early Hist. of Mankind*, p. 284.

[3] It is a pity that no really good general work on anthropology has

cursory view; but yet I think it will be found sufficient for the purpose of solving the following questions: firstly, why do many savage tribes prohibit marriage between certain relatives, and why do others prefer marriage between near kin; and, secondly, is there any innate horror implanted in mankind against incest; or is this horror only a result of custom, and brought about by various and complicated circumstances?

History, and the analogy of animal habits, seems to point to promiscuity as the first stage of relationship between the sexes.[1] Mr. Darwin seems to think

yet been written. The unfinished work of Professors Waitz and Gerland is perhaps the best we have as yet on the subject, yet this work is very imperfect, and not so well arranged as it might be. We must bear in mind, however, the difficulty in the production of these works. Few travellers take the trouble to learn how to observe. Few, again, are able to recognize nice distinctions of law and manners in the places they visit, or to qualify themselves to give a report of any value on the statistics of health and reproduction, without a long residence among the people they describe; while those whom accident or official duty compels to remain in a fair field for such observations are too often silent on subjects which especially interest us. The opportunity is in many cases now for ever lost. Many nations have entirely died out, and have left no clue to their habits. Europeans are changing the primitive savage customs all over the globe with unexampled rapidity, and where they do not introduce the most loathsome vices of our civilization, they acclimatize laws on these poor people, which are the accumulated result of ancient civilizations entirely different from theirs, and unsuited to modern times even in Europe. Where savages are not shot down or induced by Europeans to kill each other, sicknesses have been introduced among them both purposely and accidentally. They have been robbed of their lands and their food, of their children and their virtues, and then Europeans complain that it is impossible that they and the aborigines can live peacefully together! More cannot now be collected on the primitive habits of a very great number of savage tribes; we can only con over more diligently the careless statements of travellers who have written, read more correctly the codes of those nations who have left a literature behind them, generalize from the comparisons of various customs in various climes, and thus possibly make up a very great loss.

[1] M'Lennan, *Prim. Mar.*, p. 175, et seq.; Sir J. Lubbock, *Origin of Civilization*, p. 60, et seq.

polygamy is the first stage,[1] for the support of which theory he also has the analogy of the habits of animals, many kinds of which are polygamous. Promiscuosity could not indeed occur in a mere family, for a natural jealousy would lead to fighting, just as it does in polygamous animals, and perhaps with the same result that the weaker males would be turned adrift, and the stronger become polygamous. But neither could polygamy become the rule in a larger community, since it would manifestly be impossible for one man to turn out all the rest; they must live and let live ; the strongest reasons urge them to remain at peace with one another, for only their unity will enable them to withstand their enemies; from which it follows that here there must be communal marriage, with perhaps a slight touch of monogamy when some noted strong man insists on having a wife to himself, and that the community must remain in this stage until the tribe is sufficiently civilized to allow a weaker man to hold what a stronger may take away. For these reasons I am inclined to think that polygamy and promiscuosity are more dependent on the number of the community than on any particular grade in their civilization, provided always that those grades are sufficiently barbarous, since polygamy can be traced again as the direct result of a comparatively high degree of civilization. Mr. M'Lennan considers female infanticide to have been the foremost cause of exogamy ; and the cases I have been able to collect certainly seem to confirm his theory. We see a very salient example of how this may be brought about in the Koombees of the Kutree

[1] Darwin, *Descent of Man*, vol. ii. p. 362.

caste, who live on both banks of the river Myhee, and are divided by it into two classes. The higher class destroy their daughters, *in order that their sons may be forced to go to the other side of the river for wives, and receive in return a large dowry for their condescension.*[1]

The causes of exogamy are indeed manifold. Continued probably for the same reason that other tribes practise endogamy, from that pride which forbids a man to marry his daughter to her inferior in rank, or to marry her at all without trying to outdo his neighbour in the prodigality of the marriage display, it

[1] Browne, *Infanticide in India*, p. 56. Sir John Lubbock, though he acknowledges female infanticide as a cause of exogamy, does not seem to give it its full weight, but rather ascribes exogamy to the natural preponderance of male to female births, and to the desire of men to get wives for themselves (See his *Origin of Civilization*, pp. 92, 94). The last cause is a very probable factor, but not in my humble estimation likely to originate the custom alone. The natural overplus of male births is much too small to have any effect, and even in Europe it is not constant, while, as Waitz shows, it is often the other way in other parts of the world (Oosterlen, *Med. Stat.*, pp. 163, 164; and Waitz, *Anthropologie*, vol. i. pp. 126, 127). One may travel through village after village in India where for centuries no more female infants have been saved than might easily be counted on one's fingers. Among the Jarejah, of the west coast alone, it is estimated that some 30,000 are annually murdered. In Rajpootana and Malwa, when infanticide was supposed to be on the decline, thanks to the zealous endeavours of the English, "an intelligent Rajpoot chief" estimated the number of female infants murdered annually at no less than 20,000. It is among proud races, not the poor, that this crime is commonest. In the year 1852, there were in one district :—

	ADULTS.		CHILDREN UNDER FIVE.		Percentage of Boys to Girls.
	Males.	Females.	Males.	Females.	
Hindoos of all Classes.	206,452	164,735	92,777	77,818	83·87
Rajpoots of all Grades.	20,178	16,832	8,506	6,742	79·26

probably arose from the constant hostilities between neighbouring tribes. On the one side there was an artificial dearth of women caused by these hostilities;

Among the higher *clans*, this disparity between the sexes is far greater. There were of the three tribes—

Name of Tribe.	Adults.		Children under Five.		Percentage of Boys to Girls.
	Males.	Females.	Males.	Females.	
Kutoch	832	612	382	154	40·31
Pathanea ..	2,500	2,000	1,044	350	33·52
Golehrias ..	1,501	800	574	125	21·75

It must be borne in mind that the English had already put a partial stop to infanticide when these statistics were gathered. In fifty (Munhás) Rajpoot villages in the Seealkote district a careful analysis of the population produced the following result:—

Girls.					Boys.
3 years old and under.	Between 3 and 6 years of age.	Between 6 and 8 years of age.	Above 8 years of age.	Total.	
129	40	16	5	190	462

The period between eight and six years represents the time when efforts were being made by the English to suppress infanticide, and therefore we see a greater proportion of females as their age is less—the measure of the success of those efforts. Among the class of Bedees of the Jullundhur district, consisting of some three hundred families, only two daughters had been preserved under the Sikh rule; and among forty-two Rajpoot villages of the Agra districts in which the crime of infanticide was deemed to have been most prevalent, there was an average apparent increase of female births between May 1st, 1851, and January 1st, 1854, of 137 per cent.; while in particular villages this average was as much as 600 per cent.; and in one village 850 per cent! (See Browne's *Infanticide in India*, pp. 34 note, 59, 106, 185, 186, 189, 192. But the whole work should be consulted.) In China, at a place called Kea-King-Chou, about a distance of five days' journey from Canton, there were some 500 or 600 cases of female infanticide per month, and about 500 female infants were annually received into the Foundling Hospital

on the other, hostilities caused by an artificial dearth of women; both causes aggravating each other until exogamy became an institution. Under these circumstances, of course, every community would seek to strengthen itself at the expense of its neighbours. Each one would kill the females, and thus remove at once a source of weakness to themselves and of temptation to the enemy. Thus tribes who practised exogamy must have been more inured to fighting, because every man had to fight before he could obtain a wife, they must have been more compact, and the whole tribe must have been constituted more like an army than were those who habitually practised endogamy. Exogamy, however, implies an individual property in a wife; for any one who is able to distinguish himself by individual prowess is not likely to surrender the captive of his bow and spear to the will of the community; and in this way a tribe may become either monogamous or polygamous, according to their superiority over their neighbours in war. Endogamy, on the other hand, is

of Canton, and 200 into that of Shanghae. The villagers of the neighbourhood of Amoy confessed that two of every four were killed; some killed three, four, or even five out of six. Indeed, so common was the crime that the Chinese Government issued a proclamation against it in the year 1838 (Mayhew, *London Labour*, etc., vol. ii. book iii. pp. 133, 134); and lately there was another proclamation, in which the number of female infants drowned was estimated at 80 per cent. (See the *Spectator* for Aug. 23rd, 1873). The reader will readily imagine from these statistics how many female infants are annually murdered among other peoples who practise this infanticide.

Sir John Lubbock considers exogamy further due to "the advantage of crossing, so well known to breeders of stock," which " would soon give a marked preponderance to those races by whom exogamy was largely practised" (*Origin of Civilization*, p. 94), yet he says a little before: " In fact, however, exogamy afforded little protection against the marriage of relatives, and, wherever it was systematized, it permitted marriage even between half-brothers and sisters, either on the father's or mother's side" (*Ibid.*, p. 92).

essentially a mark that the tribe practising it has either been long at peace, or has had very little to fear from its enemies. We always find it as a custom of superior people, or people that fancy themselves superior to their neighbours, and who probably, at least at one time, have been so, such as the Jews, the warlike Affghans, or wherever else there is a spirit of caste. But exogamy can rarely turn into endogamy, even when an exogamous nation has long been incontestably superior to its neighbours, without leaving some permanent trace behind. They have hinged their whole system of marriage upon it; to take a wife from their own stock has so long been considered a mark of cowardice that it has become shameful for any one to do so; and the custom of taking a wife by force has become so inveterate that it is even simulated when she is taken from a friendly stock. Endogamous nations, on the other hand, can easily become exogamous, either when female infanticide has become general from some of the secondary causes already mentioned; or when the neighbours become so rich and powerful as no longer to be despised; or again when the tribe becomes more civilized, and consequently more tolerant. Of course we see apparent exceptions to both these rules. In some tribes which are exogamous it is customary to kill the males and save the females, because the former are a source of expense to the parents, while the latter have a marketable value. Many apparently endogamous nations occasionally allow marriage outside that nation to males, for the reason that men make laws, and women have to obey them; and do not allow intermarriage between certain relations, because they have formerly been accustomed to exogamy. But the general law holds good, that endogamy arises from pride of race, and exogamy from female infanticide and littleness of tribes; *not*, as Mr. Tylor and Mr. Darwin urge, from any observed bad consequences of these marriages.[1]

Mr. Tylor seems to think it probable, from the great similarity, or rather absolute identity of many customs which exist in widely separate parts of our globe, that there has been some connection between the tribes practising them until after these customs were established. Customs, such as the restrictions on the intercourse of parents and children-in-law; the cure of diseases by the pretended extraction of some foreign substance; and the custom that the husband should receive the congratulations of his friends and be dieted on the birth of a child; together with many others.[2] But I cannot admit such a deduction. It seems to me that under the same circumstances the same customs will arise independently of any connection between tribes practising them; for to deny this, would be to deny that similarity in mental structure which we have every right to assume, and which is shown in the invention or discovery of the same thing by different men independently of each other when society is " ripe " for that discovery, or in other words, when both discoverers have been led to it by the same course of circumstances. Shall we need to see an historical connection between the makers of flint knives; would not any race choose a

---

[1] Tylor, *Researches into the Early Hist. of Mankind*, chap. x.; Darwin, *Variation of Animals, etc., under Domestication*, vol. ii. p. 123.

[2] Tylor, *ut sup.*, pp. 302, 303.

flake of flint for that purpose without being taught to do so? Can we not fancy how, after generations of men have used these, they will fashion any piece of malleable metal they may find also into a flake, and use it as a knife? Even in the same tribe we see that savages are not content with doing the same thing in the same manner, from which we may fairly deduce, that they also have a genius for invention, and do not depend on some one individual to make all their discoveries for them. King found at York Sound, on the river Roe in Australia, no two huts in the village alike; and Simpson found all possible forms of tents in use among the Flatheads of North America. Can we suppose an original connection between the Bosjesmen, the Aleuts, the Australians, and the Caroline Islanders, because they have the same method for making fire by rubbing a stick perpendicularly to another piece of wood; when we see that on Radack and the Sandwich Isles they have a different method of rubbing two sticks together, and that the Algonquin obtain it by striking sparks from a flint? Mr. Tylor's further argument, that because certain customs common to many tribes are very peculiar, therefore there must have been some connection between them, is extremely difficult to reconcile with the relative positions of those tribes. Is it possible that at a time when people were civilized enough to invent that custom on which he lays so much stress, known as *faire la couvade*, it should have been taught by some original ancestor to the Basques, the Caraïbs, the people of Cassanga in West Africa, the Mandrucus, and the Abipones of South America, the people of West Yunnan, and of the

Pearl Islands, and to various other isolated communities, when we see that their neighbours have not got this custom?[1] It appears to me, indeed, as a "growth of savage psychology" arising quite independently in different tribes, just as the belief that a mother has no share in her offspring has probably arisen independently in various tribes, the rise of which idea it is easy to picture to oneself. In the same way we can account for the identity of proverbs and moral fables existing among people who have been severed from intercourse with each other even more intellectually than they have physically. Æsop's fable of the dinner given by the crane to the fox might possibly, though this is very doubtful, have been received by the negroes of West Africa from Europe, or Europe from them; but could their many identical proverbs, such as, "The apple does not fall far from the bough," "No man can serve two masters," "Fine feathers make fine birds," "A golden key will open any lock," and many others of course not in this imagery, but to the same purpose, have been received from Europe, or even from a common source? Is it not far easier to see how these proverbs may constantly arise anew?[2]

Mr. Tylor, however, insists less on the traditionary origin of the restrictions on marriage than even on these customs; and that the restrictions on marriage arose independently, I certainly see no reason to doubt—exogamy, from the common desire of getting a wife; endogamy, from the equally common desire

---

[1] Tylor, *Researches into the Early Hist. of Mankind*, pp. 300-304; Waitz, *Anthropologie*, vol. i. pp. 294-295.

[2] See Waitz, *Ibid.*, vol. ii. pp. 244-245.

of retaining one's superiority. If the former custom arose from any observed evil results of consanguineous marriages,[1] it was absurdly inadequate to its object, and would besides argue a power of observation that we civilized Europeans have not attained to, together with a power of reasoning equally incompatible with savage character. On the other hand, if exogamy arose from the mere desire to get a wife, we see no one-sided arrangement in the matter, and the whole theory rests on the sound basis that this custom has arisen from a passion, not only the deepest and most powerful in mankind, but from a passion which he shares with the whole animal and, perhaps, vegetable world. Of course there are apparent difficulties in this view also, since a few tribes forbid the intermarriage of relatives on both the paternal and the maternal sides, such as the Hindoos, the Bogo, and the people of Madagascar; but these difficulties can be explained when we reflect that it would be strange

[1] The only instances I have met with in any savage nation of any evil effect being attributed to consanguineous marriages, are that reported of the Kenai, whose custom of marrying into a different stock to that of their mother's having fallen into disuse, the old people assert that the mortality has consequently risen (Sir J. Lubbock, *Origin of Civilization*, p. 98)—an assertion like that of old people who are so fond of saying, "Ah! things are not as they used to be when *I* was young!" Probably the mortality has risen among the Kenai, just as it has among other North American Indian tribes, but not from this cause. Another instance is given by Reich (*Ehe*, etc., p. 211), who, on the authority of two Mohammedan travellers of the ninth century, says the East Indians never marry a relative, because they thought exogamous marriages improve the offspring—a natural idea when the true motive of exogamous marriage had been forgotten. A third instance is cited by Gerland (*Anthropologie*, vol. vi. p. 141), who says that the Hawaiians accounted for Tamehameha's small family by the fact that he had married a relation. I am inclined, however, to attribute this opinion to European influence, for if this had been an idea originating with the natives such marriages would long before have been forbidden.

indeed were no people to project themselves beyond the idea of one-sided relationship; and if they do this, it necessarily follows that they must extend the forbidden degrees to that side also. We see how apt mankind is to do this thing in the spiritual relationship, or rather adoption, which we find practised in spots extending all round the world, and in the idea of relationship which we find in Greenland extending to all persons reared in the same house. It would be far more just, but equally incorrect, to argue that the intermarriage of some tribes always among themselves was because they had noticed a beneficial result on the offspring; for here we have instances of an apparent good resulting from that practice in the islands of Polynesia and elsewhere, where the chiefs who practise incest are a far finer race of men than the common people who do not. I have shown reason, however, for believing that this superiority is probably not due to their incest;[1] and the reason so often urged by themselves, and which forces itself on the notice of any one who observes that this custom is commonest among nobles and superior races, that only by incestuous marriages, or marriages which exogamous nations would regard as incestuous, can they obtain wives socially their equals, is probably the sole reason for this custom.

Is there then no innate horror of incest, no relationship at which we can draw the line and say, here all humanity agree that incest begins? We are too apt to consider incest—that crime from which Milton draws his foulest picture,[2] and the practice of which

---

[1] See pp. 112-113 of this work, note.
[2] Milton, *Paradise Lost*, Book II.

Hallam describes as the vilest phase to which mankind can sink[1]—as too monstrous to discuss whether it is a crime at all; or, in other words, whether the practice of it is hurtful to the community. Now, incest was practised in the direct line by the Persians, and by all people, ancient and modern, who have no marriage, but practise promiscuity. In the first degree collateral, it was besides practised by the Egyptians, and perhaps at one time by the Jews and Peruvians. Marriage with half-sisters was practised by all these, by the Greeks, and by all nations who count relationship only on one side, and it was therefore practised by nearly the whole uncivilized world. From this we may fairly conclude that there is no natural horror of incest implanted in the human race, or it would never have become habitual among any great number of people. If the horror of incest were innate, it would be as universal as the passions, whereas there was no horror until relationship became more settled and mankind more civilized. We must therefore attribute the dislike which unquestionably exists among the greater number of people to marriage with near kin, to the practice of seizing wives from a foreign community, which made it seem cowardly for a young man to take a girl to wife in any other way. But although this is sufficient to account for the prohibition of intermarriage between relatives in the third degree and beyond, it hardly accounts for the almost universal prohibition of marriage in the direct ascending or descending line which we find in force even among endogamous tribes. In these there is apparently no reason why a father

[1] Hallam, *Middle Ages*, vol. i. p. 301.

should not marry his daughter, if it is true that there is no natural horror of incest in any degree. A father would regard his daughter as a slave, and hence add her to his hareem without more ado, were he not withheld by some feeling hitherto unexplained. That this feeling is an "innate horror" in the generally understood meaning of innate, is hardly possible, since in that case it must be universal, and must show itself by analogy among animals. Man should then shrink instinctively from an incestuous connection, even if he did not know of any relationship. Some of the older writers, indeed, have tried to show that this horror exists among the lower animals; thus Pliny relates how a horse, on the discovery that it had committed incest, threw itself over a precipice and was killed;[1] and Heywoode, after repeating the story, concludes with, "If then this sinne be so hatefull in bruite beasts and vnreasonable creatures, how much more ought it to be auoided in men and women, and which is more, Christians!"[2] Even at the present day this idea is not extinct, for Walker says, "That this aversion, however, should * * * exist among animals" is a strong proof of the impropriety of breeding in-and-in in the nearest degrees!"[3] And M. Devay asserts that M. Munaret "and other distinguished observers" noticed that animals have an instinctive repugnance to incestuous unions.[4] Perhaps M. Munaret and other distinguished observers hold with Pliny, that fowls have a sense of religion, that elephants understand the nature of an oath, and that

[1] Pliny, *Hist. Anim.*, Book VIII. cap. 64.
[2] Heywoode, *Hist. of Women*, pp. 176, 177.
[3] Walker, *On Intermarriage*, pp. 292, 293.
[4] Devay, *Du Danger*, etc., p. 65, note.

dolphins prefer the name of *Simo* to all others, because their noses are turned up?[1] Ovid, although he had a great horror of incest, as is shown in the first part of the fable of Myrrha and Cinyras, is yet at a loss to see why such marriages should be forbidden, and cites the very case of animals to show that they are not forbidden by nature.[2] Indeed, incest is constantly practised by animals, and habitually by those which are polygamous. The greater number of the mammalia are polygamous, since most deer,[3] cattle, and sheep are so. In herds of about a dozen antelopes in South Africa, says Mr. Darwin, rarely more than one mature male is seen. The Asiatic *Antilope saiga* drives away all rivals and collects a herd of about one hundred females and kids. The horse is polygamous, and so is the wild boar; it is rare to find more than one adult male with a whole herd of females. The elephant is also polygamous, but both he and the wild boar only consort with the females during the breeding season. The gorilla and several species of baboons are polygamists. Lions

[1] Pliny, *Hist. Anim.*, Book X. chap. 57; Book VIII. chap. 1.; Book IX. chap. 7.

[2] . . . . . . "Sed enim damnare negatur
Hanc Venerem pietas: coëuntque animalia nullo
Cetera delectu. Nec habetur turpe juvencæ
Ferro patrem tergo: fit equo sua filia conjux;
Quasque creavit, init pecudes, caper: ipsaque, cujus
Semine concepta est, ex illo concipit ales.
Felices, quibus ista licent! Humana malignas
Cura dedit leges: *et quod natura remittit
Invida jura negant.*"—Ovid, *Metam.*, lib. x. ll. 323-331.

[3] The roe-deer keeps even more in families than do other deer. Two generally are produced at a birth, usually a male and a female. These pair for life, and consequently the closest in-and-in breeding is practised from generation to generation. See Macdonald, *Cattle, Sheep, and Deer*, p. 643.

are, perhaps, the only polygamous terrestrial carnivora, but the *Pinnigradia*, with the exception of the walrus, are all so. Among birds, the *Natatores* are usually polygamous; the *Grallatores* are some of them; the *Rasores* are, most of them; and ostriches, the bird of paradise, the Whydah finch, and, perhaps, humming birds are polygamous.[1] Now, polygamy among animals means the closest incest. The strongest male drives away all others, often, indeed, kills them, and jealously keeps in his own hareem all the females of his family. Animals are, besides, much sooner mature in comparison with the length of their lives than are human beings; hence incest with them, as it includes more generations, is closer than would be possible in mankind. Nor does the incest cease when the old male is turned out by a more powerful young one, for a young male of the same family is far more likely to be the first in the field of combat. But enough of this question. I have met with no authority save that vaguely cited by M. Devay; and M. Devay himself cites Aubé, who indirectly shows, in accusing such unions of the production of albinoïsm, that he takes for granted animals have no horror of incest.[2]

If there is really an innate horror of incest, I have said it ought to show itself intuitively when persons are ignorant of any relationship. But does it? Can we reconcile the statements of Quintilian and Lactantius that incestuous unions often occurred in Rome from the exposure of infants who were reared by slave dealers, with this supposition.[3] Not long ago Selim

---

[1] Darwin, *Descent of Man*, vol. i. pp. 267-270; vol. ii. pp. 362-363, and others.   [2] Devay, *Du Danger*, etc., p. 54.

[3] Cited by Lecky, *Hist. of European Morals*, etc., vol. ii. p. 30, note 2.

Pasha unwittingly married his sister, like himself a Circassian slave. Selim had risen, as slaves often do rise in the East, to rank and wealth; and only found out by chance that his slave-wife was his sister.[1] The evidence afforded by stories, ballads, and other pictures of life is contradictory. We are all familiar with tales of persons being insensibly drawn towards their children, though they did not know of their relationship—such as the story of Noor ed-Deen and his son, and Shems ed-Deen and his daughter, in the *Thousand and One Nights*. It is true that here the recognition was rather one-sided, but the subject is a favourite one, and such instances are not uncommon. On the other hand there is the story of Jocasta, who unwittingly married her son; and a curious story told in the *Heptameron* by the Queen of Navarre of a double incest. The story is probably a true one; the case was brought before the University at Erfurt, and happened to a young man Luther knew. The University sanctioned the marriage, as it had already taken place; the parties remained ignorant of their relationship, but the story seems to have become widely spread, and the following epitaph was composed on them:—

> " Ci-gît l'enfant, ci-gît le père,
> Ci-gît la sœur, ci-gît le frère,
> Ci-gît la femme et le mari,
> Et ne sont que deux corps ici." [2]

In the ballad of *Eglamore*, contained in Percy's collection, Christabelle marries her own son, but happily finds out the relationship:—

---

[1] See Martineau, *Eastern Life*, etc., p. 53.

[2] See the *Heptameron*, Trans. by Kelly, 1855, note to 3rd Day, novel. xxx.; Lawrence's *Vindication of Marriage*, book i. ch. vi. pp. 111-112; Heywoode's *Hist. of Women*, p. 176; Roich, *Ehe*, etc., p. 134.

"Sir Degrabell his troth hee plight;
& Christabell, that ladye bright,
to Church they her ledd.
through the might of god he spedd
his owne mother there he wedd,
in Romans as wee reade."[1]

The Editors say that such an union is a very favourite arrangement with old romance writers, though they do not generally dare to consummate such a marriage. Boccaccio, again, tells a story of two men, Giannole and Minghino, who both fell in love with the same girl. The one tries to carry her off; the other stops him; they are both arrested; and then it is discovered that this girl is the sister of Giannole, and Minghino marries her.[2] Defoe makes Moll Flanders unwittingly marry her brother; Goethe makes Mignon the offspring of a brother and sister, who were ignorant of any relationship existing between them; Molière makes a brother fall in love with his sister, and it is only when he finds out his relationship, that he becomes indifferent to her;[3] Lessing does the same,[4] and Racine very nearly the same.[5] These are only fictions, it is true; but observers like Molière and Goethe, whose business it was to mirror nature, are not likely to harp upon improbabilities.[6]

---

[1] *Percy Ballads*, vol. ii. p. 380, lines 1061-1066.
[2] Boccaccio, *Decamerone*, Giorn. Quinta, novel. 5.
[3] Molière, *Dom Garcie*.   [4] Lessing, *Nathan der Weise*.
[5] Racine, *Don Sanche*.
[6] "In some tragedies and romances," says Adam Smith, "we meet with many beautiful and interesting scenes, founded upon what is called the force of blood, or upon the wonderful affection which near relations are supposed to conceive for one another, even before they know that they have any such connection. This force of blood, however, I am afraid, exists nowhere but in tragedies and romances." (Cited in Buckle's *Hist. of Civilization in England*, vol. ii. p. 440, note 51.)

We have abundant evidence further, in the canons enacted again and again for its prevention, that there is no horror of incest inherent, even in civilized man, sufficient to prevent that crime in the presence of a moderate amount of temptation.

From the earliest times thinkers have been puzzled to account for the prohibitions on marriage between near kin. Some ascribe them to a fear lest relationship may become too involved; others to a fear lest affection may become concentrated within too narrow a circle; because marriage would take place too early; because people would be induced to marry each other that property might be kept in the family; because they are prohibited by "God's law;" because they outrage "natural modesty;" or, in modern times, because they are supposed to prove injurious to the offspring.

Socrates objected to them that they generally took place between an old man and a young woman, as in the case of marriages between uncle and niece.[1] Plato maintained, as did afterwards Novatian, that they were contrary to the unwritten law of nature; and he adds that else everyone would marry those whom they most nearly resembled, and there would be no proper mixture of characters and property.[2] Aristotle feared that love would become immoderate if, to the usual ties of paternal or fraternal love, marital love were superadded.[3] Chrysippus and Zeno considered every prohibition, though between the nearest relatives, as

[1] John Taylor, *Elements of Civil Law*, p. 318.
[2] J. A. N. Perier, *Mém. de la Soc. d'Anthrop. de Paris*, vol. i. 1863, pp. 200, 216; John Taylor, *Elements of Civil Law*, p. 318.
[3] Amyraut, *Consid. sur les Droits par lesquels la Nat. a Reiglé les Mar.*, p. 223.

simply absurd;[1] but Philo thought the respect we owe to a father should preclude all thoughts of marriage with his wife,[2] and in this opinion he is supported by Agathias Scholast, and by Statius.[3] Plutarch took altogether a more matter-of-fact view of the question: Who would the wife have to complain to when her husband beat her, should her parents be the same as her husband's?[4] St. Chrysostom adopts the views of Aristotle: Are there not ties of love enough between relatives, he asks, that you should seek to draw them tighter by marriage to the exclusion of the rest of mankind?[5] St. Augustine took also much the same view as Aristotle: the whole idea pervading his mind is that marriage between near kin is to be avoided because kindred, and with it charity, cannot otherwise be extended. So, he considers that the marriage of Adam's children among themselves was not reprehensible, but if in the succeeding generations brothers and sisters had married each other, instead of marrying their cousins, that would have been wrong, for the families would then have become isolated. Marriages between first-cousins, he continues, are neither forbidden by the law of God nor by the law of man; yet he is glad to see it discouraged and rare, because it is so near to that which is unlawful, and that which one doth with one's cousin he almost thinketh that he doth with his sister; and besides, one person thus

---

[1] John Taylor, *Elements of Civil Law*, p. 318.

[2] Adam, *Fortnightly Review*, 1865, p. 86.

[3] John Taylor, *Elements of Civil Law*, p. 319.

[4] Amyraut, *Consid. sur les Droits par lesquels la Nat. a Reiglé les Mar.*, p. 225.

[5] J. A. N. Perier, *Mém. de la Soc. d'Anthrop. de Paris*, vol. i. 1863, p. 205.

absorbs two alliances which ought to have been employed for the increase of affinity; while lastly, there is a certain laudable natural instinct in a man's shamefastness to abstain from using that lust upon such as propinquity hath bound him chastely to respect.[1] Pope Gregory I. alleged, in a letter to St. Augustine concerning the marriages of the English, that marriages between near kin prove sterile;[2] an idea, as I have already shown,[3] he probably got from a misapprehension of the words of Moses. Thomas Aquinas gives a different reason from all the former: all persons, he says, who lived under the same roof were forbidden to intermarry, since were marriage between them permitted, this liberty would violently inflame their passions; but under the new law which is the law of the Spirit and of love, several degrees of consanguinity were forbidden, because the worship of God spreads and multiplies by spiritual grace, and not by carnal love. Men must consequently be debarred from carnal things, to the end that they may rather attach themselves to spiritual matters, that love may abound in them more and more, so that friendship may be extended to a larger number by means of affinity. The prohibitions were rightly therefore extended to the seventh degree, but afterwards the Church only

[1] St. Augustine, *Of the Citie of God, with the learned Comments of Io. Lod. Vives. Englished by J. Healey*, book xv. chap. 16.

[2] "Quædam terrena lex in Romana republica permittit, ut sive frater et soror, seu duorum fratrum germanorum, vel duarum sororum filius et filia misceantur; sed experimento didicimus, ex tali conjugio sobolem non posse succrescere; et sacra lex Mosaica prohibet cognationis turpitudinem revelare: unde necesse est, ut jam tertia vel quarta generatio fidelium licenter jungi debeat, nam secunda, quam prædiximus, a se omnimodo abstinere debet" (*Excerp. Ecgberti, Arch. Ebor.*, cxxxii.; Thorpe's *Early Inst. of England*, p. 337).

[3] See page 30 of this work.

restrained it as far as the fourth, since it was useless and dangerous to forbid it in any further degree.[1] Another argument was advanced against these marriages by Luther, who saw no actual harm in marriages beyond the third degree collateral, but considered them inexpedient on the ground that people would marry without love merely to keep property within the family, while poor women would be left spinsters.[2] Theodore Beza harps on the confusion there would be in relationship were all marriages permitted : if a father takes the daughter, and a son marries her mother, or if father and son marry two sisters, then one in the character of brother and step-father marries his brother's stepdaughter, the son marries the sister of his mother-in-law, the brother marries the daughter of his wife's brother, the son of the brother marries the mother of his paternal uncle,—and thus, since relationship gets so hopelessly confused, it is very proper that many canons should have been enacted to prohibit such marriages.[3] Robert Burton is the first author I have met with who considers that the offspring are hurt by these marriages ; and these injuries he ascribes not to created diseases, but very reasonably to an intensification by inheritance of previous disease in the parents: "For these reasons belike the Church and commonwealth, humane and divine Lawes, haue conspired to

---

[1] Dally, *Anthrop. Review*, pp. 103-104, May, 1864.

[2] "Jm vierten Grad lassen wirs zu, im dritten aber wollen wirs nicht zugeben ; nicht zwar um des Gewissens willen, sondern um des bösen Exempels willen unter den Geizigen Bauren, die würden um des Guts willen auch ihre nächsten Blutsfreundinnen nehmen. Wenn man ihnen den dritten Grad zuliesse, so gewohnten sie im andern Grad zu heyrathen. Sind doch sonst Jungfrauen gnug, warum sollen dieselben sitzen bleiben?" (Reich, *Ehe*, etc., p. 135.)

[3] *Ibid.*, p. 141.

avoide hereditary diseases, forbidding such marriages as are any whit allied ; and as *Mercatus* adviseth all Families, to take such, *si fieri possit quæ maximè distent naturâ*, to make choice of those that are most differing in cōplexion from thēs if they loue their owne, and respect the common good. And sure I thinke, that it hath beene ordered by Gods especiall providence, that in all ages there should be, as vsually there is, once in 600 yeares, a transmigration of Nations, to amend and purifie their brood, as we alter seed vpon our land, and that there should be, as it were, an invndation of those Northerne *Gothes*, *Vandales*, *Scythians*, and many such like Nations, which come out of that continent of *Scandia*, and *Sarmatia*, as some suppose, and ouerranne as a deluge, most parts of *Europe*, and *Africke*, to alter for our good, our complexions, which were much defaced with hereditary infirmaties, which by our lust and intemperance we had contracted. A sound generation of strong & able men were sent amongst vs, as those Northerne men vsually are, and innocuous, free from riot, and free from diseases. To qualifie vs, and make vs as those poore naked Indians are generally at this day ; and those about *Brasile* (as a late Writer relates) in the Isle of *Maragnan*, free from all hereditary, or other contagion, where as without help of Physicke they liue commonly 120 yeares, or more."[1] Bishop Jeremy Taylor compares marriage with a mother to a river returning to its source, or to the marriage of to-day with yesterday. Further, he says it is a contradiction of rights that any one should be at once the

[1] Burton, *Anatomy of Melancholy*, pp. 81, 82, Oxford, 1621. This last passage seems very hard on the doctors !

superior and inferior of the same person; that there is a natural abhorrence of such mixtures; that brothers and sisters would not have a chance of choosing elsewhere, and would marry while yet too young; that if there were but one sister in a large family they might all fall in love with the same sister; or they might do the same if there were several sisters, and quarrel in consequence. Yet he ridicules those who, "making the law of Nature to be a sanctuary of ignorance and an artifice to serve their end, just as the pretence of occult qualities is in natural Philosophy," would use it as a standpoint to forbid the intermarriage of first-cousins; and gives the following reasons "why the Projectors of the Canon law did forbid to the fourth or to the seventh degree." "They that were for four gave this grave reason for it. There are four humors in the body of man to which because the four degrees of consanguinity do answer, it is proportionable to nature to forbid the marriage of Cosens to the fourth degree. Nay more; there are four Elements; *Ergo*. To which it may be added, that there are upon a Man's hand four fingers and a thumb. The thumb is the *stirps*, or common Parent; and to the end of the four fingers, that is, the four generations of Kinred we ought not to marry, because *the life of man is but a span long*. There are also four quarters of the World; and indeed so there are of everything in it, if we please, and therefore abstain at least till the fourth degree be past. Others who are graver and wiser (particularly *Bonaventure*) observe cunningly, that besides the four humors of the body, there are three faculties of the Soul, which being joined together make seven, and they point out to us that men are to abstain till the seventh generation. These reasons such as they are, they therefore were content withal, because they had no better : yet upon the strength of these they were bold even against the sense of almost all mankind to forbid these degrees to marry."[1] Almost the same reasons are advanced by Amyraut. He considers there is an innate horror of incest which he accounts for in this way : our horror of marriage with a parent is due to our respect for them ; for since in a monarchy no one can be subject and monarch too, so no one can marry his mother and thus become monarch over her as her husband, and subject to her as her son ; while in our sisters we see the representative of our mother, therefore neither can we marry with them.[2] Dugard says that one argument he has heard against the marriage of cousin-germans is that they " do not thrive, and that God shows by the ill-prospering of them that such Marriages are unlawfull, and by no means approv'd of by him." One judgment " which is said often to accompany these Marriages " is " a Want of Children and a Barrenesse," which he has " taken notice of * * not only from the Mouthes of some," but has seen " in print too, and that from no less a Man than the Pope." They were also blamed for hindering the spread of relationship. The common people, as we might expect, were the most bitter against such marriages, which they believed brought ill luck—their horses died, their orchards " did not hit," and their

---

[1] Jeremy Taylor, *Duct. Dubit.*, book II. chap. ii. rule iii. secs. 18, 19, 20, 27, 34, 66.

[2] Amyraut, *Considerations sur les Droits par lesquels la Nature a Reiglé les Mar.*, pp. 258-286.

flocks were devastated by the rot.[1] Lawrence is very much against marriages contrary to the laws of nature, whatever these may be, and instances the people of Carthagena who "allow not marriage with the Sister, on this Tradition, That one who married his Sister was for that offence, carried and confin'd to the Moon, where he still remains the spot, or Man in the Moon."[2] Montesquieu places difference of age in the first rank of natural prohibited degrees. Were a man permitted to marry his mother, he says, she would be aged long before he would. Marriage between father and daughter he thinks forbidden, because it is the office of a father to bring up his daughter morally, and hence he has a natural abhorrence of corrupting her. The horror of incest in the collateral degrees he ascribes to the same source, for brothers and sisters are brought up in moral conduct at home; while uncles stand in the place of fathers. Even a teacher, or other persons living under the same roof, would for these reasons not like to intermarry.[3]

If we pass the forbidden degrees of various peoples under review, it is evident that the most universal prohibitions are those where the parties, in the course of nature, are of an unsuitable age for marriage. I do not, however, think it possible that the prohibitions of marriage between persons of unsuitable age are the result of an observed bad consequence of these marriages to the offspring, even though it be true that there is such a result, and that even

---

[1] Samuel Dugard, *The Mar. of Cousin-Germans Vindicated*, pp. 36, 50, 51, 53, 97.
[2] Lawrence, *Vindication of Mar.*, book i. chap. ii. p. 14.
[3] Montesquieu, *De l'Esprit des Lois*, book xxvi. chap. xiv.

in Aristotle's time it was already noticed,[1] for I cannot think these results gross enough to have been generally observed, or savage nations to have self-control enough not to contract marriages they would otherwise wish to contract, on this account. Indeed, were the consent of women never necessary to marriage, or had they never any choice, I firmly believe that marriages in the first degree, between old men and young women, would not be considered incestuous. But, as a matter of fact, women, even in the most savage communities, have ample opportunity of expressing their likes and dislikes in the same way as females have among animals; they do not like marrying old men, and young men do not like marrying old women; hence the natural law against ill-assorted marriages, and hence the prohibitions against marriage into a different generation, of which we see further evidence in the general fact that marriages between old male relatives and young females are more common than the reverse, since marriages between uncles and their nieces are more common than marriages between aunts and their nephews; and in the fact that it is far more generally lawful for brothers to marry their younger sisters than their elder: I say, this is a confirmation of the supposition, because females have less control over their marriages than have males.

As far then as a deduction may be trusted from the general customs of men, no marriage is prohibited by nature unless the parties are of an age unsuited to each other.

---

[1] Aristotle, *Hist. Anim.*, book v. xii. 1.

CHAPTER IV.

OBSERVATIONS ON SOME ISOLATED COMMUNITIES WHO HAVE CONSTANTLY MARRIED AMONG THEMSELVES.

IN the preceding chapters we have seen that no deduction as to the harmfulness of marriages between near kin can be drawn from the fact that nearly all mankind have certain prohibited degrees, and it remains for us to discover by observation whether these marriages are really harmful. We shall therefore consider a set of cases where such marriages have occurred constantly in communities for so long a time that the effects for evil, if any, must show themselves; at the same time, however, we cannot ascribe every disease observed in these isolated communities to consanguineous marriage, because we cannot point to one particular factor where many are present, and say, this is the cause, and no other; but if we see people constantly contracting these marriages and find they do not suffer from disease more than other communities who seldom marry near kin, then we may justly reason that these marriages are in fact harmless.

In the year 1789, a mutiny occurred on board the royal ship *Bounty*, which has now become famous. The *Bounty* had been sent out, under the command of

Lieutenant Bligh, to make further observations among the South Sea Islands, and to obtain the bread-fruit tree from the natives; but, in consequence of Bligh's temper, the beauty of the climate, combined with the prospect of an idle life, and the contrast of such a life with the hardships a sailor is compelled to undergo, so enthralled a part of the crew that they determined to seize the ship at all hazards. How some of the crew escaped to Otaheite, some were killed, and some were taken and brought to England, is unnecessary to relate—the *Bounty* was carried away from Otaheite by nine of the mutineers, who brought with them six men and twelve women, natives of Otaheite and Tabouai, and landed on Pitcairn Island, which before their arrival was uninhabited. But they soon began to quarrel. One of the mutineers having lost his wife by an accident, took away the wife of one of the natives, who, indignant at such treatment, laid a plot to murder all the whites. This plot, however, was discovered by the women, and the injured husband with another native were treacherously slain. Cowed, but thirsting for revenge, another plot was laid about a year or two after by the natives to free themselves from their slavery, which proved only too successful: five whites were killed, only four escaping. Then the women rose, and murdered all the natives; and thus out of nine white men, six native men, and twelve native women, who landed early in 1790, only four white men and ten or eleven native women remained alive in October, 1793. From this time for about five years things went on smoothly, excepting that one of the men killed himself in a drunken fit; but in 1799, another man lost his wife, and insisted on taking the wife of one of his companions. The husband resisted, the widower sought to murder both him and his wife, whereupon the husband with his remaining companion killed the would-be murderer to secure their own safety. A year after one died of asthma, and the other, Adams, remained the only man on the island.

Thus, in the year 1800, the population consisted of one man, five women, and nineteen children, the eldest nine years old, in all twenty-five persons. In 1808, Mr. Folger makes the total thirty-five. In 1814, Sir Thomas Staines makes the *adult* population as much as forty; probably a mistake, for only fourteen years before, nineteen of the twenty-five then existing were mere children. In 1825, there were according to Captain Beechey, thirty-six males and thirty females, or a total of sixty-six persons. In 1830, Captain Waldegrave makes the total seventy-nine, showing an increase of thirteen in five years, yet only three strangers had joined the little community, nor had these joined long before.

As to their health, all observers agree nothing could be better, that they are strong, and average six feet in height. Both sexes are well formed and handsome, and their children uniformly enjoy good health, while the women are almost as muscular as the males, and taller than the generality of females. Parturition was easy, seldom lasted over five hours, and had never proved fatal. Captain Waldegrave saw but one defective person, a little one-eyed boy.[1]

[1] *The Eventful Hist. of the Mutiny of H.M.S. "Bounty,"* etc., London, Anon., 1831, pp. 302-307, 333, 334, 326-328. Their present energy was strikingly shown on the visit of the "*Pearl*." See *The Times* for November 21st, 1874.

There is a community established eastward of Surabaya, Java, on the Tengger Hills, and near the so-called Sandy Sea, consisting of some 1,200 persons, distributed in about forty villages, who still profess the ancient Hindoo religion. These curious people seem to live in a state of Arcadian purity. The chief of every village is appointed by election; and four priests, intelligent, though uncultivated men, have charge of the sacred writings. There is no pœnal code, for there is hardly any occasion for it—a reproof from the village chief meets every case. They live frugally, peacefully, and happily; are proud of their institutions and themselves; and therefore take care never to marry out of their own community; yet they are bigger and stronger than any other race in Java.[1]

A very remarkable case is given by Dr. Thibault of a Portuguese slave-dealer, named Souza, who died in the year 1849, at Widah, Dahomey, leaving behind him 400 disconsolate widows and about 100 children. By order of the king, the whole of this family was interned in a particular part of the country, where reigned the most complete promiscuosity. In 1863, there were children of the third generation; and Dr. Thibault, who verified the fact himself, asserts that at that time, although these people were born from all degrees of incestuous unions, there was not a single case of deaf-mutism, blindness, crétinism, nor any congenital malformation. Nevertheless it is probable that they are now decreasing in number, for misery, debauchery, and syphilis co-operate to their destruction.[2]

[1] Waitz, *Anthropologie*, vol. i. p. 482.
[2] Thibault, *Mar. Cons. dans la Race Noire*, in the *Archiv. de Méd. Navale*, Ann. 1864, vol. i. p. 310. Cited by J. A. N. Perier.

But besides these remarkable cases, there are many isolated populations even in the midst of crowded countries. Dr. Beddoe gives many instances of isolated communities in Great Britain. At Brighton, the fishermen are said to be a separate caste; some assert them to be of Spanish origin. They intermarry among themselves, and are as robust and healthy as the agricultural population.[1] At Itchinferry, Southampton, the people are little given to intermarriage with their neighbours. Some intermixture there has probably been with the Channel Islanders, but it does not appear to have been very great. They seem of average height, and if anything rather above the average weight.[2] The inhabitants of Portland Island, a hardy, tall, primitive race, seldom if ever intermarry with any on the mainland.[3] At Bentham and its neighbourhood, in the West Riding, Yorkshire, there has been until quite lately continual close intermarrying; yet the race is stalwart and long-lived, and very free from idiocy and insanity.[4] In Cornwall there are villages so situated that marriages among blood-relations are necessarily frequent. The villages of Mousehole and Newlyn, in Mountsbay, are instances; yet these people are by no means degenerate. Dr. John Davy asserts that there is only one case on record of any Stuart of Glenfinlass marrying out of the glen, and yet all the families were healthy. Nor in the Scot-

---

[1] Beddoe, *Stature and Bulk of Man in the British Islands*, pp. 91, 171; and Sharp's *Gazetteer*, 1852, Art. *Brighton*.

[2] Beddoe, *ut sup.*, pp. 92, 93.

[3] Brand, *Pop. Antiq.*, vol. ii. p. 87; Sharp, *ut sup.*, Art. *Portland Isle*.

[4] Beddoe, *Stat. and Bulk*, etc. p. 55.

tish inlets, where marriages between near kin must necessarily occur, are the inhabitants at all degenerate, while consumption is rare.[1] This last is confirmed by the *Medical Times and Gazette*.[2] "We confess," it says, "that our experience—founded on a tolerably intimate acquaintance with the inhabitants of the fishing villages along the coast of Scotland, who are altogether a race apart, seldom if ever intermarrying with the peasants of the surrounding country, but to such an extent among themselves that it is nothing unusual to find only one or two family names in a village, the deficiencies being made up in nick-names—leads us to agree with Dr. Voisin,[3] for a hardier or more robust set of men and women than these people do not exist." At Boulmer, Alnwick, Northumberland, there is a population of about 150 persons, of which about twenty are fishermen. How closely they intermarry may be inferred from the fact that with five exceptions all are named Stephenson, Stanton, or Stewart; and withal, they are rather above the average height and weight of the district.[4] The people of Burnmouth and Ross, Berwickshire, form a colony about 120 years old, first founded by smugglers with the sole view of security in their trade, and hence by no means favourably situated for health. The houses, though clean and orderly, are damp from their situation close to the sea at the foot of the cliff; the population of these two connected villages is estimated at

[1] Child, *Essays on Physiological Subjects*, pp. 50-53.
[2] Nov. 24th, 1866, pp. 573, 574.
[3] Who gives an account of the community of Batz. See p. 179 of this work.
[4] Beddoe, *Stature and Bulk*, etc., pp. 44, 45, 171.

420 persons; the men are tall and strong, swarthy complexion, prudent, and sober; the women are also tall, stout, and high-featured. Formerly they married strictly among themselves, but now they occasionally marry the daughters of the agricultural population who are in their service. At the time of Dr. Mitchell's investigation, five of the married women were not born in the place, and "careful inquiry only brought to light seven marriages between full cousins," and no case between second-cousins; but there were many marriages between persons whose relationship was recognized, though distant. The result of these seven marriages was as follows:—

	No. of Children Born.	No. of these Dead.	No. yet Alive.	Remarks on the Children.
I.	6	3	3	Those alive were big, strong, sound, and healthy. Those dead were said to have been sound.
II.	6	—	6	"Never had a headache."
III.	9	1	8	All are sound and sane.
IV.	1	—	1	Sound. Newly married.
V.	3	2	1	The living one is not robust.
VI.	5	1	4	All were sound.
VII.	5	—	5	All were sound.
Total.	35	7	28	

Three out of the twenty-eight living children were married, and to persons distantly related to them, with the following result:—

	No. of Children Born.	No. of these Dead.	No. yet Alive.	Remarks on the Children.
I.	7	1	6	All of them healthy.
II.	6	2	4	All the living ones are sound. The two deaths were in infancy.
III.	—	—	—	No children born.

There were, in the whole population, only two imbeciles, both of whom were self-supporting, and neither was the child of blood-relations. Besides these, there were two cases of *acquired* insanity, both women. The disease in both cases resulted from grief and shock to the nervous system, caused by the sudden news of the death of their husbands by drowning. The parents of one of these women were distantly related. There was one case also of mild epilepsy, in a child whose parents were not related; but there was no case of a lame, deformed, blind, dumb, or paralytic person to be heard of; and in the school which was twice visited, and where nearly all the children of the village were assembled, no strumous sores were found, nor were any of the children puny, pale, or languid, but on the contrary, merry and active, though their teacher thought them slower and duller than other children he had had under his care. Dr. Mitchell adds, that though none of the children of cousins were found defective, there was in the whole population a greater proportion of "unsound" persons than the average for the whole of Scotland.[1] He does not, however, explain the nature

---

[1] Mitchell, *Mem. read before the Anthrop. Soc. of London*, vol. ii. 1866, pp. 438-441.

of their "unsoundness;" and I presume that the unhealthy position of the villages must have some effect. This much is however plain, that the greatest grievances alleged to follow consanguineous marriages are here conspicuous by their absence, and that the people were generally healthy and strong—a fact which Dr. Beddoe confirms, and adds that the Eyemouth fishermen, of the same county, are also handsome and stalwart in the main, though they are a very isolated community.[1] At Boyndie, in Banffshire, and Rathen, in Aberdeenshire, the fishermen are a very closely interbred community, somewhat shorter and lighter than the landsmen, and their heads are not quite so big.[2] The fishermen of Buckhaven, Fife, are also closely interbred, but differ less from the agricultural community than do their brethren of Rathen and Boyndie. They are supposed to be the descendants of a Dutch ship's crew wrecked in the time of Queen Mary.[3]

The population of the two villages of Portmaholmack and Balnabruiach, Easter Ross, is estimated at 1,548,[4] of which 710 are married. Of these 355 couples, 62 are first-cousins, and 20 second-cousins, or persons more distantly related. These 82 marriages between relations produced 340 children, or an average of a little more than 4·1 per marriage. Of these children, 90 died, most of them under the age of ten years, making nearly 26·5 per cent. In

---

[1] Beddoe, *Stature and Bulk*, etc., pp. 42, 43, 171.
[2] *Ibid.*, pp. 27, 171.
[3] *Ibid.*, pp. 31, 73, 171; Sharp's *Gazetteer*, 1852, Art. *Buckhaven*.
[4] In 1841, the population was only 646, which shows an increase of 902 in twenty-five years (Sharp's *Gazetteer*).

the whole number of 340 children there were 2 imbeciles, 1 idiot, and 2 cripples, or nearly 1·5 per cent. malformed or diseased. There were besides 4 imbeciles, 2 idiots, 2 insane, and 2 cripples among the children not born from marriages between near kin.[1]

Dr. Mitchell gives his observations on another village, which he does not name, but which he says is situated " on the north-east coast of Scotland." The fishing population is estimated at 779, and contains 119 married couples, of which 11 couples are first-cousins, and 16 are second-cousins. Of these 27 marriages between near kin 3 are barren, and from the remaining 24, 105 children were born, or nearly 4·4 for every fruitful marriage, and nearly 4 for all the marriages between near kin, including the barren ones. Of these 105 children 38 are dead (of which 35 deaths were in childhood), or 33·4 per cent; 4 are deaf-mutes, or nearly 6 per cent.; 4 are imbecile, and

[1] Dr. Mitchell attempts to show from these data "that such calamities fall on the offspring of blood-related parents with greater frequency than on the offspring of parents not related," by means of a computation from the data of the productiveness and mortality of consanguineous marriages, to supply the data wanting concerning the non-consanguineous marriages. To this end, he puts the probable number of children born from 273 non-consanguineous marriages at 1,160, a mistake for 1,132, nearly; whence the probable number dead would at nearly 26·5 per cent. be about 300, leaving 832 alive. Of these it was known that 10 were malformed or diseased, making, if the data assumed are correct, 1·2 of the children born from non-consanguineous marriages malformed or diseased, as compared to 1·5 per cent. of those born from the consanguineous marriages, or a difference of nearly 0·3 per cent.; a small fraction indeed, and besides quite an unreliable deduction, seeing that many of the data are conjectural, and that the number altogether is so small, that if a mistake was made in the collection of the facts, which is not at all improbable considering that the enumeration was not official and that the diseased are more likely to have been among the 300 dead, the balance might be on the other side. See *Mem. read before the Anthrop. Soc. of London*, vol. ii. 1866, pp. 444-445.

4 more "slightly silly;" 1 is paralytic, or 1·5 per cent.; and 11, or a little over 16·4 per cent., were scrofulous and puny. The children of those who are first-cousins are described as "*all* of them neither strong in mind nor in body;"[1] but Dr. Mitchell says nothing about the results of the marriages of those who were not related, without which the above facts are of no value whatever; except to show, as Dr. Mitchell observes, that there must be more than the average causes of idiocy at work in this community.

Berneray, situated at the west of the Isle of Lewis, contains a population of 427, in which there are 74 married couples. Of these 74 couples, 2 are first-cousins, and 6 are second-cousins. From the first two, 10 children were born, of which 8 are yet alive; from the other six, 20 children were born, of which 18 are still living; making an average of 3·75 children per marriage, and a mortality of 13·4 per cent.; and yet "not one of these is either insane, imbecile, idiotic, deaf, dumb, blind, lame, deformed, or in any other way defective in mind, morals, or body." "We have thus a population where every ninth marriage is between blood-relatives; yet, instead of finding the island peopled with idiots, madmen, cripples and mutes, not one such person is said to exist in it."[2]

[1] Mitchell, *Mem. read before the Anthrop. Soc. of London*, vol. ii. 1866, p. 442.

[2] Mitchell, *Ibid.*, p. 434. His informant was Mr. J. Macdonald, land-steward at Berneray, a resident, and intimately acquainted with every family, who was himself astonished at the paucity of consanguineous marriages. "Were I not certain of it," he says, "as I am now by a minute search, I would doubt the fact, from the island being inhabited by the present race from time immemorial" (*Ibid.*). If this is the case, the above facts are much more important, for in a population of nearly 427, marriages between near kin must undoubtedly be very frequent, even excluding those between first and second-cousins.

The rugged island of St. Kilda, barely six square miles in area, with a population, in the year 1860, of 33 males and 45 females, was the subject not long ago of a series of letters to *The Times* [1] on the great mortality among the infants born there. The second letter which was signed *C. A. W.*, attributed it to consanguineous marriage ; and perhaps reasonably, according to the accepted theory that every ill that human flesh is heir to must be due to consanguineous marriage if it cannot be clearly proved due to something else. Nor is *C. A. W.* alone in his theory, for both Mr. Macdonald and Dr. Mitchell partially ascribe this undoubted mortality to the same cause.[2] The population in 1700 was 180 persons, in 27 families. In 1820, it was 103 persons in 20 families. In 1851 it was 110 persons in 19 families ; while in 1860 it was 78 persons in 14 families. Of these 78 persons, 4 are below the age of five years, 6 between the ages of five and ten, 9 between ten and fifteen, 13 between fifteen and twenty, 12 between twenty and thirty, 9 between thirty and forty, 12 between forty and fifty, 11 between fifty and sixty, 1 between sixty and seventy, and 1 between seventy and eighty. Not one of the 14 marriages is between first-cousins, but 5 are between second-cousins; while only one of the married people is a stranger on the island—a woman from Lochinver who married a native, and had 14 children, of which only 2 are alive, both unmarried. Indeed, strangers rarely settle on the island, and the children of those who do are always the first to

[1] Aug. 23rd-31st, 1871.
[2] Mitchell, *Mem. read before the Anthrop. Soc. of London*, vol. ii. 1866, p. 429 ; Macdonald, *Cattle, Sheep, and Deer*, p. 411.

emigrate.[1] The mortality among the infants, says Mr. Macdonald, was excessive when his father visited the place fifty years ago; but now it is still greater, and may be estimated at 80 per cent.[2] From the five marriages between second-cousins, 54 children were born, or an average of 10·8 per marriage; and of these 37 died of *Trismus nascentium*, the Icelandic *Gin-Klófi*, or 68·5 per cent.; while to the whole 14 marriages 125 children were born, or nearly 9 per marriage, and of these 84 died within the first fortnight, or 67·2 per cent.,[3] a much smaller percentage than that estimated by Mr. Macdonald.

We find, therefore, that the population is small, isolated, and dwindling fast; that the infant mortality below the age of five years is just about ten times as great as the infant mortality for England and Wales—what more natural than to attribute this to marriages between near kin?

Unfortunately for this theory, Mr. Corfield shows that whatever the cause of the mortality is, it certainly is not the consanguinity of the parents; since infants brought from the mainland are subject to the same mortality.[4] *Trismus nascentium* is, besides,

[1] Mitchell, *Mem. read before the Anthrop. Soc. of London*, vol. ii. 1866, pp. 425, 427.

[2] I have adopted Dr. Mitchell's figures in preference to those of Mr. Macdonald, as the former seems to be much better informed. Macdonald says that the population was 108 in the year 1822; in 1851 it was 100, and now (apparently in 1871) it is 72. "We have not the least hesitation," he says, "in attributing it to the continued intermarrying of a population already too closely related" (see his *Cattle, Sheep, and Deer*, etc., p. 411). Now this was written after the letters appeared in *The Times*, and taking into consideration the fact that the example of St. Kilda is put forward to serve as a *proof* that in-and-in breeding is injurious to cattle, his conclusion is hasty, to say the very least of it.

[3] Mitchell, *ut sup.*, pp. 425, 428.

[4] Letter to the *Times*, Aug. 28th, 1871.

caused solely by the irritation to the umbilical wound of foul air; while convulsions and other infantile disorders are not more frequent, considering the wretched food they get, than they are in other communities.[1] The natives, says Rear-Admiral Otter, from whom Dr. Mitchell had much of his information, live chiefly on sea-birds, a very oily food; and this he thinks the chief cause of the mortality, for a mother fed on cocoa, meat, and biscuit, was able to rear her child successfully.[2] This inference is countenanced by the fact that the mortality among Kalmuck children is found to be greatest in the eastern districts, where there are few pastures, and the chief food is fish; while it is least in the western districts, which contain large cattle pastures.[3] Here the factor consanguinity, as we have seen,[4] is absent, and therefore we have no right to say the similar mortality in St. Kilda is due to consanguinity of the parents. Only one person on the island, a woman upwards of fifty years of age, is insane; while of the 17 children born from the consanguineous marriages, not one is defective in body or mind. The inhabitants are strong, healthy, and robust, and of a particularly good and clean complexion; intelligent, sharp, cautious, sober, and moral.[5]

Between the years 1750 and 1846, the population of Iceland ranged between a minimum of 38,142, in the year 1786, and a maximum of 58,619 in 1845. So scanty a population as this, in a country like Iceland, cannot fail to contract very many consanguineous

[1] See below, pp. 173, 174, on Iceland.
[2] Letter to *The Times*, Aug. 31st, 1871.
[3] *Pall Mall Gazette*, May 29th, 1874.
[4] See p. 103 of this work.
[5] Mitchell, *Mem. read before the Anthrop. Soc. of London*, vol. ii. 1866, pp. 425, 427.

marriages; for it is practically shut out from all communication with the rest of the world; it is almost roadless, and travelling is so dangerous that many persons are annually drowned in the attempt to cross those torrents which constantly intersect their lonely path; while the villages are small and far apart, to the end that sufficient pasture may be found for their flocks and herds.[1] Were we to believe all that has been said as to the evil of these marriages, we must expect to find a dwindling population, dwarfed, and bowed down with phthisis, twisted with rickets, loathsome with swollen glands and unhealed sores, unable to communicate to each other for their deaf-mutism any thoughts that might be spared to them by idiocy and insanity, even were they able to distinguish whether they were speaking to a human being or to a post. And certainly they do not live under conditions so healthful that they might hope to escape from these. They eat their food generally cold, often putrid, and always at irregular times. They have no artificial means of warmth, and therefore allow no ventilation in their miserable hovels, which are built of damp earth, and where the whole family remains huddled up, not only at night but the greater part of the day also, during six months in the year, with their cattle, sheep, dogs, and all the live stock they may happen to possess. Indeed, the air in these dwellings becomes so poisonous from the breath of the inmates, their refuse, and the fuel they use composed

---

[1] M'Culloch says that only one-third of the island, or 10,000 square miles, has vegetation of *any* sort; while the cattle, estimated at 36,000 to 40,000, their 50,000 to 60,000 ponies, and 500,000 head of sheep, have to live as best they may on grass, moss, and seaweed alone (*Geographical Dict.*)

of dung, rotten bones, and anything that can be got to burn, that it becomes extremely dangerous to women after parturition and to new-born infants. Nevertheless, the population, though subject to great fluctuations in consequence of the great liability to epidemic diseases, is steadily increasing.[1] The number of legitimate births in Iceland is 16·3 per cent. higher than in Denmark for women between the ages of twenty and twenty-five; while the illegitimate births are 22·9 per cent. higher than in Denmark; and while the number of children per marriage in Denmark averages 5, a high average compared to Europe, in Iceland it averages 6·8, and the percentage of still-born is below that of Denmark.[2] The Icelanders average 5 ft. 8·5 in. in height, and 156 lbs. in weight;[3] are robust, and very rarely deformed, even in the one district where rickets is known, for this disease generally disappears at the age of five or six.[4] As in St. Kilda, however, the infant mortality is exceedingly great, almost double that in Denmark at the same period of life. The children are nearly always put out to nurse, and are hardly ever suckled, because the women almost destroy the bust by a close-fitting knitted garment they are accustomed to wear, which renders them physically incapable;[5] and hence the children are brought up by hand on very unsuitable food. The foul air of their houses also frequently poisons the wound on the navel, which suppurates and brings on *trismus nascentium*, the most fatal infant

---

[1] See Thomsen, *Ueber Krankheiten*, etc., *auf Island und den Färöer-Inseln*, Table II. and pp. 12-15, 17.

[2] *Ibid.*, pp. 147, 148.

[3] Beddoe, *Stature and Bulk*, etc., p. 167.

[4] Thomsen, *Ueber Krankheiten*, etc., p. 24.

[5] *Ibid.*, pp. 146, 26, 27, 16.

disease perhaps known, and which is often combined with bowel complaints brought on by their bad food.

| On every 100 deaths ||Were caused by|
In Denmark 1840–1844	In Iceland 1827–1837	
2·0	—	Hydrocephalus.
1·5	0·6	Whooping-cough.
2·4	—	Scrofulous disease of the glands.
13·7	30·0	Other diseases of childhood, including convulsions, cramp, trismus, etc.[1]

It is curious that rickets is found in only one district, the only part of Iceland where Iceland-spar is found, and this is the only part of the island where scrofula is otherwise than extremely rare.[2] In the year 1845, there were 110 idiots in Iceland, of which 66 were males, and 44 females; of mad, there were 10 males and 34 females, making a total of 154 mentally deranged, or 0·26 per cent. In Denmark the mentally deranged were 0·31 per cent.; but while here idiocy composed only 57 per cent of all mental diseases, in Iceland it composed 71·4 per cent.[3] The Icelanders are given to hard drinking: indeed, *delirium tremens* forms 0·08 per cent. of all diseases in Iceland;[4] and when we consider that drunkenness, illegitimacy, and unskilful midwives are fruitful causes of idiocy, causes which all three are most powerful in Iceland, we can only wonder there is so little idiocy there, not that there is so much. Moreover, it is probable that

---

[1] Thomsen, *Ueber Krankheiten*, etc., pp. 115, 116.

[2] *Ibid.*, pp. 11, 31. Sir Thomas Watson, quoting from Schleisner, on whom Thomson chiefly relies, says that scrofula is entirely unknown in Iceland (*The Principles and Practice of Physic*, vol. i. p. 227).

[3] Thomsen, *Ueber Krankheiten*, etc., pp. 112, 113.

[4] *Ibid.*, pp. 111, 44.

the disproportion between Iceland and Denmark is not so great, since many persons are returned in Denmark as insane and under the age of ten years, who must probably be idiots. The Icelandic insanity is mostly of the gloomy-religious kind,[1] and, as we have seen, taking all kinds of mental derangement together, there is less in Iceland in proportion to the population than there is in Denmark.

Since the year 1842, all the deaf-mutes in Iceland have been registered. Thomsen picks out the years 1844 and 1846, and says that the number of deaf-mutes at that time was 36.[2] In the year 1845, the population was 58,619, and therefore the deaf-mutes were 6·14 per 10,000 of the population. But of these 36 cases, 24 were congenital only, and of 5 no particulars are supplied, which gives the deaf-mutes of congenital origin a proportion of 4·6 per 10,000 of the population, counting 3 of the 5 whose origin was unknown.[3] That is, there was 1 case of deaf-mutism, congenital or acquired, on every 1,628·3 of the population: while in Ireland there was in 1851, 1 deaf-mute on every 1,380; in 1861, 1 on every 1,176; and in 1871, 1 on every 1,222;[4] or an average of 1 on every 1,259. On a total population of 1,700 at Asprières, in France, where many consanguineous marriages occurred, there were only 2 deaf-mutes, and these were from non-consanguineous marriages.[5]

[1] Thomsen, *Ueber Krankheiten*, etc., p. 113.
[2] His statement is not very clear, for he only says, "Nach dem Verzeichniss für die Jahre 1844 und 1846, fanden sich damals 36 Taubstumme." Does this mean the average for the years 1844 to 1846 *inclusive*, or is it the average for only these two years?
[3] *Ibid.*, p. 114. Two small districts are omitted in the returns.
[4] *Report on the Status of Disease*, 1871, Table II.
[5] Devic, *Gazette Méd. de Paris*, March 7th, 1863, p. 158.

Phthisis, says Thomsen, is extremely rare. Indeed it is doubtful whether it occurs at all, since from the close similarity of the symptoms of hydatis and phthisis, and the frequency of the former disease, it is probable that physicians have often been deceived.[1] Dr. Hjaltelin, a distinguished physician of Reykjavik, says that among 30,000 patients who have passed through his hands during a practice of fifteen years, he has never found a case of indigenous consumption or tuberculosis; while Dr. Skaptason, the oldest and most experienced physician in Iceland, says that during thirty-two years of practice, though he has seen a great many diseases of the lungs, he has never once met with a single case of *phthisis tuberculosa*, either in his practice or his autopsies.[2]

In the little island of Westmannoë, the infant mortality is even greater than in Iceland. For while in Denmark 189 boys and 160 girls of every 1,000 born do not survive the first year, and in Iceland 326 boys and 281 girls of every 1,000 do not survive the first year, in Westmannoë the death-rate reaches the enormous figure of 762 boys and 722 girls. This mortality is chiefly due to *trismus nascentium*, shown statistically to be entirely due to the horrible state of their dwelling places,—worse, if conceivable, even than those of Iceland; at times crowded excessively, and made still worse by the damp and cold of the place.[3] As in Iceland, women are rarely able to suckle their babes, and this is the more unfortunate since they do not

[1] Thomsen, *Ueber Krankheiten*, etc., pp. 32, 41.
[2] Sir Thomas Watson, *The Principles and Practice of Physic*, vol. i. p. 227.
[3] Thomsen, *Ueber Krankheiten*, etc. pp. 15, 89, 100, 91-96.

keep so many cows as the Icelanders, live more on oily birds, and less on vegetable food. What cows there are, have to be partly fed on dried fish.[1]

Notwithstanding their wretched way of life, which is much the same as that in Iceland, the Faroë islanders enjoy an average of life, even including the still-born, of forty-four years and two-thirds; while in Denmark it does not amount to more than thirty-six, not including the still-born. Scrofula and tuberculosis are very rare, indeed Sir Thomas Watson says they are entirely exempt from the latter, although their climate as regards cold and damp is exactly that which has been thought most likely to produce it. But on the other hand the proportion of those mentally deranged is very great. Panum estimates them at 1 per cent.; and as in Iceland, their madness assumes a gloomy-religious tendency which often degenerates into hopeless imbecility, and which Panum ascribes to the melancholy look of the islands, their imposing character, and the general low pressure of the atmosphere. Congenital idiocy also occurs sometimes, though excessive indulgence in alcoholic liquors is not very common. But Panum believes that of those who do indulge in it, a greater relative number go mad than is the case in other countries.[2]

M. Devay gives a deplorable account of an Irish colony now supposed to be settled in Sligo and County Mayo, of people driven out of Armagh and Down in the years 1641 and 1689, and therefore subjected to great hardship. For nearly two centuries they have been, in consequence of always inter-

[1] Thomsen, *Ueber Krankheiten*, etc., pp. 96, 97.
[2] *Ibid.*, pp. 160, 156, 157, 158; Sir Thomas Watson, *ut sup.* vol. i. p. 228.

marrying among themselves, a diminutive, pot-bellied, crook-legged, and generally miserable people, who form a race easily distinguishable from the other Irish of the neighbourhood.[1] Dr. Beddoe, in noticing it, declares the passage to be an entire libel on the inhabitants of eastern Sligo, and though he has never visited Mayo, he does not think that any diminution of stature has taken place there. To Connemara, where the people, though small, are well built and well favoured, it certainly does not apply; nor to Joyce's and O'Flaherty's country, near Galway, where the people are notoriously tall.[2]

Again, the *Pall Mall Gazette*,[3] in a biography of the late Mr. Augustus Smith, for a long time owner of the Scilly Islands, says that he removed some of the inhabitants of the outlying islets to better neighbourhoods: " On some of these the scanty resident householders—never forming connections out of their own island—had, it is said, degenerated into a condition approaching that of imbecility." Yet Dr. Beddoe remarks on these same islanders, their " proportions certainly give the lie to the current notion that men and quadrupeds must degenerate in small islands,"[4] and McCulloch calls them a healthy people.[5]

In France, also, there are many communities which

---

[1] Devay, *Du Danger*, etc., p. 190, who gives the authority of Quatrefages, but does not point out the place in his works, which, according to M. Godron (*De l'Espèce*, etc., vol. ii. p. 316) is in the *Moniteur des Cours Publics*, 1857, p. 64. Quatrefages probably took it from Prichard, who, in his turn, took it from the *Dublin University Magazine*, No. 48, p. 658. See Waitz, *Anthropologie*, vol. i. p. 63.
[2] Beddoe, *Stature and Bulk*, etc., p. 190.
[3] Aug. 5th, 1872.
[4] Beddoe, *Stature and Bulk*, etc., p. 163.
[5] McCulloch, *Geographical Dict.*

do not intermarry with their neighbours. The commune of Batz, near Le Croisic, says M. Voisin, is situated on a peninsula, bounded on one side by precipitous rocks, bathed by the sea, and shut off from the mainland by a salt marsh. The inhabitants number 3,300, and have but a very limited intercourse with the rest of their department (Loire-Inférieure). They seem to be a very simple race, are intelligent, but reserved to strangers. Drunkenness, prostitution, concubinage, and crime are unknown; nor was there a single individual afflicted with any malformation, or suffering from any disease of the mind, or from deaf-mutism, albinoïsm, blindness, or *retinitis pigmentosa*, though they have been in the habit of closely intermarrying among themselves from time immemorial. At the time of M. Voisin's visit, he found 46 consanguineous marriages, of which 5 were between first-cousins, 31 between second-cousins, and 10 between cousins of the fourth degree. The 5 marriages between first-cousins produced 23 children, an average of 4·6 per marriage, while the average for all France, according to M. Husson, is only 3.[1] All these children were healthy, but 2 died from acute diseases. The 31 marriages between second-cousins produced 120 children, or 3·87 per marriage, none of whom were affected by any congenital malformation or infirmity, but 24 of them died of acute diseases. The 10 remaining marriages produced 29 children, all healthy, but 3 of whom died of acute diseases. Of the whole 46 marriages only 2 proved barren, or 4·3 per cent.; while the average of barrenness is 11·7 or 15 per cent.[2] The

---

[1] Duncan, *Fecundity*, etc., p. 110.
[2] *Medical Times and Gazette*, July 20th, 1867, p. 76.

average number of children for all the marriages was 3·7 per marriage.[1]

M. Broca's results obtained in Bretagne, says Dr. Beddoe, seem at first sight to show that in-and-in breeding has a tendency to cause the race to dwindle in size and in numbers. The Bretons, for instance, of the central cantons, where little admixture of blood has taken place, are far smaller men than their compatriots in general. But the greater elevation of stature in some islands, such as that of Ushant, and Cape Clear Isle, and the evidence of other secluded districts where the population, while far from being purely Armorican, has not recently been crossed, points to the essential character of the race rather than its freedom from admixture as the cause of the remarkably low stature, and such indeed is the opinion of M. Broca himself.[2]

There exists, between St. Armand and Bourges, says Dr. Revillout, a village inhabited by a community of foreign origin, supposed to be descendants of Irish prisoners of war, settled there by a king of France. They live on the sale of the produce of their orchards, which they sell at all the markets of their province (Cher). The Forèatines, as they are called, are all descended from marriages between near kin, for they never marry except among themselves, and form one of the handsomest races of France.[3]

[1] Voisin, *Mém. de la Soc. d'Anthrop. de Paris*, vol. ii. 1865, pp. 433-459. This account is confirmed by Dr. Revillout (see Art. No. 6818, p. 53, vol. xxxvi. of the *Journal de Méd. et de Chir. Prat.*, 2nd Series, note). The full particulars of these cases are given in the Appendix to this work; see Cases Nos. 254-299.

[2] Beddoe, *Stature and Bulk*, etc., pp. 170 and 163, note.

[3] Revillout, extract from the *Gazette des Hôpitaux*, in the *Journal de Méd. et de Chir. Prat.*, 1865, p. 53, note, vol. xxxvi., 2nd Series.

Pauillac (Gironde), says Dr. Ferrier, contains 1,700 inhabitants, most of them robust, vigorous, and well-made sailors; while the women are renowned for their beauty and the clearness of their complexion. Yet, he continues, there is perhaps no other part of France where consanguineous marriage is more common, nor is there any other where exemptions from military service, on account of physical defect, is more rare. It is much the same at Granville, where the maritime population is quite distinct and isolated. At Arromanches, a little village containing less than 100 fishermen, and Portel, a village near Boulogne, containing a few hundred inhabitants, the maritime population always marry among themselves, and never among the agricultural population, whom they disdainfully call " shepherds ;" yet they are healthy and robust.[1]

M. Gubler also bears witness to the great beauty of the inhabitants of Gaust, in the valley of Assau, Pyrenees. The custom of marrying in their own commune, though it barely numbers 200 people, is so well established, that should any young man wish to marry out of it he is obliged to ask permission from the village elders.[2] The Bas-Bretons and Basques, who live at the foot of the Pyrenees, are handsome, hardy and fertile; yet they never intermarry with other races: the Basques always marry Basques, and the Bas-Bretons always Bas-Bretons.[3]

The people of the nominal Republic of Andorra, on

[1] Dally, *Anthrop. Review, etc., of London*, May, 1864, p. 98.
[2] *Ibid.*
[3] Châteauneuf, in the *Ann. d'Hygiène*, vol. xxxv. pp. 43, 44.

the south side of, and comprising some of the wildest and most picturesque valleys of the Pyrenees, number some 7,000 or 8,000 inhabitants, who are divided into six communes. They live as their fathers lived before them; the eldest of each family is its chief, and every man chooses a wife from a family of equal consideration with his own, particular care being taken that no one marries below his rank, and very little attention being paid to fortune. No one leaves the paternal roof till he marries, nor may any one till married have a share in the public affairs. When there are only daughters in the family, the eldest becomes heiress and marries the youngest son of another house, who adopts the name and family of his wife. By this arrangement the principal families have remained for centuries unchanged. They are a strong and well-proportioned race, and mental disease and vice are almost unknown.[1]

M. Devay gives an instance of a little village near St. André and Rives, in the Department of Isère, called Izeaux, very isolated, and situated on poor and barren soil. The inhabitants, he says, necessarily intermarry among themselves, and frequently in their own family. Towards the end of the last century all of them, men and women, had an extra finger on each hand, and an extra toe upon each foot, but about the year 1860 these extra digits had disappeared, through crossing.[2] This very terrible accusation against consanguineous marriage I shall have occasion to return to, and therefore shall say no more about it here.

[1] M'Culloch, *Geographical Dict.*
[2] Devay, *Du Danger*, etc., pp. 95-97.

The Cagots, Capots, Gahets, Cassati, or Chrestiaa, formed a widely distributed race in France from the time of Charlemagne to the beginning of this century, when the last vestiges of them were finally absorbed. Their origin is doubtful, but from the researches of M. Michel it is most probable that they were Spanish and Goth refugees from the vengeance of the Saracens, after Charlemagne's unsuccessful incursion into Spain.[1] This much is certain, however, that they were looked upon with no friendly eye by the inhabitants near whom land was granted them: whether because of the jealousy to foreigners inherent in every ignorant people, to the favour shown them by the Carlovingian dynasty, or to a taint of the Arian heresy,[2] they were treated in a way that would speedily have caused them to emigrate to a more hospitable country had there been the chance of such a country's existence. As it was they fell gradually lower and lower. They were not strong enough, distributed as they were in small detachments, to hold their own against a hostile population. They were gradually forbidden to carry arms, to carry on any of the nobler professions, to move beyond certain bounds, or to mix at all with their neighbours. These disabilities again brought on greater. The people forgot the origin of the Cagots, if they ever knew it, but remembered their hatred. They were tainted with heresy, and the mob in consequence fancied they were tainted with diseases sufficient to account for the regulations which forbad all intercourse with them. It might be that they were confused with the lepers, who were also an out-

[1] Michel, *Hist. des Races Maudites*, vol. i. p. 293, *et seq.*
[2] *Ibid.*, pp 302, 305-311.

cast lot, or it might be as Michel thinks, that their original name of *Gavacho*, taken from *Gabali*, or mountaineers, and which, by what Michel calls a philological accident, had also formed the term *Gafo*, first used in the sense of dirty or stinking, and then for leper,[1] caused them after a time to be looked upon as really tainted with leprosy, and that they were said to stink, and to infect everything they touched. For this reason they were obliged to wear a distinctive mark, a piece of red stuff, since yellow was already appropriated by the unfortunate Jews; they were not allowed to mix in the village sports, to be laid in the village churchyard, or even to mix with the rest of the faithful in the village church—nay, to such extremes was this repugnance carried, that they had a separate door, a separate holy-water dish, and the priest handed them the consecrated wafer at the end of a stick.[2]

As may be imagined, it was rarely indeed that they intermarried with their neighbours. Mixed unions of this kind did occasionally take place; but, as a rule, a man would rather see his daughter turned on the streets than married to one of the hated Cagots; nor would popular prejudice even leave the offspring of these mixed marriages in peace, but taunted them for generations with their Cagot blood.[3] The marriages which they contracted among themselves must have been very close indeed, since bounds were generally set to the peregrinations of every Cagot community; and they were forbidden ever to enter a hostelry, for fear lest they might infect the place.[4]

[1] Michel, *Hist. des Races Maudites*, vol. i. pp. 345-350.
[2] *Ibid.*, vol. i. pp. 9, 11, 74, 103, 182, 183.
[3] *Ibid.*, vol. i. pp. 7, 17, 21, 27, 44, 221, 227, 3, 4.
[4] *Ibid.*, pp. 165, 166, 181, 183. Compare also p. 170.

Under these circumstances, we are indeed surprised to find that the Cagots, far from being cast down, as almost any people would have been, by the general hate of what to them was the world, nobly struggled for their freedom. Kings and parliaments were importuned again and again. But in vain. For though justice was often nominally granted them, it was never carried into practice; and at last when, tired out with these misfortunes and cruelties they attempted to compel respect, they were soon disarmed, and forbidden ever again to carry anything more formidable than a pointless stick.[1] Their wretched habitations outside the walls of cities, their oppression, and the manifold disabilities under which they laboured, seem only to have had the effect of making them more industrious, and not less healthy than their neighbours.[2] De Buziet describes them as indistinguishable from the other people, except that some families had a fair and fresh complexion, and were tall and lithe; while others again were dark and strong, of middle height, and well-knitted frames; and Làa, Minvielle, and Zamacola, who, like De Buziet, had exceptional means of observation, confirm this description.[3] It was a popular idea that they had very small or no ear-lobes; which may be true of some of them,[4] or it may be that those people who had small ear-lobes were believed to be Cagots. Of course if any family should produce a "sport" of this kind or of any other, such as six fingers,

---

[1] Michel, *Hist. des Races Maudites*, vol. i. chap. ii.
[2] *Ibid.*, vol. i. pp. 214, 14, 17, 28, 37, 40, 41, 48, 53, 103, 217, 218.
[3] *Ibid.*, pp. 266, 267.
[4] *Ibid.*, chaps. i. and ii.

consanguineous marriage must necessarily tend to fix it.¹

In short, the Cagots are a people who have married continually among themselves for centuries, who have been terribly oppressed for centuries, who have been as healthy as their neighbours despite this ill-treatment, and who have now, through the advance of civilization, broken through the barriers which opposed their intermarriage with their neighbours, and have been absorbed.²

The paper-makers of the Angoumois, Limousin, and Auvergne, are a separate race, so much attached to their villages that they never quit them, and consequently never marry except among themselves. Their children are always brought up in their own profession, which from its nature is very unhealthy. Living as they do in a damp and marshy country, their factories full of water, and having to pass twelve or fourteen hours at a time

---

[1] The Chaouia Berbers of the Aures Mountains also have lobeless ears. Waitz, *Anthropologie*, vol. i. p. 97; Godron, *De l'Espèce*, etc., vol. ii. p. 260.

[2] M. Devay, with his usual disingenuousness, chooses to repeat a mistake of the Middle Ages, which even at that time was hardly excusable, and confound the Cagots with the lepers, crétins, and goîtreux. He has read nothing except Michel's work on the subject, yet he chooses to ignore Michel's deductions founded on an intimate and comprehensive knowledge of his subject, and put up instead some chimerical views of a physician who shared in the prejudices of an ignorant age, and who has been contradicted by more enlightened physicians of the same time. I must refer my readers to Devay, *Du Danger des Mariages Consanguines*, p. 186, and to Michel, *Histoire des Races Maudites*, chaps. iii. and v. vol. i. But perhaps the whole of the volume should be read to form an adequate idea on the subject. Compare also Hecker, *Die Grossen Volkskrankheiten des Mittelalters*, pp. 65, 75, 96-100, who in his account of the Black Death shows how ready the mob are to fix on any outcast race as the disseminators of disease.

surrounded by vapour in their paper vats, it is not surprising that they seldom attain even the age of sixty-five, and frequently suffer from varix, dropsy of the legs, chronic rheumatism, and ulcers of the leg and ankle joints, from tertian fevers in autumn and spring, and that they suffer greatly from catarrh, their knees bend inwards, and their teeth fall out early in life. They are said also to be subject to scurvy,[1] which would account for their teeth going. It is especially noteworthy, however, that of all these diseases not one is to be found in the list of those said to be caused by the marriage of near kin.

The Marans, of Auvergne, a race of Spanish converted Jews, were for their origin left particularly isolated. The Christians of Spain would not intermarry with them, nor would either the Jews or the Moors. It is probable that they were expelled with the Moors from Spain by Philip III., and that they then settled in France. They have been accused of having introduced syphilis into France, an accusation of doubtful truth, since they always intermarried among themselves, and were generally hated as strangers. No other accusation of ill-health seems to have been made against them.[2]

The Hautponnais and Lyzelards, of St. Omer, form populations absolutely distinct from their neighbours, and which have preserved themselves without a cross from any other race, since their first establishment about A.D. 449. Some say they were Saracens, others say Saxons, others Moors, Flemings, or the descendants of German lanzknechts. They have a separate

[1] Michel, *Hist. des Races Maudites*, vol. ii. p. 30, and note.
[2] *Ibid.*, vol. ii. pp. 52-54, 94-96.

language, a sort of Flemish patois, and have always lived at peace with the city, and fought with the other community.[1] No accusation of ill-health has been brought against them as far as I am aware, and they certainly have not " died out."

There are various communities scattered over Sermoyer, Arbigny, Boz, and Ozan, communes of the Department of Ain, and the Arrondissement of Bourg-en-Bresse, Canton Pont-de-Vaux. An immemorial tradition ascribes their origin to the Saracens, but Reinaud has shown that this is impossible. Whatever be the origin of their isolated state and the hatred shown them by their neighbours, it still exists according to the Curé of Boz. They cannot get even the commonest girl in marriage, and hence always intermarry among themselves. Yet what a difference is there, says the Curé, between these Burins, industrious and rich, whose active labour makes the earth put forth her riches in abundance, and their neighbours who hold them in contempt, yet often rest in idleness and poverty! The Burins have been labourers, cattle-breeders, and butchers, for centuries. Some among them are remarkably fine men, the women pretty, with quick black eyes, fair, and rather inclined to roundness. They have a strange air, which the curé ascribes to their isolation. The Sermoyers, like the people of Boz and Uchizy, are hated by their neighbours, and pass for an avaricious and bad people. They have had great disputes concerning the pastures of the Saône, and many have been killed on either side. Now they fight in the law courts; but these quarrels, adds the curé,

[1] Michel, *Hist. des Races Maudites*, vol. ii. pp. 102-104.

do not prevent them being a very sensible lot of people.[1]

Another isolated community are the Vaquéros, who, notwithstanding their isolation, have probably the same origin as the rest of the Asturians. Their little villages are perched on the Asturian mountains, in positions where they are protected by others higher still. Their sole occupation is the breeding and sale of sheep; and every year they leave their habitations for the higher mountains of Leon, where they remain from the middle of June to the end of September for the sake of the pastures. As merchants, they are sharper than those who occupy themselves with agriculture, and also more dishonest. The consequence is they are hated by the other Asturians, a sentiment which the Vaquéros cordially return, and both avoid as much as possible any intercourse, and especially the ties of relationship. But if, in spite of all this, interest or violent love lead to a mixed marriage, it can never occur without great scandal, and the greatest expression of repugnance from the Asturian family. The Vaquéros pay more money to Rome for dispensations for marriage within the forbidden degrees than all the rest of the principality together. They are all plebeians alike, with the exception of one family, a very extended one now, since it has married among the others for more than half a century.[2]

The Chuetas, of Majorca, were another isolated community of Jewish converts, who never intermarried with their neighbours. Of course they were

[1] Michel, *Hist. des Races Maudites*, vol. ii. pp. 108-110.
[2] *Ibid.*, vol. ii. pp. 42, 43.

treated brutally, both by the people and by the Inquisition, which tribunal sentenced them to be burnt in hecatombs on suspicion of a relapse to the faith of their fathers. Notwithstanding this treatment, so far from dying out, there were in the year 1782 more than 300 families left, who were still generally hated for their origin, and still forced to intermarry among themselves.[1]

The upper classes of the Azore islanders, owing to the general desire to hoard wealth, are in the habit of intermarrying very closely among themselves, so that marriages even in the third degree are very frequent. They are very ignorant, and contented with their ignorance; the strain of Moorish blood in their veins tinges their habits and customs, and they are mean and parsimonious in their way of life. Consul the Hon. E. Monson finds an unnatural frequency of idiocy and enfeebled constitution among them, which he considers a consequence of their consanguineous marriages.[2] M'Culloch, however, points out that they are also morally debased by a corrupt clergy,[3] and this of course is likely to lead also to physical debasement.

The descendants of the original French settlers on the island of Réunion, known as Petits-Créoles, or Petits-Blancs, are described by Dr. Yvan, who visited them in the year 1844, as the aristocracy of the island. These families, he says, have acquired, under the influence of the most salubrious climate in the world, a remarkable degree of beauty. The men

---

[1] Michel, *Hist. des Races Maudites*, vol. ii. pp. 38-41.
[2] *Consular Reports*, 1871, No. 4.
[3] *Geographical Dict.*

are straight and vigorous, their skin a delicate sunburnt shade, the forehead broad, and their countenances noble and dignified. The women are also well formed and beautiful, with chestnut hair, long eyelashes, and large brown eyes, regular features, and if anything too proud and energetic an expression of countenance. The manners of these people are simple and peaceful; the families live in the closest intercommunion; they commit few offences, and crime is almost unknown among them. It is worthy of remark, adds Dr. Yvan, that in spite of their poverty, these people will never marry a half-breed—nothing would induce them to defile the purity of their race by one drop of mixed blood.[1]

In the year 1835 the settlers of Tasmania, finding the aborigines rather inconvenient, organized a regular battue, and massacred all but 210 persons, men, women, and children, who were transported to Flinder's Island, Bass's Straits. Among these were one or two adult Anglo-Tasmanian half-breeds. The island is about thirteen leagues in length by seven in breadth; the British Parliament ordered that they should be kindly treated and supplied with food; but this was after the news had slowly travelled to England, undergone a debate, and the order had slowly travelled back. When Count Strzelecki visited them in 1842, only fifty-four Tasmanians remained alive, and during the whole time which elapsed between their deportation and his visit, not more than fourteen children had been born.[2]

[1] J. A. N. Perier, *Mém. de la Soc. d'Anthrop. de Paris*, vol. i. 1863, pp. 191, 192.
[2] Broca, *Hybridity*, etc., translated by Blake, p. 46, note; J. A. N. Perier, *Mém. de la Soc. d'Anthrop. de Paris*, vol. iii. 1870, p. 242.

On some other isle of Bass's Straits was a small colony of half-breeds, between English seal-fishers, and Australian and Tasmanian woman. These people have continued to intermarry among themselves, are tall, quick, and of good intelligence, while they are free from nearly all the vices of civilization.[1]

There is some doubt whether the old Samaritans were a half-bred race, or whether they were a pure race, removed to Samaria by Esarhaddon in place of the captives of Shalmaneser. If they were a pure race they were probably an Assyrian colony, on whose fidelity the king was able to rely. If a mixed race they may have been remnants of the ten tribes, or a fusion of these, Phœnicians and Syrians, or a colony from a great distance placed there by the Assyrians. But whatever their origin, they were, by the force of circumstances, a peculiarly isolated race. Like the Chosen People, they spurned all connection with the neighbouring races, and were spurned in their turn by the Jews themselves. Stung by this contempt—a contempt which naturally followed from the exclusive ideas and laws which clung to the Jews even in their bondage—the Samaritans arrogated to themselves that title which the Jews refused to share with them, and thenceforward a hate burned between them, only possible between neighbouring nations holding different and bigoted religious views. They were exclusive then, they are an exclusive family now. Though they had been slaughtered in thousands by Alexander, by

---

[1] J. A. N. Perier, *Mém. de la Soc. d'Anthrop. de Paris*, vol. iii. p. 241. He, however, rather doubts the authority, which is derived from an Australian paper, whence it was copied in Petermann's *Geographische Mittheilungen*, and thence into the *Moniteur Univers* for June 9th, 1863, p. 852.

Pilate, and by Vespasian, Epiphanius in the fourth century considered that the Christians had more to fear from the Samaritans than from all their other adversaries. In the fifth century they justified this prediction by an outrage on the Christian population, which was so severely punished that they sank into an obscurity scarcely broken until the sixteenth century. A short time ago they numbered 150 or 200 in all; an islet among a tempestuous sea of Mohammedans, Christians, and Jews, while now they are reduced by constant ill-treatment and persecution to 135 individuals, of which 28 couples are married, 10 persons are widows, 49 are unmarried men and boys, and 20 are girls.[1] According to a memoir by Mr. Glennie, in which he describes the heads of the seventy families, whom he saw all together in their synagogue in the year 1862, " Every man of them was full six feet or upwards, erect, and well proportioned, with very fine, though, of course, Jewish features; beautifully clear, fair complexions, and dark, lustrous eyes."[2]

[1] Smith, *Dict. of the Bible*, Art. *Samaria*. See also a Letter to *The Times*, April 4th, 1874, from Yacoub esh Shallaby.

[2] See *Fraser's Magazine* for August, 1863, an article entitled *Mr. Buckle in the East*, p. 182. Mr. Glennie adds: "We (*i.e.*, Mr. Buckle and himself) had no gallop for a long time that morning; for the physiological laws of breeding in-and-in, the influence of race, the worth of phrenological indications, and related subjects, occupied us and gave our horses a rest." On the publication of Mr. Buckle's *Common Place Book*, I had hoped to have found the results of his reading on this subject, together perhaps with some brilliant flashes of his genius to help me on my way. I cannot describe my disappointment on finding that Miss Taylor, who so generously and conscientiously edited these remains, has been induced to leave out certain articles for their coarseness. This I cannot but think a mistake, just as the opinion I have heard expressed that more should have been left out to make the book more readable. I see in the Index:—"Incest: Persians, 66; Athenians married their sisters, 157; Of near relations not contrary to Nature, 125." The first

The Jews, before all others, are the chosen people against whom the attacks of Parasyngeneiasts have been directed, yet M. Boudin, the most pertinacious champion of this class, has himself written a most charming eulogy on the strength, viability, and wonderful endurance of this much-maligned nation. There are some races, he says, who seem marvellously able to adapt themselves to variations of climate, while there are others who can scarcely support the least change. The Jews, and perhaps the Gypsies, are instances of the former, for the Jews are now spread over every quarter of the world; they are scattered in Europe, from Norway to Gibraltar; in Africa, from Algeria to the Cape of Good Hope; in Asia, from Cochin China to the Caucasus, and from Jaffa to Pekin; while in America they are to be met with everywhere, from Monte Video to Quebec. Only fifty years ago they invaded Australia, and have already given proof of their ability to flourish and multiply in a climate which other people of European origin have never yet been able to withstand. Though they are not generally found at any great elevation above the level of the sea, possibly because such places are rarely suited to their usual occupations, there is yet no reason to suppose that they are at all unsuited to such places;

article is there, but it is not of much value. The second has under No. 157, an article on the word "Deist;" while Article No. 125 is entirely expunged. Again, under the head "Marriage," I see "Of cousins (see *Incest*), 411," which article only refers to the origin of the Puritans. The book is a work for the student, not for the every-day reader; a most valuable work, and the editing of which was peculiarly laborious. It is to be hoped that in the event of a second edition the expunged parts may either be restored to their places or be printed in a separate volume, and by this means please both those who wish to read the *Common Place Book* for amusement, and those who would use it as a work of reference.

while on the other hand they have lived for centuries, and still continue to live, at the only spot of the globe situated 400 mètres below the level of the sea, a part where it is very doubtful whether any European could live and multiply. Wherever, continues M. Boudin, the Jewish race has hitherto been studied, it has shown itself governed by statistical laws as to births, deaths, and the proportion of the sexes, entirely different from those which govern the surrounding communities.[1] Nor is this astonishing when we recollect that they always marry among themselves, except under the supposition that marriages between near kin are harmful, for by marriages of this kind every animal is far more easily acclimatized than by continual crosses. If however M. Boudin's declared opinions are correct, we should expect this stiff-necked generation to have a lesser viability, a greater percentage of still-born, of early mortality, and of other evils, as befits a people who will not submit to laws laid down by pontiffs and sanctioned by emperors, excepting when they applied to themselves. Yet, strangely enough, M. Boudin informs us that the Jewish population has doubled in fifty years, while the infant mortality is less among the Jews by more than one-third in some countries, and by one-half in others, than in the Christian populations.[2] In Prussia, says Oesterlen, one-fifth of all children born in wedlock, and including the still-born, die before they reach their fifth year, while only two-thirteenths of the Jewish infants die before reaching that age, including, besides the still-born, also all the children born out of

[1] Cited by Devay, *Du Danger*, etc., p. 180, from Boudin's *Géographie Médicale*.
[2] Dally, *Anthrop. Review*, London, May, 1864, p. 93.

wedlock. How great a difference this really is may be seen by the fact that while the average mortality for almost the whole of Europe was 21·8 per cent. for legitimate, it was 32·5 per cent. for illegitimate children.[1] Neufville shows that in Frankfurt, while the average duration of life among the Christian population was 36 years and 11 months, that of the Jewish population was 48 years and 9 months; and yet, in comparison to other towns, the Christian population was long lived.[2] The following two tables will show this difference more clearly:—

There died between the age of	Out of every Hundred Deaths among the Christians.	Out of every Hundred Deaths among the Jews.
1—9 years	26·4	13·3
10—19 ,,	4·5	4·5
20—29 ,,	12·4	8·8
30—39 ,,	10·6	9·5
40—49 ,,	11·0	9·9
50—59 ,,	10·3	9·9
60—69 ,,	11·4	16·7
70—79 ,,	9·7	20·5
80—89 ,,	3·5	6·5
90—100 ,,	0·2	0·4

[1] Oesterlen, *Med. Stat.*, pp. 146-147. If a great proportion of male births is a sign of vigour in a nation, as appears to be the case from the Scotch Census Report of 1871, the Jews seem pre-eminent. Thus for every 100 girls there were born in—

Country.	In the Years	Male Jews.	Male Christians.
Prussia	1820—1834	111·0	106·0
,,	1849—1852	106·9	105·9
Austria	1851	121·0	105·9
Algiers	1836—1851	106·5	103·0

Ibid., p. 164. And Waitz gives the number of male births for every hundred female births among the Jews as 208 in Berlin, and 120 in Leghorn (See his *Anthrop.*, vol. i. p. 127).

[2] Neufville *Lebensdauer*, etc., pp. 18, 19.

Of the comparative number living :—

Jews per Cent.	Christians per Cent.			
86·7	73·6	reach their	10th	year.
82·2	69·1	,,	20th	,,
73·4	56·7	,,	30th	,,
63·9	46·1	,,	40th	,,
54·0	35·1	,,	50th	,,
44·1	24·8	,,	60th	,,
27·4	13·4	,,	70th	,,
6·9	3·7	,,	80th	,,
0·4	0·2	,,	90th	,,
0·0	0·04	,,	95th	,,
0·0	0·0	,,	100th[1]	,,

Consanguinity of the parents of course does not affect the mortality one way or the other. The cause of the greater mortality among the Christians may be partially due to the danger from too early exposure for baptism,[2] but this can have but a very slight effect. We must look for the cause of infant mortality in that of the greater viability of the adults. The Jews are an abstemious race, and their professions are chiefly mental, rarely mechanical. Hence they are less subject to the influence of harmful trades, and the good health of the parent cannot but affect the viability of the child. It is for the same reason that the Jewish population seems comparatively more prone to mental derangement. There were, in the year 1847 in Denmark, 5·85 per 1,000 Jews suffering from some form of mental disease, while there were only 3·34 per 1,000 Catholics thus affected;[3] and M. Legoyt gives the following table, showing the relative

[1] Neufville, *Lebensdauer*, etc., pp. 111, 113.
[2] Oesterlen, *Med. Stat.*, p. 148, note 3; and p. 309, note 3.
[3] Boudin, *Ann. d'Hygiène*, vol. xviii. 2nd Series, p. 15.

frequency of mental disease among Catholics, Protestants, and Jews:—

	Number for one Insane of		
	Catholics.	Protestants.	Jews.
In Bavaria	908	967	514
,, Hanover	528	641	337
,, Silesia	1355	1264	644
,, Wurtemburg	2006	2028	1544
Total	4797	4900	3039[1]

Hence there were four insane on every 4,797 of the Catholic population; four on every 4,900 of the Protestant population; and four on every 3,039 of the Jewish population. "Must we see in this frequency of lunacy among the Jews," says M. Legoyt, "an influence of race, or merely a consequence of the fact that they inhabit those towns, and exercise those professions the most exposed to economic crises? Must we see there, like Dr. Martini, the influence of the fact that marriages between near relatives are commoner among the Jews than among the Christians? Or ought we to admit the concurrence of all three causes?"[2] MM. Boudin and Chipault consider it solely due to the last cause, and even fancy they can detect a direct relationship between the frequency of insanity, and the closeness of the degrees within which marriage is permitted by the Jews, the Protestants, and the Catholics.[3] How such a deduction

[1] Legoyt, *Journal de la Soc. de Stat. de Paris*, 1863, p. 90.
[2] *Ibid.*, pp. 90, 91.
[3] Boudin, *Mém. de la Soc. d'Anthrop. de Paris*, vol. i. 1863, p. 526; Chipault, *Etudes*, etc., p. 74.

is supported by the table given above I am at a loss to see. In the first place the Catholics, who are forbidden marriage with a first-cousin, appear to be more subject to insanity than the Protestants, who are not; and in the second place, how can we say that the greater proportion of insanity among the Jews is not due to their professions, when we do not know the relative number of Jews and Christians who live in towns, and the relative number who live in the country? How important, nay, how absolutely necessary it is to know this before any deduction can be made is shown by the difference of about 25 per cent. between madness in the country and madness in towns.[1] Hence, to make the table really useful, it would be necessary not only to compare the proportion of insane of each faith, but also the proportion who lead a town life; and further, the proportion who are engaged in professions which tend to cause constant anxiety. It is well known that a far greater proportion of Jews than of Christians are occupied in professions which may be classed as mental, and not mechanical, such as teachers, professors, merchants, and speculators in money, etc.; and it is also well known that just these professions give the highest percentage of lunacy; indeed, it is nearly certain that it is these professions which make madness so much more frequent in towns than in the country. There can be very little doubt but that it is this which brings up the proportion of lunacy among the Jews, a theory which is confirmed by the character of that madness,

---

[1] Oesterlen, *Med. Stat.*, p. 522. On every 100,000 inhabitants of Hanover living in towns, there were 222 mad; on the same proportion in the country there were 161.

generally an acute or raving mania, and seldom hypochondriacal or despondent.[1] But according to M. Legoyt, idiocy also is commoner among the Jews. Thus there were in—

	Number for one Idiot of		
	Catholics.	Protestants.	Jews.
Silesia (1856)	4113	3207	3003
Wurtemburg	580	458	425[2]

and this is not so easily explained as the former. The Jews are not drunkards, and hence we should expect, if anything, a smaller proportion of idiots among them than among the Christians. It is not certain, however, owing to the smallness of these numbers, that any deduction is reliable; and further, the causes of madness in the parent will doubtlessly lead to imbecility in the offspring, while, at all events, continued residence in towns must largely influence it. We must therefore await a more elaborate table before we may venture to reason upon it.[3] M. Boudin further asserts that deaf-mutism is more frequent among the Jews

[1] Oesterlen, *Med. Stat.*, p. 522.

[2] Legoyt, *Journal de la Soc. de Stat. de Paris*, 1863, p. 91.

[3] It is amusing how, in the absence of reliable statistics, attempts to fix upon marriages of consanguinity as the cause of idiocy refute themselves. "It is remarkable in England," says Esquirol, talking of mental derangement, "*especially among the Catholics*, who always ally themselves in marriage, with those of their own denomination" (*Mental Maladies*, etc., translated by E. K. Hunt, p. 49). It is more frequent in England and Scotland, according to Dr. Stark, because the Protestants marry their cousins, than in Catholic Ireland (*Journal of the Stat. Soc. of London*, vol. xiv. p. 62). How baseless all these thoughtless off-hand statements are, may be seen from subjoined Table, compiled from the *Report of the Census Commissioners for* 1871 *on the Status of Disease in Ireland*, for in every one of the diseases here given, and which are supposed to be peculiarly

than it is among Christians; that on examination of the records of the Imperial Institution for Deaf-mutes at Paris, he found three Jews on a total of 200 congenital cases; whereas the Jews are, to the rest of the population of France, as 1 to 350.[1] Unfortunately, however, for this theory, M. Boudin's premises do not happen to be beyond doubt. The Chief Instructor at that Institution, M. Vaisse, says it has only brought up a very inconsiderable number of Jews; it has sometimes contained one, and sometimes none at all. M. Dally, who examined the records very carefully, utterly denies that M. Boudin's three Jewish cases were either congenital, or derived from marriages between near kin.[2] M. Isidore, Grand Rabbi of Paris, estimates the Jewish population of that town at 25,000, in which number, he says, there were a few weeks before the date at which he writes, only 3 Jewish deaf-mutes in Paris; of which 2 came from Bordeaux, and 1 from Rhenish Prussia, while the latter has now left.[3] Counting all three, this gives a pro-

due to consanguineous marriages, the Roman Catholics have a far greater proportion of sick, whatever may be the cause. Thus:—

There were in Ireland in 1871.	A total number of	Proportion to the population of Roman Catholics of 1 in every	Proportion to the population of other religious professions of 1 in every	Proportion to the total population of 1 in every
Deaf-mutes and Mutes ..	5554	968	996	974
Blind	6347	795	1117	852
Lunatics and Idiots ..	16505	321	347	328
Malformed.. ..	2931	1615	3766	1846

[1] Boudin, *Ann. d'Hygiène*, vol. xviii. pp. 14, 15.
[2] Dally, *Anthrop. Review of London*, pp. 92, 93.
[3] *Comptes Rendus*, vol. lv. 1862, pp. 128-129.

portion of 1 on every 8,334. Now according to official reports, says M. Dally, there is 1 deaf-mute on every 4,694 inhabitants of the department of the Seine; hence, so far from giving a greater percentage of deaf-mutism, the Jews give a considerably lesser.[1] Yet Dr. Liebreich, the celebrated oculist, gives his support to M. Boudin's supposition, and adds that *retinitis pigmentosa* is also much commoner among the Jews than among the Christians, since in the establishment for deaf-mutes at Berlin, of 14 patients who also suffered from this disease, 8 were Jews.[2] Whether this disease is really caused by consanguineous marriage is discussed below; but are his statistics on the deaf-mutes who are Jews trustworthy? He says that on a total number of 341 deaf-mutes in the asylum at Berlin, 42 were Jews; while of 223 out of those who were born in Berlin, 23 were Jews.[3] Now M. Boudin gives the Jewish population of Prussia about that time as one-seventieth of the total population;[4] hence, if we are to accept the results taken from this asylum as an accurate proportional reflection, the Jewish deaf-mutes are very greatly in excess of what we should expect. But here is the weak point. Asylums are not, and cannot be, a correct mirror to the general population. We have seen how very widely M. Boudin was misled by deductions from the Paris asylum; and no one can accept a theory without hesitation which is so slenderly supported by statistics. Even should we grant that he has proved the Jewish deaf-mutes greatly to

[1] Dally, *Anthrop. Review of London*, p. 93.
[2] Chipault, *Etudes*, etc., p. 53.
[3] *Ibid.*, p. 54; Dally, *Anthrop. Review of London*, p. 91.
[4] Boudin, *Mém. de la Soc. d'Anthrop. de Paris*, vol. i. 1863, p. 528.

exceed the Christian, we have no evidence to show that this is due to consanguineous marriage; on the contrary, we have far more reason to believe that those causes which lead to mental disease in general, and which in all probability work more on the Jewish than on the Christian population, will also partially show themselves in deaf-mutism and other mental maladies.

I shall not notice the assertions of Dr. Elliotson, that from his personal experience the rich Jews of England are afflicted with more than the ordinary share of squint-eyed, stammerers, peculiarity of manner, imbecility, insanity, and nervousness; of Lallemand, that they have degenerated; of Dr. Pruner, of Cairo, who considers that there are a greater proportion of deaf-mutes among the Cairene Jews than there should be; or of Grellois and Furnari, that hydrophthalmia in Algiers is almost exclusively confined to the Jews.[1] These statements are wholly unsupported; and until they are supported by evidence of some sort they do not deserve a reply. Dr. Elam, who apparently is unable to accuse them of all the disease so freely lavished on them, and unwilling seemingly to allow that consanguineous marriages are harmless, hints that the Jews are the chosen people of God, and may perhaps on that account marry as their fathers did before them, a permission denied to the Gentiles![2] The fact is, that the Jews are not influenced, as far as our evidence as yet goes, either for good or for evil by their consanguineous

---

[1] Boudin, *Mém. de la Soc. d'Anthrop. de Paris*, vol. i. 1863, pp. 524, 526, 528.
[2] Elam, *A Physician's Problems*, p. 73.

marriages, unless these are the cause of their greater ability to withstand the variations of climate. They transmit their type unchanged, as do all races of men and animals when they constantly breed in-and-in, but they have probably been somewhat affected by the centuries of persecution they have endured, and still endure in many parts of the world. The history of the Jewish community of one Christian town is very much the same as that in any other; and a sufficient idea of the miseries they have endured, and the marvellous elasticity which has preserved them from extinction when so many other outcast races have been utterly stamped out, may be conveyed by the history of the Roman Jews, probably the oldest unbroken community of that nation in Europe. The quarter that has been assigned to them in that town, known as the Ghetto, is the most foul, most crowded, and most evil-smelling part of that evil-smelling city. The position is so low that the whole quarter is very frequently under water, an inconvenience which the inhabitants remedy by having their houses communicating with each other by the upper floor, so that this forms, as it were, an upper street. Yet they have inhabited this den from the time of Paul IV., A.D. 1555, with occasional short-lived mockeries of liberty, and long spells of diabolical persecution. During all this time it is far from probable that they have ever received the slightest amount of fresh blood in their veins. What inducement was there to go to the centre of persecution ? But they have constantly emigrated. As soon as ever a member became rich enough, he left this wretched place ; and thus much property and much mental capital was constantly withdrawn, and a disproportionate amount of poverty left behind. Under Claudius the Roman Jews numbered 8,000; about the year 1667 there were some 4,500; not long before the Italians entered Rome there were only 4,000; but under more humane laws they have since then risen to 4,500 again. And yet, despite their persecution, despite their bad drainage, their dirty habits and chronic state of overcrowding, the Ghetto seems always to have been one of the healthiest places in Rome. The deathrate has been small, fever rare, fewer died when cholera was devastating Rome in 1837 of that disease in their quarter than in any other quarter of the city. They thrive and multiply, paid heavy taxes to the Papal government, and support their own poor; they founded a university as well as schools, and they support a synagogue. Truly, as Mr. Story says, "In a people thus oppressed" must there be "immense vitality and energy, or they would long ago have ceased to exist."[1]

[1] Story, *Roba di Roma*, pp. 306, 310, 322, 327, 332.

CHAPTER V.

THE VALUE OF STATISTICS HITHERTO COLLECTED CONCERNING
MARRIAGE BETWEEN NEAR KIN EXAMINED.

THE Parasyngeneiasts have advanced many figures in support of their theories taken chiefly from collected cases, the deductions from which, if trustworthy, may almost be said to be conclusive as to the harmfulness of marriages between near kin. But in making use of any statistics on the proportion of diseased from consanguineous and non-consanguineous marriages, it is necessary first of all to know the proportion that one class of marriages bears to the other. This unfortunately we do not know.' Of course some attempts have been made to determine the proportion. Thus M. Boudin attempts it by means of the following table from the French official returns:—

Year.	Between Nephews and Aunts.	Between Uncles and Nieces.	Between First-Cousins.	Total of Consanguineous Marriages.	Total of all Marriages.
1853	38	107	2,309	2,454	280,609
1854	36	106	2,427	2,569	270,896
1855	48	141	2,592	2,781	283,335
1856	58	147	2,738	2,943	284,401
1857	48	136	2,892	3,043	295,510
1858	66	173	2,806	3,076	307,056
1859	35	111	2,108	3,045	298,417
Total	329	921	17,872	19,911	2,020,224

	Between Nephews and Aunts.	Between Uncles and Nieces.	Between First-Cousins.	Total of Consanguineous Marriages.
Per cent. of all Marriages	0·016	0·04	0·88	0·91

From another official return of marriages in France in the years 1863–1865[2] there were—

Marriages between nephews and aunts	179
,, ,, uncles and nieces	552
,, ,, first-cousins	10,810
Total of consanguineous marriages	11,541
,, all marriages in France	900,197
On every 100 mixed marriages there were consanguineous	1·28

These results therefore do not agree with those obtained by M. Boudin; and the difference is so great that even were there no other evidence it would be impossible to accept either as conclusive. But M. Dally supplies further evidence as to their untrustworthiness: "The regulations of the préfecture," he says, "prescribe the registration of marriages between first-cousins, uncles and nieces, nephews and aunts, sisters-in-law and brothers-in-law; but these registrations are incomplete in the towns, and entirely neglected in the communes. In the offices of the *mairie* at Paris, the statistics of marriage are registered monthly with great exactness, with the individual relationship, and it is from these that the clerks register the degree of cousinship. The future parents[3] are not directly questioned, and

[1] *Ann. d'Hygiène*, Paris, vol. xviii. 2nd Series, 1862, pp. 7, 8. The returns of the year 1853 do not include the departments of Crouse, La Manche, Seine, and Vaucluse.

[2] *Loc. cit. Medical Times and Gazette*, May 15th, 1869, p. 520.

[3] Evidently a mistake of the translator. It should be "the contracting parties."

their relationship is not an object of particular registration, either in the record, or the registers. One can understand, then, how many omissions must be made in the long and uninteresting work of abstracting from numerous records (1000 per month in the 8th district) a page of statistics in which are comprised thirty or forty questions. But it is not in towns that marriages of relations are most numerous: it is certainly in the country. Now, in a great many country places (I myself know of three communes), no account is taken of the relationship of parents, excepting in the case where legal dispensations are necessary (uncle and niece, aunt and nephew). Most people know that, in general, the communal schoolmaster fulfils the duties of secretary to the mayoralty; these *employés* have usually a manual recommended by the Minister of the Interior,[1] and according to the instructions found therein they draw up their records; one can read there the enumeration of 'eleven declarations common to all the records of marriage,' and in these there is not a single question on the subject of relationship. These instances alone would suffice, I think, to authorize us to consider that the official number of marriages between cousins is very much under the reality, and that they do not comprise the country population, in the centre of which the statistics are regulated. The total number of marriages is, on the contrary, rigorously exact everywhere. Hence it follows that when M. Boudin values by official documents, the proportion of 'relationship marriages' at 0·9 per cent., this number has, in my eyes, notwithstanding its official origin, no scientific authority whatsoever, because I know that it rests

[1] Hallez-d'Arros, *Guide du Maire et du Secrétaire du Mairie*, 1858.

upon the authority of incomplete data. And that which was at first my opinion only is become a certainty, since I inquired of M. Legoyt, the head of the statistical office of France, the manner in which the numbers published by M. Boudin had been obtained; this official has authorized me to declare that he cannot answer for any statistics, except those which have reference to the legal dispensations necessary for marriage between uncle and niece, aunt and nephew; so, since marriages between first-cousins do not require this permission, M. Legoyt is convinced himself that the extracted numbers are incomplete, and he has prepared a circular destined to remedy the various mistakes already noticed. Future parents[1] will be henceforward directly questioned about their relationship; mention will be made in the records of their answers, and there is reason for hoping that, in a few years, we shall learn the real proportion of consanguineous marriages.[2] In

---

[1] Contracting parties.

[2] Dr. Mitchell quotes what is probably the text of M. Legoyt's letter; but he only states that he got it from the *Medical Times and Gazette*:— "Sir, the question so warmly debated in learned bodies as to the influence of marriages of consanguinity upon the physical aptitudes of the generations which are the result of these, gives quite a special importance to the table which the annual movement of the population should furnish me with respect to the number of marriages. Now, information derived from trustworthy sources authorizes me to believe that these indications are remarkably incomplete as regards marriages between cousins-german. Omissions of this kind are very easily explained when we bear in mind that the marriages in question not being, as are those contracted between brothers-in-law and sisters-in-law, uncles and nieces, aunts and nephews, the objects of legal prohibition, the local authorities have no means of recognizing them. I beg of you, then, to issue special instructions inviting the mayors to make direct inquiries in the case of all future marriages, when the papers laid before them do not contain the necessary information whether the parties are related in the degree of cousins-german or even of cousins the issue of cousins-german" (*Mem. read before the Anthrop. Soc. of London*, vol. ii. 1866,

the meanwhile I have examined, at the *mairie* of the eighth district of Paris (formerly the first), the monthly records of marriages celebrated during a period of ten years, from 1853 to 1862, and I have obtained from them the following results:—

Total number of marriages	.	.	.	.	10,765
Marriages between first-cousins		.	.	.	141
,, ,, uncle and niece	.	.	.	.	8
,, ,, aunt and nephew	.	.	.	.	1
Total consanguineous marriages				.	150

(These numbers may vary from 146 to 152, on account of three figures which are uncertain.) These numbers give us a proportion of 1·4 per cent. And it appears to me impossible to admit otherwise than this, that, in a district of Paris which is inhabited by foreigners, showing a considerable floating population, there are many less marriages between cousins than in the midst of small towns and in the country. This is why, finding here 1·4 per cent., I am authorized to say that 0·9 seems to be three or four times too small a percentage for the whole of France. But our criticism does not stop there. Starting from the incorrect proportion of 0·9 per cent. of marriages between first-cousins, M. Boudin wishes to value the number of marriages between cousins, *children of first-cousins*, so as to be able to comprise in the morbid cases, due according to him to consanguinity, those which are observed in children who are the issue of such marriages. Unfortunately, here all the elements of statistics are completely at fault; in such a case, it is not worth while giving up a value which is neces-

p. 413, note). This letter is very instructive as to the trustworthiness of M. Boudin's deductions. See *Gaz. Méd de Paris*, p. 806, 1863.

sarily arbitrary. M. Boudin does not understand it thus: he wishes to comprise in his statistics the cousins who are themselves children of first-cousins (and even, as we shall see, as far as cousins of the seventh degree), and he believes that by adding 1·1 for these last he has sufficiently valued the proportion in relationship marriages:[1] we thus obtain 2 per cent. (0·9 + 1·1) as the number around which to group a large number of deductions. Now, as for this second fraction, M. Boudin is not more fortunate about it than he was with the first; for while he fixes at 1·1 the proportion of marriages between children of first-cousins (and others), I can myself fix it at 5, 10, or 15 per cent. We are here speaking of a matter of pure hypothesis. Every one can choose his own; and since we ought to find three or four times more children who are the offspring of first-cousins than first-cousins themselves, my first number, however exaggerated it may seem, will be much nearer the truth than that of M. Boudin. In whatever way we regard it, it is impossible to agree with M. Boudin that marriages between first-cousins are in the proportion of 2 per cent."[2]

Dr. Down made careful inquiry into the family history of 200 persons who were sane and healthy, lived in different districts, and who belonged to different families. He found only one of these was the offspring of first-cousins, and concludes therefore from this that consanguineous marriages form 0·5 per cent. of all marriages.[3] Dr. Mitchell estimates them at about

[1] Boudin, *Ann. d'Hygiène*, vol. xviii. p. 8, 1862.
[2] Dally, *Anthrop. Review*, pp. 73-75, May, 1864.
[3] Down, *London Hospital Reports*, vol. iii. 1866, p. 226.

1·5 per cent.: a rough guess on the basis of his inquiries.[1] From an inquiry into the parentage of deaf-mutes, by the Census Commissioners of Ireland, there were

In the Year	Out of Marriages.	Consanguineous Marriages.	Degree.				Per Cent.
1871	3,005	201	Between cousins up to the sixth				6·7
1861	3,523	242	,,	,,	,,	fourth	6·9
1851	3,415	170	,,	,,	,,	third	4·9[2]

But this, of course, is only the parentage of deaf-mutes, and would only apply if these correctly represent the community.

We have therefore absolutely no basis from which to start a statistical inquiry as to the effect of consanguineous marriage on the offspring; and can only proceed on very imperfect estimates indeed.

Among the many evils that these marriages are supposed to bring upon a population, crétinism occupies a conspicuous position. This disease is one which forces itself upon the notice of everybody, and like other widely spread and little understood diseases, has been ascribed to every cause that the ingenuity of man could devise. The very list of supposed causes is so great that it occupies three pages of Dr. St. Lager's luminous work on the subject; and as every one of these alleged causes has found several supporters, it is not astonishing that so determined a blackener of these

[1] Mitchell, *Mem. read before the Anthrop. Soc. of London*, vol. ii. 1866, pp. 415, 423.

[2] *Report on the Status of Disease*, 1871, table x. and p. 21; *Ibid.* 1861, p. 20; *Ibid.* 1851, p. 17.

marriages as Devay should have accused them of causing crétinism.[1]

Now, it is very nearly certain that goître is an early stage of crétinism, or rather that the same conditions which show their presence by goître, will in time lead to that disease. For we find that the greater number of crétins also have goîtres; and that even those who have no apparent goître nevertheless have got a tumefaction of the thyroïd gland, which either develops internally or merely impedes respiration and thickens the neck generally. Wherever there are crétins there are goîtreux; but the reverse is not always the case, and hence goître is probably the first stage. Thus in descending a valley, first a few cases of goître will be noticed, then more, and lower down frequent cases of goître mixed with crétinism.[2] It is said, says Sir Thomas Watson, that a couple of generations of goîtreux are sufficient to produce crétinism in the third generation. Suspicion for a long time pointed to the drinking water as a source of goître; and further and wide investigations as to the nature of the soil over or through which these waters ran, chiefly undertaken by M'Clelland, induced Sir Thomas Watson to ascribe their disastrous effects to lime rocks.[3] Dr. St. Lager goes a step further. As the result of his researches, he declares that the most constant mineral in those regions where goître and crétinism are common, is iron pyrites, whose presence is betrayed by crystals of sulphate of calcium in calcareous rocks, by sulphate of magnesium in dolomite,

[1] Devay, *Du Danger*, etc., pp. 7, 8, 110; and *Un Mot sur le Danger*, etc., p. 25.
[2] St. Lager, *Etudes sur les Causes du Crétinism*, etc., pp. 22, 23, 25.
[3] Sir Thomas Watson, *Lectures on the Principles and Practice of Physic*, vol. i. p. 831.

and by sulphate of iron and aluminium in other rocks. Second in frequency, he found copper pyrites, and the double sulphate of copper and iron.[1] But whatever may be the particular poison, it is certain that drinking water is its vehicle. This is manifest from the fact that after different drinking water had been laid on at Bozel, in the valley of Doron, Tarentaise, no more crétins were born in those families who exclusively used the new water, notwithstanding that before, in a community of only 1,472 persons, there were 900 goîtreux and 109 crétins; and only in those parts where the people were still obliged to drink the old waters did new cases of goître and crétinism occur.[2]

To those who urge that the people should cross as a means to wipe out crétinism, Dr. St. Lager observes: "I have seen, and whoever has visited crétinous countries has seen, as I have, strong and healthy girls who have been forced by poverty to marriages of this sort, give birth to hideous crétins." Lombroso asserts that this disease is deplorably frequent at Artogne, the only village of the valley of Camonica, where it is customary to seek wives from abroad, and "everyone assured him, with one voice, that those families were most affected with crétinism who were most accustomed to intermarry with strangers." M. Billiet observes that at St.-Alban and St.-Georges-des-Hurtières, which are the principal foci of endemic crétinism in Maurienne, it has long been customary to seek healthy wives from the opposite side of the mountains; but that these soon get goîtres, and their children are often crétins. And Fabre de Meironnes

---

[1] St. Lager, *Etudes sur les Causes du Crétinism*, etc., p. 444.
[2] *Ibid.*, pp. 197, 198.

says, that in the commune of Condamine-Châtelard, Basses-Alpes, although it is customary to intermarry with the neighbouring population of Meyronnes, l'Arche, and St. Paul, where there are no crétins and but few goîtreux, the people are yet very subject to madness, idiocy, and crétinism. Dr. Trombetto indeed heard that the Valaisans accuse their intermarriages with the inhabitants of the valley of Aosta, and these their intermarriages with the Valaisans, of engendering crétinism amongst them. Lombroso further adds, that the people of Quistello, Mantua, where crétinism is unknown, all belong to the same family, the Valvassini:[1] indeed this kind of evidence is given in abundance in the preceding chapter. As far as statistics go concerning consanguineous marriages in Valais, Maurienne, Tarentaise, and Aosta, they point to the paradoxical conclusion that mixed marriages produce more crétins than do others; and for this reason: in healthy districts the young men are not obliged to go out of their village to seek health and beauty in a wife; moreover, the healthiest population live on the higher Alps, and are rather isolated, but the population lower down, where the water has had more time to linger among the rocks, are less healthy, and here it is easier to contract mixed marriages, because, as a rule, the ground is not so difficult, and intercommunication is consequently easier.[2]

It is not a little remarkable that the greatest proportion of idiots and deaf-mutes is found precisely

---

[1] St. Lager, *Etudes sur les Causes du Crétinism*, etc., pp. 110, 111, 115.
[2] *Ibid.*, p. 116.

where are the greatest number of goîtreux. The four following States of North America—

	Idiots.	Deaf-mutes.
State of New York	1,739	1,307
Pennsylvania	1,448	1,004
Ohio	1,399	947
Virginia	1,285	711

are precisely those where goître is commonest. Florida is credited with only 37 idiots, and this State is entirely free from goître. Hence Dr. St. Lager supposes that most of the above cases are not idiots, but are crétins; while the 37 cases of Florida are really idiots.[1] This connection of deaf-mutism with goître is well worthy of further investigation, and may explain M. de Watteville's observation that there are about half as many deaf-mutes again in the mountainous parts of France compared to those in the plain;[2] and M. Chazarain also noticed that where deaf-mutes were commonest, there also were most crétins;[3] an observation to the same effect was also made by M. Menière.[4]

Crétinism has often been mistaken for idiocy, but the two are essentially different, as may be seen from the following comparison:—

CRÉTINISM.	IDIOCY.
I. It is endemic in particular parts.	I. Spread indiscriminately all over the world.
II. There is an arrest of development which affects the whole system.	II. Many idiots are perfectly well formed.

[1] St. Lager, *Etudes sur les Causes du Crétinism*, etc., p. 26; also p. 52, *et seq.*
[2] Chipault, *Etudes sur les Mar.*, etc., p. 84.
[3] Chazarain, *Du Mariage*, etc., *comme Cause de Dégénérescence*, etc., 1859, p. 31.
[4] *Gaz. Méd. de Paris*, vol. xi. 1856, p. 304.

CRÉTINISM.	IDIOCY.
III. Over two-thirds of the crétins also have goitres.	III. Neither idiots themselves nor their immediate ancestors have goitres.
IV. The teeth appear late, are irregular, and decay prematurely. A premature ossification of the cranial sutures (except in cases of hydrocephalus). Occasional ossification of the cartilages. A malformed, narrow, *foramen magnum*. The cerebral hemispheres are not symmetrical; and some parts of the cerebral substance is thickened.	IV. Various injuries.
V. The sole cause is the nature of the soil through which the drinking water percolates.	V. Multiple causes:— *Congenital Idiocy* is caused by the vices of the parents: drunkenness, senility, accidents, and inheritance. *Acquired Idiocy* is due to different diseases, such as convulsions, epilepsy, typhoid fever, drunkenness, etc.[1]

The causes of idiocy are not sharply defined; they are multiple and obscure, and it has been found convenient, therefore, to shift the responsibility of its production upon the broad shoulders of consanguineous marriage. The sources of every congenital disease must be looked for in the pathological history of the family, in the state of both parents before the conception, in the history and health of the mother during gestation, and in the accidents of birth. Now, Dr. Carpenter believes that a continued state of anxiety and nervous shocks during gestation may cause idiocy in the offspring. Thus, at the siege of Landau, in the year 1793, in addition to a violent cannonading, which kept the women in a constant state of alarm, the

---

[1] St. Lager, *Etudes sur les Causes du Crétinism*, etc., p. 10.

arsenal blew up with a terrific explosion, which few could hear with unshaken nerves. Out of 92 children born in that district within a few months afterwards, 16 died at the instant of birth; 33 languished for from eight to ten months, and then died; 8 became idiots, and died before the age of five years; and 2 came into the world with numerous fractures of the bones of the limbs, probably caused by irregular uterine contractions.[1] It may be caused during parturition, by a bungling midwife, or the use of surgical instruments.[2] It seems to be connected with immaturity or senility of the mother: for Dr. Mitchell found that out of 443 idiots he examined, 138 were first-born, or 31·1 per cent.; 89 were the last birth, or 20·1 per cent.; making in all 227 idiots, out of 443, who were born at the extremes of married life, or more than half.[3] Now, about every sixth idiot is illegitimate in Scotland, and these as a rule are either first-born or last-born,[4] hence Dr. Mitchell collected 85 cases of idiocy where the subjects were all born in wedlock, and all congenital cases, although he excluded

---

[1] Carpenter, *Human Physiology*, p. 864; see also Reynolds' *System of Medicine*, vol. ii. p. 38; and the *Census Report of* 1871 *on the Status of Disease in Ireland*.

[2] Down, *London Hospital Reports*, vol. iii. pp. 233, 234.

[3] This is confirmed by the *Census Report of* 1871 *on the Status of Disease in Ireland*, for in 1,216 cases out of 3,087 of deaf-mutism, the deaf-mute was the first birth, the last birth, or both.

[4] Dr. Howe says that many idiots have become so by attempts at abortion, and these are usually made when the child is illegitimate (*Journal of Psych. Med.*, July, 1858, pp. 394, 395). M. Perier also considers that there are a greater proportion of malformed children among foundlings than among other classes; and Morel seems to be of the same opinion. In the years 1865-1866, 22 per cent. of the bastards were rejected from the conscripts for want of height; 33 per cent. of the foundlings; and only 16 per cent. of the legitimate children (See *Mém. de la Soc. d'Anthrop. de Paris*, vol. iii. 1870, pp. 217-218.)

all in whose family there was more than one case of idiocy, and all whose mother was not already beyond the age of childbearing when the inquiry was instituted, with the result of confirming his previous theory, that the children born at the extremes of married life are more likely to be idiots than are others.[1]

The effect of the habitual abuse of alcohol by parents is unfortunately only too evidently a cause of idiocy in the offspring. Dr. Elam states that on the removal of the spirit duty in Norway, insanity increased 50 per cent., and congenital idiocy by 150 per cent.[2] Dr. Lannurien, of the establishment for mental diseases at Morlaix, in Bretagne, says, "I do not hesitate to attribute the greater number of cases of idiocy in this establishment to that cause."[3] Dr. Ruez noticed that idiocy was very common among the miners of Westphalia, who, living apart from their wives, only came home, and generally got drunk, on their holidays; and M. Devay says that Morel and Demeaux also observed the same result from the same cause.[4] Dr. Delasiauve instances the village of Carême, whose riches were its vineyards, the inhabitants of which place were forced to be a little more sober in consequence of ten years' vine disease. This, he says, had a sensible effect in diminishing the cases of idiocy.[5] From an inspection of the Report of the Commissioners appointed by the Legislature of Massachusetts in 1846, to inquire into the condition

---

[1] *Loc. cit.*, Duncan, *Fecundity*, etc., pp. 392-393, note.
[2] Elam, *A Physician's Problems*, p. 84.
[3] St. Lager, *Etudes sur les Causes du Crétinism*, etc., p. 114.
[4] Devay, *Du Danger*, etc., pp. 10-12.
[5] Devay, *Ibid.*, p. 111. M. Devay says *crétinism*, but he is probably mistaken.

of the idiots of that State, Dr. Howe finds that out of 359 cases in which the parentage could be ascertained, 99 were the children of notorious and habitual drunkards. By pretty careful inquiry as to the number of idiots of the lowest class derived from parents known to be temperate in their habits, it was found that not one quarter were so derived.[1]

Nor does the evil influence of drunkenness end here. Demeaux assured himself that of 36 epileptic patients he had had under his observation during twelve years, and whose history he was able to trace, 5 were conceived in drunkenness. He observed 2 children of the same family suffering under congenital paraplegia, whose conception also took place while the father was drunk. A youth tainted with insanity, and an idiot five years of age, were engendered under the same circumstances.[2] Dr. Bennett says that phthisis may be caused by an abuse of alcohol, because it prevents the proper assimilation of food, though Dr. Anstie asserts that unless there is a previous hereditary tendency this is not the case; and Dr. Maudsley assures us that insanity, epilepsy, hysteria, syphilis, tuberculosis, alcoholism, and even neuralgia in the parent predisposes the offspring in many cases to insanity. Alcoholism is also induced by these diseases; a tendency to alcoholism is also inherited; and thus they act and react on each other. " Perhaps the most frequent causes of an arrest of mental development are those which operate after birth up to the third or fourth year: they are epilepsy, the acute exanthemata,

---

[1] Howe, *On the Causes of Idiocy*, in the *Journal of Psych.*, etc., July, 1858, p. 388.

[2] Devay, *Du Danger*, etc., p. 10.

perhaps syphilis, and certainly conditions of bad nutrition, such as produced by overcrowding, dirt, and want," to which may be added hydrocephalus, convulsions, and their causes.[1] As to the question whether epilepsy is hereditary, Moreau concludes that it is; Tissot thinks cases of inheritance exceptional, and in this he is supported by Gintrac; Beau found in 232 cases which he examined 22 cases which were inherited; and Delasiauve found in 133 cases, 13 inherited, since 3 had mothers who were epileptic, 1 had a brother and 1 an aunt thus affected, while in 8 other cases 2 had uncles who were idiots, 1 a brother an idiot, 1 a mother subject to convulsions, 1 a brother subject to convulsions, 2 had mothers who were hysterical, and 1 an aunt who was insane.[2] Out of 95 cases of epilepsy enumerated in the Irish Census Report for 1871, 19 were said to be inherited.[3] Dr. Reynolds says the "large majority of cases owe their malady to other causes than inherited tendency, a certain number of those whose parents exhibit a like affection to their own may have become morbid independently of hereditary taint. It is well known that many of the children of epileptic parentage are free from the disease, and it is quite clear that many epileptics, descended from epileptic stock, have been exposed to causes of the malady which would, of themselves, have been held sufficient to have produced the malady independently of any constitutional taint." He considers that perhaps one-third of the cases may be

[1] Reynolds' *System of Medicine*, vol. ii. pp. 12, 15, 37, 38, 164; vol. iii. pp. 549-550.

[2] Chipault, *Etudes sur les Mar.*, etc., p. 15.

[3] *Report of the Census Commissioners on the Status of Disease in Ireland*, 1871, p. 77.

hereditary; of other causes there were in 63 cases, 29 due to fright, grief, or overwork; 16 to irritation, such as indigestion, dentition, dysentery, venereal excesses, etc.; 9 to general organic changes, such as fatigue, pregnancy, rheumatic fever, and acute diseases; and 9 to accidental hurts, such as blows, cuts, etc. In about half of these cases convulsions had occurred in infancy; and it is extremely probable that convulsions, from the organic lesions of the brain which they often indicate, are frequently followed by idiocy and epilepsy.[1] Dr. Howe finds that the children of parents addicted to intemperance are often scrofulous, and their children again are apt to be weak in body and mind.[2] Indeed, since the abuse of alcohol is a fruitful source of degeneracy in the parent, we should expect that it also frequently led to degeneracy in the offspring.

Since then, any cause such as syphilis, alcoholism, or other diseases or habits of the parents tending to weaken them are apt to produce an arrest of mental development in the offspring, since most of these causes are hidden to all but the parents, and sometimes doubtlessly are ignored even by them; since, moreover, an idiotic child may thus be born from apparently healthy parents, and even in many cases from really healthy parents, quite irrespective of any consanguinity between them, it behoves us to be especially cautious how we accept the evidence tendered to us by those who advocate the theory that marriages between near kin lead to degeneracy in the offspring.

[1] Reynolds' *System of Medicine*, vol. ii. pp. 294, 295, 297, 298, 262-263.
[2] Howe, *Journal of Psych.*, etc., July, 1858, p. 388.

Now Dr. Voisin states, as the result of a careful examination of 1,077 of his patients at Bicêtre and Salpétrière, that in no one instance could healthy consanguinity be regarded as a cause of idiocy, epilepsy, or insanity;[1] yet Dr. Howe evidently considers his case proved when he says that he found, among 359 idiots, 17 who were known to be the produce of such marriages, and thinks, that from collateral evidence, 3 more cases should be added, making in all 20 out of the 359 whose parents were related. I do not quite understand what he means here, for he goes on to say, " The statistics of the 17 families, the heads of which, being blood-relatives, intermarried, tell a fearful tale. Most of the parents were intemperate or scrofulous; some were both the one and the other; of course, there were other causes to increase chances of infirm offspring, besides that of intermarriage. There were born unto them 95 children, of whom 44 were idiotic, 12 others were scrofulous and puny, 1 was deaf, and 1 was a dwarf! In some cases all the children were either idiotic, or very scrofulous and puny. In one family of 8 children, 5 were idiotic."[2] From which I presume that each of the 17 families had one representative idiot in some institution from which the returns were culled. Hence, dismissing the first comparison, we find that among 95 children, the produce of 17 consanguineous marriages, 44 were idiots, or nearly half. Dr. Mitchell examined every case of idiocy in nine counties of Scotland, namely, Aberdeen, Bute, Clackmannan, Fife, Kincardine, Kinross, Perth,

---

[1] *Loc. cit., Medical Times and Gazette*, Oct. 10th, 1868, p. 436.
[2] Howe, *Journal of Psych.*, etc., July, 1858, pp. 393, 394.

Ross and Cromarty, and Wigtown, with the following result:—

<table>
<tr><td>Whole number of idiots and imbeciles examined</td><td>.</td><td>711</td></tr>
<tr><td>Of these were illegitimate . . .</td><td>108</td><td></td></tr>
<tr><td>,, parents unknown .</td><td>. 84</td><td></td></tr>
<tr><td></td><td>Total</td><td>192</td></tr>
</table>

Total number whose parentage was known . . 519
Of these the parents were related in cases numbering. 98
,, ,, not related 421

Of the 98 idiots whose parents were related:—

The parents were first-cousins in 42 cases.
,, ,, second ,, 35 ,,
,, ,, third ,, 21 ,,

He also gives an analysis of 59 cases where he found that more than one child in the same family was an idiot, but with this we have nothing to do here, as such cases do not represent the whole. Taking then the idiots born in wedlock whose parentage was known, on a total of 519 we have 98 who were born in consanguineous marriage up to the third degree, or 18·8 per cent.[1] Dr. Down, on the other hand, found that on a total of 852 idiots, from which all cases in which an element was doubtful have been excluded, 60 were the produce of consanguineous marriages up to the third degree, thus:—

The parents were first-cousins in 46 cases.
,, ,, second ,, 6 ,,
,, ,, third ,, 8 ,,

Of the marriages between first-cousins, two were a little more nearly related, the parents of these cases being

[1] Mitchell, *Mem. read before the Anthrop. Soc. of London*, vol. ii. 1866, pp. 414-417.

themselves the issue of consanguineous marriages.[1] Dr. Down therefore derives 7 idiots in every hundred from consanguineous marriages. Dr. Bemiss found among 192 children, born from 34 consanguineous marriages, in all 4 idiots, or only 2 per cent.[2] And on a total of 833 consanguineous marriages, he found 7 per cent. of the children were idiots, while on a total of 125 marriages between persons in no way related by blood, he found 0·7 per cent. of the children were idiots.[3] Yet, notwithstanding these facts, and the difficulty which he acknowledges there is in obtaining an insight into the parentage of patients in asylums, he asserts that he feels he is authorized to assume that 15 per cent. of the idiots in the United States asylums are of consanguineous parentage![4] The commissioners appointed to report on idiocy to the General Assembly of Connecticut, in the year 1856, found that of 160 cases in which the question was answered whether there was any relationship between the parents of the idiot in question, 20 were found to be so related, or 12·5 per cent.[5]

Now let us compare the results of these observations:—

[1] Down, *London Hospital Reports*, vol. iii. 1866, p. 225. He has evidently made a mistake in the total number of male idiots from which he takes the number of those born from consanguineous unions—having taken the total of those where the doubtful cases have not been eliminated, while he takes the right total of the females. I have ventured to correct this from the context.

[2] Dr. Bemiss says *191 children* on one page, and 192 children on the next (*Journal of Psych. Med. and Ment. Path.*, p. 370, April, 1857).

[3] Bemiss, *Trans. of the Amer. Med. Assoc.*, vol. xi. for 1858, pp. 420-423. See also the Appendix to this work.

[4] Bemiss, *Ibid.*, p. 330.

[5] Mitchell, *Mem. read before the Anthrop. Soc. of London*, vol. ii. 1866, pp. 420, 421.

Observers.	Total Number of Idiots examined.	Total Number derived from Consanguineous Marriages.	Percentage.
Howe	359	17 (or 20)	4·7 (or 5·5)
Down	852	60	7·0
Commission of Connecticut	160	20	12·5
Bemiss	"from researches."		15·0
Mitchell	519	98	18·8

We see fluctuations here between 5·5 and nearly 19 per cent.; a difference which points conclusively to the fact that these statistics are entirely worthless. Indeed, the disturbing causes are so many, that a few hundred cases like these cannot rise above them. When governments see the force of making these returns compulsory in the census; when this is carried out with a due regard to other questions necessary to elucidate the matter, and when we thus have returns by the million so that a false report or two on either side may be but as a drop in the ocean; then, and not till then, can we venture to deduce with safety from observations of this kind.

Of 35 cases of insanity and idiocy reported on by Dr. Mitchell, in the Island of Lewis, there were born

```
Of parents known not to be related .              16
Of parents known to be related :—
    First-cousins .           .  .   .   . 2
    Second   ,,   .           .  .   .   . 3
    More distant relatives    .  .   .   . 6
    The issue of cousins .    .  .   .   . 1
                                    Total —  12
Of parents concerning whom nothing was known
    (this includes illegitimates)    .  .   .  7

                                    Total    35
```

Also, on a total of 260 cases reported on from the counties of Ross, Cromarty, and Wigtown, Dr. Mitchell obtained information concerning the parentage of 177, of which 41 were born from marriages between relations. Of these 41, 37 were idiots, and 4 laboured under acquired insanity. Of the remaining 136 not descended from consanguineous marriages, 101 were idiots, and 35 laboured under acquired insanity.[1] In the Island of Lewis, therefore, nearly 42·7 per cent. of the cases of insanity were derived from consanguineous marriages; and in Ross, Cromarty, and Wigtown, there were nearly 23·2 per cent. The two do not agree: nor should we expect them to do so, since the numbers are so small. But for the same reason they are worthless as evidence, and especially so from the way this evidence is deduced.

Chorea, another alleged result of consanguineous marriages, is brought on by anything that makes a person what is called "nervous"—by inheritance, therefore, of a tendency to epilepsy, paralysis, hysteria, insanity, etc.; or it may be caused by fright, or irritation such as dentition, worms, etc.; and it is nearly always combined with cardiac disease, which is, in its turn, induced by rheumatism. Of 104 cases of chorea examined, 15 only were free from cardiac murmur, and had not suffered previously from articular rheumatism. It follows, therefore, that a rheumatic diathesis, when inherited, may be numbered among the causes of chorea, just as it may, also, of epilepsy.[2] The statistics adduced[3] to prove the

---

[1] Mitchell, *Mem. read before the Anthrop. Soc. of London*, vol. ii. 1866, pp. 418, 433.

[2] Reynolds' *System of Medicine*, vol. ii. pp. 189, 198-201, 206, 207, 221.

[3] See Appendix of this work.

origin of this disease in consanguineous marriage, are disproved in the refutation of other statistics of the same class.

M. Chazarain, however, has attempted to show that these marriages tend to the production of deaf-mutism in the offspring;[1] and in France, MM. Boudin, Devay, and Chipault have caught up and enlarged upon the theory; it has been specially investigated in the United States of America; and, after being the subject of much desultory inquiry in Great Britain, it is gradually getting to be a subject of official investigation by means of the census. It behoves us, however, before we compare the results of all these inquiries, to examine as far as possible what other causes may produce congenital deaf-mutism, in order that we may eliminate disturbing factors from our problem, and start as nearly as possible on clear ground, just as we have already done in our inquiries on insanity, and on goître. Now, these various causes are tabulated in the Irish Reports,[2] from which it appears that, whatever the immediate, the remote cause of congenital deaf-mutism must lie in the transmission of nervous disease, and must therefore be looked for primarily in the general causes of nervous disease, and secondarily in accidental causes. This nervous disease may affect both the tongue or the ear, or the one organ alone, and may be inherited in such a variety of ways, that we may well despair of ever ascertaining, without the help of a general census, how consan-

[1] Chazarain, *Du Mariage entre Cons. considéré comme Cause de Dégén. Org., et plus particulièrement de Surdi-mutité Congéniale.*
[2] *Census of Ireland, Status of Disease.*

guinity of the parents affects it. For it may be caused by hydrocephalus, chorea, convulsions, paralysis, or epilepsy directly, and by the causes of these and of other nervous affections indirectly: we have seen that deaf-mutism is influenced probably as idiocy, by the position in the family of the child, whether a first or last birth, or the only child; it is also caused by an inherited tendency to inflammation of the lining membrane of the ear, with a gouty, rheumatic, or scrofulous constitution; by amaurosis of the nerve from inherited syphilis; or it may be caused in a variety of ways so soon after birth that it is impossible to say whether the disease was congenital or not. It may very possibly be sometimes caused by drinking goîtrous waters;[1] and is certainly hereditary like other chronic diseases.[2]

Seeing, then, that trustworthy statistics are wanting, we should expect the question to remain in abeyance until there was some safe ground to go upon. Other men, however, have thought differently, and have asserted as a truth, deductions from statistics which could not possibly yield the truth; and since scientific men have been found who accept these assertions, it will be my duty to point out more clearly where they are fallacious. In the first place, their method of procedure is obviously untrustworthy, for they either inquire into the parentage of as many deaf-mutes as they can find, and then compare the result with what they suppose to be the relative numbers of consan-

[1] See p. 216 of this work.
[2] Reynolds' *System of Medicine*, vol. i. pp. 740, 757; vol. ii. pp. 112, 188, 262, 263, 375, 569, 573; Campbell, *Deafness and its Causes*, etc.; Pennefather, *Deafness*, etc.; and the *Census Reports on the Status of Disease in Ireland*.

guineous and non-consanguineous marriages, or they
collect as many cases of consanguineous marriage as
they are able, and then analyze the result. Now, in
the first place, suppose that in 400 marriages between
near kin, 200 produced 2 deaf-mutes a-piece; and that
400 marriages between persons not related produced
each 1 deaf-mute; we should say that all the non-consanguineous marriages produced unhealthy offspring,
while only half the consanguineous marriages turned
out badly. If, however, we had all these deaf-mutes
together in an asylum, and inquired into their
parentage, consanguineous marriages generally, would
appear to produce double as many deaf-mutes
as the non-consanguineous. In other words, this
method of research presupposes three things to be
true, all of which we have more evidence to believe
untrue, to wit: that an equal proportion of deaf-mutes
are born to each consanguineous as to each non-consanguineous marriage; that an institution (generally
charitable) is a true mirror of the state of deaf-mutism
in the general population;[1] and, thirdly, that we
know the proportion of consanguineous marriages to
the non-consanguineous, which I have already shown
that we do not.[2] The uncertainty of any one of these
premises is sufficient to damn any conclusion based on
the whole. The uncertainty of all three premises
makes the worthlessness of the deductions overwhelm-
ingly decisive. The second method of research—by the
collection of cases of consanguineous marriage—I have

[1] The fallacy of which is strikingly exemplified in the seventeen cases
of idiocy reported by Dr. Howe (see p. 223 of this work), and in the facts
concerning the number of Jews in the Deaf-mute Asylum at Paris (see
p. 201 of this work).
[2] See pp. 206-212 of this work.

shown in the Appendix to be utterly untrustworthy except as a proof of the harmlessness of these marriages; because there is a greater likelihood that more than the natural proportion of consanguineous marriages which have turned out badly will be noticed.

But let us see what has been done. M. Boudin says he examined the books of the Imperial Institution for Deaf-mutes at Paris. At the time of his visit, which occurred in January, 1862, there were 200 deaf-mutes, of which 95 were congenital cases; but the parentage of 67 only was satisfactorily ascertained. The parents of 48 of these 67 were not related, while the remaining 19 were derived from marriages between relatives, or 28·35 per cent. were of consanguineous origin. Hence, he argues, there were fourteen times as many deaf-mutes of consanguineous origin as there should have been were these marriages as harmless as others;[1] because he values these marriages at 2 per cent. of all marriages, a figure which, as we have seen, is founded on mere guesswork. M. Dally went to the same institution, but instead of restricting himself to the deaf-mutes then present in the asylum, he carefully analyzed the records of all the cases which had passed through that institution, with the following result:—
Out of 315 cases, 124 were congenital, of which 18 only were derived from consanguineous marriages; thus—

Derived from marriages between first-cousins	6[2]
From second-cousins and others up to the seventh degree	11
From marriages between aunt and nephew	1

[1] Boudin, *Ann. d'Hygiène*, vol. xviii. pp. 8-10.

[2] Of these one has a relation of the mother's also deaf-mute; another has a first-cousin of the father deaf; in another case the mother had her first child when she had attained the age of thirty-four years; and in another the mother was ill during gestation (Dally, *Anthrop. Review*, p. 78, May, 1864).

Nor, perhaps, were all these 18 cases really derived from consanguineous marriages, for 4 of them were only supposed to be so from the similarity of the parents' names. Dropping, therefore, these 4 from our calculations, we have a total of 14 deaf-mutes, half of whom were born from parents very distantly related, and the other half from parents who can really be considered as related, on a total of 120 cases; or 11·7 per cent. if we count the distant relationship, and 5·8 per cent. if we only count the nearer. Now M. Dally, with quite as much authority as M. Boudin, estimates the proportion of marriages between first-cousins alone at figures varying between 1·4 in a division of Paris, where the population is constantly shifting, to 8 per cent. in some provinces where the population is stationary.[1] From which it follows that the number of deaf-mutes derived from consanguineous marriages, which chance has sent to the institution at Paris, is not out of proportion to the estimated number of consanguineous marriages, as compared to the non-consanguineous.

M. Boudin further cites M. Landes, to the effect that in the Asylum for Deaf-mutes at Bordeaux there were 24 congenital deaf-mutes on a total of 79, who were derived from marriages between near kin.[2] But M. Dally points out that the number 24 refers not to the deaf-mutes, but to their families. On a total of 287 families, who had sent deaf-mutes to the asylum between the years 1839–1859, the parents were related, in 24 cases, up to a distance of the fourth degree, or in the proportion of about 8·4 per cent., which ought

---

[1] Dally, *Anthrop. Review*, pp. 75-78, May, 1864.
[2] Boudin, *Ann. d'Hygiène*. vol. xviii. p. 9.

to be reduced if we only take into account marriages between first-cousins and nearer relatives.[1]

M. Dally adds, concerning the statement of M. Boudin that Piroux found 21 per cent. of the deaf-mutes at the asylum of Nancy were the offspring of consanguineous marriages: "I had the pleasure of speaking to M. Piroux last year on the subject of his remarkable institution at Nancy. At that time M. Piroux had made no inquiries, and could give no information about the question I was studying. Since then M. Piroux has published a statement which referred to 612 cases of deaf-mutism received at his establishment from 1828 to 1863. He arrived at his conclusions by retrospective inquiries, comprising in them the most distant degrees of relationship, and obtained 15 or 17 per cent. as a maximum. Why then does M. Boudin speak of 21 per cent.? But if we take from M. Piroux's documents the relations beyond the third degree, we only find, in thirty-five years, 42 who owe their origin to parents who are related, or a little more than 6 per cent., the number at which we can approximately value the marriages between these relations for all France."[2]

"I begged Dr. Jantet," continues M. Dally, referring to a further statement of M. Boudin's that Dr. Perrin found at Lyons 25 per cent. of the deaf-mutes resulted from these marriages,[3]—"I begged Dr. Jantet, a distinguished physician of that town, to ask him about the statements mentioned in M. Devay's book, and

---

[1] Dally, *Anthrop. Review*, p. 79, May, 1864.
[2] Boudin, *Mém. de la Soc. d'Anthrop. de Paris*, vol. i. 1863, p. 509; Dally, *Anthrop. Review*, pp. 79, 80, May, 1864.
[3] Boudin, *Mém. de la Soc. d'Anthrop. de Paris*, vol. i. 1863, p. 509; Devay, *Du Danger*, etc., pp. 123-124.

quoted in M. Boudin's essay. Now here is M. Perrin's answer: 'I have never made any statement on the subject of deaf-mutism caused by consanguineous marriages. They were merely some verbal data which I gave to M. Devay. I can hardly remember the fact myself. Besides, no register of this establishment will show whether cases of deaf-mutism result from consanguineous marriages or not.'"[1]

There remain two more reports on which M. Boudin leans for his statistics. During a space of fifteen years, according to M. Brochard, physician to the asylum at Nogent-le-Rotrou, there have been 55 cases of congenital deaf-mutism under his care. Of these 55 there were 15 children of first-cousins, and 1 child of second-cousins,[2] or a total of 16 children born from near kin out of 55, which makes 29 per cent. The other is a report from M. Chazarain, that at the asylum at Bordeaux, on a total of 66 congenital deaf-mutes, 15 were derived from consanguineous marriages, and these 15 had between them 12 brothers and sisters who were also deaf-mutes, but who were not in the asylum. Of 50 deaf-mutes, out of 51 not derived from consanguineous marriages, 8 each had 1 brother or sister, and 1 had 3 brothers or sisters also deaf-mutes.[3] Hence, if there be no confusion between this account and that of M. Landes, we have a percentage of 30·4 deaf-mutes of consanguineous parentage. The exact

---

[1] Dally, *Anthrop. Review*, p. 80. I have put these quotations into slightly better English.

[2] Brochard, *Comptes Rendus*, vol. lv. 1862, pp. 43, 44; and Boudin, *Mém. de la Soc. d'Anthrop. de Paris*, vol. i. 1863, p. 509.

[3] Chazarain, *Du Mariage entre Cons.*, etc., p. 34; Devay, *Du Danger*, etc., pp. 122, 123; Boudin, *Mém. de la Soc. d'Anthrop. de Paris*, vol. i. 1863, p. 509).

relationship is not given; but judging from 18 cases, of which a detailed account is given, M. Chazarain seems to give himself a very wide scope.[1]

The director of the asylum for deaf-mutes at Rome sent out a circular to the parents of the inmates, at M. Balley's request. Only 33 were answered, and of these only 13 were congenital. Two of the 13 resulted from consanguineous marriages, or 15·4 per cent.[2]

Loubrieu, cited by M. Mantegazza, found 43 in 500, or 8·6 per cent., of the deaf-mutes resulted from consanguineous marriages. From the reports from Italian asylums it appears that only 12 out of 306 were derived from marriages between first-cousins or second-cousins, and none at all from marriages between aunt and nephew, making in all only 3·9 per cent.[3]

Dr. Bemiss found that out of 833 marriages between blood-relations up to the degree of third-cousins, and including cases of incest, 3·6 per cent. of the children

[1] See Chazarain, *Du Mariage entre Cons.*, p. 36, where he describes the relationship in one case as *assez éloigné*. The 18 cases are given in full in the Appendix to this work. (Cases Nos. 30 to 48.)

[2] Balley, *Gaz. Méd. de Paris*, Dec. 5th, 1863, p. 804. M. Balley says there were three, but he gives the following history of one of the three:—A certain young lady had a child before marriage, which she sent to the foundling hospital. On her subsequent marriage, she only had one child, a boy, and therefore prevailed on her husband to adopt a daughter from the foundlings. "Comme on le *pense*," she chose her own child, who afterwards became the wife of her half-brother, with the result of four still-born, then the deaf-mute in question, sixthly a dwarf, and seventhly an apparently healthy child, now eleven years old. But there is no proof whatever that the girl was half-sister to her husband. Indeed nothing was known of her beyond the fact that she was "d'une rare beauté;" while the father of her husband seems to have been decidedly unfertile, and of the girl's parents nothing was known. A man who condescends to an argument like this is unworthy of attention.

[3] Cited by Mantegazza, *Studj sui Matrimonj Consanguinei*, pp. 36, 37, note.

were deaf-mutes. He was also informed by a principal of a deaf-mute asylum that 21 out of 139 pupils, whose history had been inquired into, were known to be derived from consanguineous marriages, or 15 per cent.; and another principal of an asylum informed him that 28 out of 183 cases, or 15·3 per cent., were derived from these marriages.[1] From a late report of the Kentucky Deaf and Dumb Asylum, says Dr. Allen, 10 to 12 per cent. of the deaf-mutes are the children of cousins. He adds that it has "been reported" that in some parts of Kentucky there has been an unusual number of such marriages in certain families for several generations, thus intensifying the hereditary effect. That moreover a Dr. Mulligan, of Dublin, found that 100 out of 154 children born from consanguineous marriages were deaf-mutes, or 0·65 per marriage! That Dr. Buxton, of Liverpool, found that 170 marriages between cousins produced 269 deaf-mutes, or an average of 1·58 for the 170 marriages.[2] If Dr. Allen's Dr. Buxton is identical with Dr. Mitchell's Dr. Burton, he belongs to the Liverpool Asylum for Deaf-mutes, and he found 10 per cent. of the inmates of that asylum to be derived from consanguineous marriages.[3]

Dr. Mitchell, dissatisfied with former investigations, wrote to the superintendents of the various institutions of Great Britain, and obtained the following results from ten institutions:—

---

[1] Bemiss, *Trans. of the Amer. Med. Assoc.*, vol. xi. 1858, pp. 330, 420, 421.

[2] N. Allen, *Intermarriage of Relations*, pp. 17, 18. It is a pity no references are given to his authorities.

[3] Mitchell, *Mem. read before the Anthrop. Soc. of London*, vol. ii. 1866, p. 421.

	No. of Pupils.	No. of Families represented.	No. of Pupils the Off-spring of Consanguineous Marriages.	No. of Families represented.
I. SCOTCH INSTITUTIONS: Glasgow, Dundee, Aberdeen, and Donaldson's Hospital	201	181	12	9
II. ENGLISH INSTITUTIONS: Bath, Newcastle-on-Tyne, Swansea, Exeter, Doncaster, and Brighton	343	323	16	15
Total . .	544	504	28	24

He deducts from this total 25 per cent. for the cases which were not congenital, and thus obtains 21 deaf-mutes out of 408, as derived from consanguineous marriages, or 5 per cent.[1]

The Irish Census Commissioners of 1871 also inquired into the parentage of deaf-mutes, with the following result, which I have tabulated:—

Degree of Relationship.	No. of Marriages.	No. of Congenital Deaf-mutes.
First-cousins	85	128
Second-cousins . . .	63	89
Third-cousins	32	40
Fourth-cousins . . .	7	11
Fifth and sixth-cousins . .	14	19
No relationship . . .	2,804	3,216
Total . .	3,005	3,503[2]

[1] Mitchell, *Mem. read before the Anthrop. Soc. of London*, vol. ii. pp. 422, 423. According to the *Irish Census Report for* 1871, 22·4 is the percentage of non-congenital cases. Dr. Mitchell says that for the United States and for Germany it is respectively 42 and 52 per cent.

[2] This table is obtained from tables i. x. and xi. in the *Census Report on*

Comparing these results with those obtained in the Census of 1851 and that of 1861, we have—

From the Irish Census Reports.	Census of 1871.	Census of 1861.	Census of 1851.
Total number of congenital deaf-mutes derived from consanguineous marriages.	287	362	242
Total number of congenital deaf-mutes derived from non-consanguineous marriages.	3,216	4,096	3,885
Average number of congenital deaf-mutes per consanguineous marriage	1·427	1·496	1·423
Average number of congenital deaf-mutes per non-consanguineous marriage	1·146	1·248	1·197
Percentage of congenital deaf-mutes derived from consanguineous marriages.	8·192	8·120	5·863

We see here the effect of intensification of a family tendency to deaf-mutism, in the average of 1·4 for consanguineous marriages, as compared to 1·2 for non-consanguineous marriages; but, as I have already pointed out, this by no means proves that a greater proportion of deaf-mutes are born from consanguineous marriages than from non-consanguineous marriages.

M. Devic found at Asprières, where to his knowledge many consanguineous marriages take place, only 2 deaf-mutes on a total population of 1,700; and even these 2 were not born from consanguineous marriages.[1]

the Status of Disease, pp. 20, 22. The number of congenital deaf-mutes from consanguineous marriages is from table xi., less those who were not both deaf and dumb. The number of congenital deaf-mutes from non-consanguineous marriages is the difference of the total of congenital deaf-mutes given in table i., and those found to result from consanguineous marriages.

[1] Devic, *Gaz. Méd. de Paris*, March 7th, 1863, p. 158.

It has already been shown that these statistics are worthless from the way in which they have been collected. If they be now compared, this will be shown beyond a doubt by the figures themselves:—

Name of Observer.	Total No. examined.	No. found to be Congenital.	No. of Consanguineous Origin.	No. derived from Second-Cousins.	No. derived from First-Cousins.	Total Percentage from Consanguineous Marriages.
Chazarain	—	66	15	—	—	30·4
Brochard	—	55	16	1	15	29·0
Piroux	—	612	—	42		16·0
Balley	33	13	2	—	—	15·4
Bemiss	—	183	28	—	—	15·3
,,	—	139	21	—	—	15·1
Dally	315	120	14	—	6	11·7
Allen	—	—	—	—	—	11·0
Buxton	—	—	—	—	—	10·0
Loubrieu	—	500	43	—	—	8·6
Report of the Irish Census for 1871	4,467	3,503	287	89	128	8·2
Do. for 1861	4,930	4,458	362	—	—	8·1
Do. for 1851	4,747	4,127	242	—	—	5·8
Mitchell	544	408	21	—	—	5·1
Report from Italy	306	—	12	11	1	3·9

Thus, we see a difference between the results obtained by various observers of from 4 to 30 per cent., a difference utterly inconsistent with trustworthiness.

Since children born from consanguineous marriages are no more exempt from diseases of the eyes than those born from any other kind of marriage, this also forms an article in the charge against them. But here again no satisfactory statistics in support of such a theory have been adduced sufficient to prove that these marriages are the cause and not, for instance, inherited syphilis, which may cause iritis, or retinitis,

or injuriously affect the optic nerve, as it may other
nerves of special sense, and thus cause complete congenital blindness; nor chronic hydrocephalus, which
may cause atrophy of the optic nerves, or strabismus;
nor other diseases of the brain and nervous system,
such as idiocy and other disease of obscure origin, all
of which we know may arise independently of any
consanguinity of the parents.[1] Yet Dr. Liebreich, the
well-known oculist, has advanced the theory that that
particular disease known as *retinitis pigmentosa* is the
result of these marriages, and attempts to substantiate
his theory by the help of some statistics from the
Asylum for Deaf-mutes at Berlin. His attention
being first called to the matter on finding that one of
his patients was the child of first-cousins, he made
inquiries, and found that out of 35 persons suffering
from *retinitis pigmentosa*, 14 were deaf-mutes, 3 idiots,
and 18 could hear; and out of 26 whose parentage
was traced, 14 were derived from marriages between
relations. The whole 14 were distributed in six
families, three of which had 1 a-piece, one had 2,
another 4, and another 5 affected. To these belonged
5 of the deaf-mutes affected with retinitis, 3 who could
hear, and the other 5 he failed to trace the parentage
of. If we may venture to form general conclusions
from a total of 26 cases, in which alone the parentage
was traced, 53·8 per cent. were the produce of consanguineous marriages, or more than half the cases of
retinitis pigmentosa! In Paris, on a total of 329
deaf-mutes, he found 11 also suffering from retinitis,
of which 4 were the offspring of consanguineous mar-

---

[1] Reynolds' *System of Medicine*, vol. i. pp. 737, 740, 763; vol. ii. pp. 38, 262, 264, 412.

riages, 2 of non-consanguineous marriages, and the parentage of the remaining 5 could not be traced. Of 66 people who were otherwise healthy, but suffered from *retinitis pigmentosa*, 25 were the offspring of consanguineous marriages, 38 of non-consanguineous marriages, and the parentage of the remaining 3 could not be traced. Taking all these together, we have a total of 95 cases where the parentage was traced, 43 of which were of consanguineous origin, or 45 per cent.

Now, at Bicêtre, Dr. Liebreich told M. Chipault he found no case of retinitis among 89 idiots; but he found 1 case among 69 idiots at Salpêtrière.[1] Taking the mean alleged percentage of these 158 idiots, that ought to be derived from consanguineous marriages if the observations collected above[2] could possibly be correct, we find that 16·7 ought to be so derived, of which only 1 was afflicted with retinitis, instead of about 8; nor do we know that that 1 was of consanguineous origin. It is evident, however, that *retinitis pigmentosa* is an uncommon disease, and even more evident that 95 cases are much too few to generalize on. There was not one case of this disease in the commune of Batz, in which, as we have already seen, there was a population of 3,300, among whom were 46 consanguineous marriages,[3] certainly a sufficiently large number to find some cases in if Dr. Liebreich's theory is true.

Let us now see how far the statement, that mar-

[1] Chipault, *Etudes sur les Mar.*, etc., pp. 50-59.
[2] See p. 226 of this work.
[3] Voisin, *Contribution à l'Hist. des Mar. Cons.*, in the *Mém. de la Soc. d'Anthrop. de Paris*, vol. ii. 1865, p. 446. See also pp. 179-180 of this work, and the Appendix, Cases Nos. 254-299.

riages between blood-relations are more frequently sterile than others, is born out by fact. According to Oesterlen, 20 per cent. of all marriages in Great Britain were sterile in the year 1851;[1] Simpson found 11·7 per cent. of the marriages were sterile; Dr. West found the average about the same;[2] and Dr. Duncan puts it at 15 per cent.[3] Taking an average from the last three we get 12·8—certainly a low average, for the sterility of marriages in England—on which to base our comparisons. As for the prolificness of marriages, there were born in Scotland, which is a very fertile country, an average of 4·64 legitimate children on every marriage in 1861; in England, only 3·89; and the average is generally about the same; in France, according to M. Husson, the average is decreasing, in 1866 it was only 3·1 per marriage for all France, and only 2 per marriage for Paris, while fifty years before it was 3·7 for all France.[4]

Now, M. Devay refers to 121 marriages between relations, of which 22 were sterile, or 18 per cent.[5] From 10 marriages in his own family, cited by M. Seguin, between first-cousins and nearer relatives, 61 children were born, or an average of 6·1 per marriage, and more may have been born since.[6] Of 46 marriages between near kin reported by M. Voisin, 2 proved barren, or 4·3 per cent., and 172 children were born, or an average of 3·7 for every marriage barren and fertile.[7]

[1] Oesterlen, *Med. Stat.*, p. 196.
[2] *Med. Times and Gaz.*, July 20th, 1867, p. 76.
[3] Duncan, *Fecundity, Fertility, and Sterility*, p. 193.
[4] Duncan, *Ibid.*, pp. 105, 106, 109, 110, 185, 186.
[5] Devay, *Du Danger*, etc., p. 93.
[6] Seguin, *Comptes Rendus*, vol. lvii. 1863, pp. 253, 254.
[7] Voisin, *Mém. de la Soc. d'Anthrop. de Paris*, vol. ii. 1865, p. 447.

Of 833 cases of marriage between relatives collected by Dr. Bemiss, 53 proved barren, or 6·4 per cent.; the remaining marriages produced 3,942 children, or an average of 4·7 per marriage barren and fertile;[1] and the following table shows that marriages between blood-relations are not the less fertile the closer the relationship:—

10 Cases of incest between parent and child, or brother and sister, gave an average of children per marriage of	3·10
12 Marriages between uncle and niece, or aunt and nephew, gave an average of	4·42
56 Marriages between blood-relations who were themselves born from consanguineous marriages, gave an average of	4·18
27 Marriages between double first-cousins, gave an average of	5·70
580 Marriages between first-cousins gave an average of	4·80
112 Marriages between second-cousins gave an average of	4·58
12 Marriages between third-cousins gave an average of	4·92
24 Marriages between first-cousins irregularly reported gave an average of	5·0 [2]

M. Cadiot collected 54 cases of marriage between relations, of which 14 were barren, or nearly 26 per cent.; but he does not give the total number of children produced.[3] Of the 299 cases of consanguineous marriage given in the appendix to this work, which includes some of those already cited, there were 16 barren marriages on the total, or 5·3 per cent.; 283 fertile marriages, which give an average of 4·08 per marriage, if 1 is counted for every sign of fertility, and 3·95 per marriage if the abortions are not counted.[4]

---

[1] Bemiss, *Trans. of the American Med. Assoc.*, vol. xi. 1858, pp. 420, 421.
[2] Bemiss, *Ibid.*, pp. 420, 421. I have corrected these averages: see the Appendix to this work.
[3] Cadiot, *Comptes Rendus*, vol. lvii. 1863, p. 978.
[4] See Appendix: Index to Cases.

From these figures it appears that consanguineous marriages, so far from being less prolific than others, are, on the contrary, more prolific, and a much smaller percentage are barren. But the fact is that these statistics are not strictly reliable, although they are advanced chiefly to show that these marriages are harmful; for they have been collected mostly by the Parasyngeneiasts for observed evil results, and consequently they have unconsciously chosen fertile cases to report upon. I may add, however, that as far as I can judge from these cases, there appears to be no foundation for the theory that they are more likely to be barren. On the contrary, the probability is that they are rather more fertile, since cousins generally marry younger than do strangers, and Dr. Duncan finds that such marriages are the most prolific,[1] provided always that the contracting parties are not too young.

In spite of this fact, obvious to any one who has at all looked into the subject or has made any pretence to a statistical inquiry, M. Devay, and after him M. Boudin, argue that because noble families generally marry in-and-in, they die out; and then continuing their argument in a circle, they assert that because noble families die out, therefore consanguineous marriages are less prolific than are others.[2] The idea has not even the merit of novelty to atone for its untruth, for, as we have already seen, it probably took its origin from the letter of Pope Gregory, who most likely made his mistake because he did not understand the sense of Moses' words.[3] Esquirol says,

[1] Duncan, *Fecundity, Fertility, and Sterility*, p. 144.
[2] Devay, *Du Danger*, etc., pp. 198-204; Boudin, *Ann. d'Hygiène*, 1862, vol. xviii. pp. 46-52.
[3] See p. 30 note 2, and p. 151 of this work.

talking of mental degeneration as caused by consanguineous marriages: "The same may be said of the great lords of France, who are almost all parents. What a lesson for fathers, who in the marriage of their children, consult rather their ambition, than the health of their descendants!"[1] Moheau asserts that when in the centre of the French *noblesse*, one seems to be among a crowd of sick people.[2] According to the *Revue Britannique*, it is said that when a grandee is announced in Spain, everyone expects to see a lump of deformity.[3] Salvandy remarks that one can see at once if there has been a *mésalliance* in a noble's family by the improvement which results.[4] Knight says, "Amongst ancient families, quick men are abundant; but a deep and clear reasoner is seldom seen. How well and how readily the aristocracy of England speak! how weakly they reason!"[5] But although these statements may, in a few cases, be partially true, it is one thing to notice a disease, and quite another to assign its cause. Though at certain epochs, and in certain countries, a part of the nobles may waste their bodies and substance in riotous living, we never see that their consanguineous marriages lead to their degeneration. M. Devay cites the family of the Chevalier Bayard as an example of a great family which has died out: since but five persons now remain out of an original total of sixty-one. To strengthen his argument, he adds the case of an idiot boy, aged twelve years, who was descended from a noble family, the members of which

---

[1] Esquirol, *Mental Maladies*, trans. by E. K. Hunt, p. 49.
[2] Cited by Chateauneuf, *Ann. d'Hygiène*, vol. xxxv. 1846, p. 40.
[3] Loc. cit. *Ibid.*
[4] St. Lager, *Etudes sur les Causes du Crétinism*, etc., p. 112.
[5] Cited by Walker, *On Intermarriage*, p. 434.

were in the habit of closely intermarrying among themselves. His elder brother, though perfectly healthy and successful in his studies, was of a most violent temper. There had also been epilepsy in the family.[1]

If these marriages are sterile, says M. Mantegazza, this at least must be due to consanguineous marriage, and to no other cause, for most certainly it cannot be hereditary.[2] The probability, however, is that it is hereditary. Mr. Galton points out that the heirs of noble houses often retrieve their fortunes by marriage with an heiress, though she be of plebeian extraction. For a girl to be an heiress, she can have no brothers, hence there is every probability that she will inherit the comparative sterility of her parents. Mr. Galton brings forward some evidence to corroborate this belief;[3] and Mr. Macdonald also recommends the farmer to " select his rams and ewes from among twins, or the progeny of fertile parents, fecundity being well ascertained to be hereditary."[4] It is therefore the cross, if anything, which brings sterility into noble families, and not any intermarriage with near kin. But there does not seem to be any marked sterility or loss of viability. It was shown at a meeting of actuaries, in April, 1861, that the families of the English peerage are an unusually long-lived class.[5] Chateauneuf shows that the Boileau family lasted scarcely two centuries, although they had sixteen male children; Racine's family only lasted three generations; Molière died childless; Corneille

---

[1] Devay, *Du Danger*, etc., pp. 111, 112, 201; *Ibid., Un Mot.*, etc., pp. 33-35.

[2] Mantegazza, *Studj sui Matrim. Cons.*, p. 35.

[3] Galton, *Hereditary Genius*, pp. 130-140. The evidence is not very strong, but his theory is highly probable in itself.

[4] Macdonald, *Cattle, Sheep, and Deer*, p. 470.

[5] Letter to the *Lancet*, signed "Genesis," Nov. 15th, 1862, p. 553.

never married; Crébillon's lasted but one generation. Where, he asks, are now the descendants of Juvénal des Ursins, of de Mêmes, of Lhospital, L'Huillier, Lemaître, Pothier, Harlay, or de Thou?[1] Did these families die out from marrying in-and-in? Malthus found that out of 487 families admitted to the rights of citizenship between the years 1583 and 1654, by the sovereign council of the Canton of Berne, but half existed one century, and only 168 still existed in the year 1783; while of 112 families who belonged to the council of Berne, in 1653, only 58 still existed in 1796. In fine, Chateauneuf concludes that seven or eight consecutive generations, lasting about three centuries, is the average life of a nobleman's family in France; and shows that so far from being less prolific, they are perhaps even more prolific than ordinary families.[2] We cannot compare the viability of a nobleman's family two or three centuries ago with the present viability of even an ordinary family. Many of the sons could not marry, since the law of entail forced all younger sons either into the service of the Church, or into the army. Those who chose the former were of course obliged to live unmarried; while those who entered the army were killed in wars, crusades, in internecine strifes, and duels, which latter not only concerned the principals, but often included a long train of noble seconds. When war ceased to be the every-day occupation of a gentleman, they spent their time and money in debauchery, soon lost all their substance, and were forced to have recourse to selling themselves to the highest bidder among the

---

[1] Chateauneuf, *Ann. d'Hygiène*, 1846, vol. xxxv. pp. 54, 55.
[2] *Ibid.*, pp. 55, 56, 32-34.

daughters of the *Bourgeoisie*, a state of things very much deplored by Count Boulainvilliers in his writings,[1] and, as we have seen, thus introduced sterility into naturally fertile families. M. Bourgeois further points out that since neither vaccination nor the treatment of children was understood, and since the education of boys at that time necessarily subjected them to very rough treatment, their viability was naturally far lower than it is in modern times, while we have no standard in plebeian families of that time to compare them with.[2]

Various ancient authors have noticed the curious fact that certain persons are barren together, who are both perfectly fertile with other people.[3] This was the case with Augustus and Livia, and this it is which has

---

[1] Chateauneuf, *Ann. d'Hygiène*, 1846, vol. xxxv. pp. 32-36, 38, 44-50.

[2] Bourgeois, *Quelle est l'Influence des Mariages Cons.*, etc., pp. 22, 23. "M. Benoiston," (de Chateauneuf,) says Devay, "constate *l'effrayante mortalité des enfants en bas âge chez les nobles;* il voit dans ce fait une cause puissante de leur décadence ; je lui responds : ' Très-bien ; mais ce qui pour le savant académicien est une *cause*, devient un *effet* pour le physiologiste. Où git donc la cause de cette effrayante mortalité des enfants nobles ? On ne saurait alléguer, ici, la misère et les privations ; il y a donc là un fait de l'ordre vital, un phénomène organique dans la déchéance de ces grands familles. Qui doit mieux en rendre raison que la consanguinité dont les effets connus, bien constatés, (!) sont d'introduire un principe léthifère dans les races comme dans les familles ? Et puis d'ailleurs toutes les familles nobles, ou privilégiées, n'ont pas guerroyé, toutes n'ont pas suivi l'état ecclésiastique, et toutes ont eu le même déclin.'" (Devay, *Un Mot*, etc., p. 33.) M. Devay, however, in the heat of his defence against the attacks of M. Dally, takes no notice of the unimportant fact that Chateauneuf does not say the viability of noble families was less than that of commoner's families. Chateauneuf merely says that nobles then lived a shorter life than is usual in our times, that the mortality of children was among all classes greater than it is now, and that there were many causes to extinguish ancient families.

[3] Pliny, Book vii. chap. xi., translated by Bostock and Riley, who also cite Hippocrates, Aristotle (*Hist. Anim.*, Book vii. chap. vi. sec. 2), Lucretius, Book iv. chap. 1242, *et seq.*

induced certain authors to attribute entire barrenness to consanguineous marriages. They see a case where two relatives marry and are childless; the one dies, and the other marries again and has children. Yet although this case is more partial than those noticed by Pliny, the observer immediately attributes it to consanguinity, and not to natural barrenness in one of the parties.[1] We know little of the causes of this barrenness, and hence a great many causes have been assigned to it. One that is often overlooked is obesity. Aristotle notices this cause in his *History of Animals*,[2] and Drs. Hewitt[3] and Power[4] both agree with him; while Mr. Youatt asserts that the English export trade for breeding purposes has been much spoiled by the cattle shows; for now it is essential that the animals should be much fattened; and Mr. Laverack recommends that the bitch selected for breeding should not be too fat, "she should be rather lean to ensure fruitfulness."[5] We shall see that unacclimatized animals are often sterile, and that races of animals and plants are generally sterile with each other in proportion to their difference. The most usual cause is probably disease in the female of which obesity is only one form; but brutal treatment may also lead to barrenness.[6] That it is due to consanguineous marriage, is however obviously untenable, as we have already seen that even in the cases collected by the Parasynge-

[1] See the Appendix, Cases Nos. 60, 194.
[2] Aristotle, *Hist. Anim.*, book iii. chap. xiii. sec. 4; book v. chap. xii. sec. 9.
[3] Hewitt, *Diseases of Women*, p. 689.
[4] Carpenter's *Human Physiology*, p. 863, note.
[5] Youatt, *On the Pig*, edited by Sidney, pp. 47, 48; Laverack, *The Setter*, etc., p. 32; Stephens, *The Book of the Farm*, vol. i. p. 258.
[6] Waitz and Gerland, *Anthropologie*, vol. vi. p. 141; Gerland, *Aussterben der Naturvölker*.

neiasts, the average of barrenness is apparently below the normal.

Another common accusation against these marriages, is that they cause congenital malformation in the offspring. But congenital malformations are in nearly all cases due to a morbid state of the cerebro-spinal nervous system, acting through the muscles to distort the bones. Any irritation before birth, therefore, may cause it; it may occur of course in anencephalous infants, and also in cases of *spina bifida*. Like other diseases of nervous origin it may be inherited directly; but it may also be inherited through other diseases which react on the nervous system, such as a rheumatic or scrofulous affection of the joint, which may either directly affect the nerve and cause disease in it, or cause deformity in the limb, because the patient finds it too painful to move it. In this last way again, almost any lesion may produce deformity; or a mere habit may do the same. It is possible that it may be caused occasionally by irregular uterine contractions, which may in their turn be caused by a sudden alarm to the mother; and in cases of extreme obesity where the pressure interferes with due nutrition, or where the state which produced fat in the parent also interferes with the nutrition of the fœtus; but this is probably only very rarely a cause. It may be caused by an arrest of development, since Wagner has shown that club-foot is natural to the fœtus up to the fourth or fifth month, and even later, and the same is the case with cleft-palate and hare-lip. It may be caused at birth, especially when parturition is assisted by a clumsy midwife, or by the use of instruments, or when the passage is narrow between the bones of the pelvis: probably many infants have received a life-long injury through pressure on their soft and yielding heads; but in these cases it is rarely that the upper extremities are deformed.[1] It is also connected with the period of life of the parent in which the child is born, and with the position in the family of the latter, whether it is a first or last birth or not. For Dr. Duncan found that twin bearing was most frequent at the extremes of the child-bearing age, and he found further, a connection between twinning and the production of idiocy and bodily deformities; indeed, the greater number of idiots are twins; and anencephalous monsters hardly ever, perhaps never, occur unless in plural births, in which also miscarriages and hydramnios are more frequent.[2] Aristotle considers

---

[1] Brodhurst, *The Deformities of the Human Body*; Salt, *On Deformities, etc., of the Lower Extremities*; Adams, *Club-foot, its Causes*, etc.; Carpenter, *Human Physiology*, p. 866.

[2] Duncan, *Fecundity, Fertility, and Sterility*, pp. 76, 78, 80, and 68, note. The connection between twinning and monstrosities is shown in the production of hermaphrodites when twin calves are born. The Hindoos regard hermaphrodites with the utmost abhorrence, and exclude them from caste and inheritance, for they are considered to be persons who in their former life have committed the greatest sins. It is the same with other congenital malformations. (Steele, *Hindoo Castes of the Dekhun*, pp. 224, 61, 439 note). Many African tribes, says Waitz, consider twinning a monstrosity. In Bonny, a village of Benin, twins with their mother are always killed. In Fetu, on the Gold Coast, if twins are perfect and strong, both are allowed to live; if the sexes are different, the male alone is allowed to live. The Hottentots do the same if the twins are female; if male they are generally delighted, and bring them both up. In the east part of Madagascar, one of twins is killed. The Ibu are accustomed to expose one of twins, while the mother has to dwell away from the community till she is considered pure again, during which time she often suffers great hardships. The Salivas of South America kill one of twins, for they fancy them to be the result of unfaithfulness, thus unconsciously admitting the doctrine of superfœtation. So do the Lules and

premature marriage produces puny and dwarfed offspring,[1] and Dr. Duncan corroborates this statement as far as it concerns the weight and size of infants.[2]

It is necessary, therefore, in gathering statistics as to any supposed relation between consanguineous marriages and the production of malformations, to consider the age of the parents, the number of the birth, whether the malformed was a twin, whether the parents suffered from any nervous affection or disease belonging to the *degenerative* class, such as scrofula, tubercles, syphilis, etc., whether they were always sober, and many other like considerations equally important and impossible to collect, and yet withal to leave a margin for accidental causes. Of course, if the statistics were taken from the whole population, and were numbered by millions of cases instead of a few hundreds, and if we also knew the proportion which consanguineous marriages bear to others, the thing might be done; at least as far as the influence of consanguineous marriages on malformations is concerned—but even then it would be impossible to say whether any evil effect was caused by the intensification of a previous family taint; or, as many authors would have us believe, by the mere fact of consanguinity of itself.[3]

Moxos, for the same reason. The ancient Peruvians fasted if twins were born to them, to ward off the evil of which this was considered an omen; while in some provinces of the empire, one was always killed. The Indians of the Orinoko also always kill one of twins. Are we dogs, say they, that our women should bear us a whole litter? In Tahiti, they did the same (Waitz, *Anthropologie*, vol. ii. p. 441; vol. iii. pp. 394, 480, 537; vol. iv. pp. 417, 461; Reich, *Ehe*, etc., p. 308, 323, 350, 419).

[1] Aristotle, *Hist. Anim.*, book v. chap. xii. sec. 1.
[2] Duncan, *Fecundity, Fertility, and Sterility*, p. 64.
[3] Boudin, *Ann. d'Hygiène*, vol. xviii. p. 17, *et seq.*; Devay, *Du Danger*, etc., p. 143; Chazarain, *Du Mariage entre Cons.*, etc., p. 13, *et seq.*; Her-

Out of 146 children, the produce of 37 consanguineous marriages, Dr. Mitchell found 3, or 2 per cent., were deformed, and 6, or 4·1 per cent., were lame;[1] and out of 3,942 children, the produce of 780 consanguineous marriages, 2·4 per cent. were deformed, according to Dr. Bemiss.[2] I have already remarked on the untrustworthiness of statistics collected in the way in which these are, and will only refer the reader to the history of certain isolated communities given above, where consanguineous marriages are common; and to the history of certain individual families where marriage in-and-in was customary, given in the appendix to this work. As an instance, however, of the loose way in which these charges are made, I will quote M. Devay, who says that an abnormal number of digits is "of all malformations, that which I have most frequently noticed."[3] To support this wonderful theory, he cites the instance of a boy, the child of first-cousins, whose toes were undivided; and says he found 17 cases of these malformations in the progeny of 121 consanguineous marriages. He further gives an instance of a little village near St. André and Rives, in the Department of Isère, called Izeaux, very isolated, and situate on barren soil. The inhabitants, owing to the nature of the roads, necessarily intermarry much among themselves, and form in reality but one family. Now comes the dreadful part of his story. Towards the end of the eighteenth century, all of them, men,

bert Spencer, *Biology*, vol. i. p. 290; Darwin, *The Variation of Animals and Plants*, etc., vol. ii. p. 116, etc.

[1] Mitchell, *Mem. read before the Anthrop. Soc. of London*, vol. ii. 1866, p. 403.

[2] Bemiss, *Trans. of the American Med. Assoc.*, vol. xi. 1858, pp. 420-423.

[3] Devay, *Du Danger*, etc., p. 94.

women, and children, had an extra finger on each hand, and an extra toe on each foot! They gradually resumed their normal form about the year 1847, through crosses.[1] It is, perhaps, a pity that the people of Izeaux got rid of their so-called deformity—the sixth digit might in time have revolutionized the world. Who can say what advantages a sixth finger might not confer on a race by dint of careful breeding and exercise! If, however, M. Devay means to assert that in a space of about fifty years, there were no inhabitants in all Izeaux with more than the normal number of fingers and toes, after every person in the place had had them, the fact is one of the most extraordinary on record, for polydactylism is most strongly hereditary. If, again, M. Devay means to assert that this abnormal growth was originated by consanguineous marriage, I have no hesitation in flatly denying it; for all observation hitherto has shown that crosses are far more likely to produce abnormal developments, through imperfect atavism; while in-and-in breeding is perfectly well known to have a tendency to fix the type so that it never changes.

Rickets, and rickety malformations, have constantly been attributed to marriages of near kin; and here again, as in so many instances, it is only because the causes of this disease are not yet accurately determined, that persons have been able to attribute it to these marriages, not only unchallenged, but with applause.

Rickets, together with phthisis, tabes mesenterica, tubercular meningitis, hydrocephalus, spina bifida,

[1] Devay, *Du Danger*, pp. 95-97, 105.

and a few others, with their consequent mental and nervous diseases, are at present considered merely different members of the great scrofula family of disease, which are, as we know, brought on by so many causes that it is simply impossible to say whether consanguineous marriage can do ought but intensify a previous scrofulous diathesis.[1] Besides the theory of Küttner, that the intermarriage of relations is a cause of rickets, and that of Schönlein, that it is caused by marriages between immature persons, it has been supposed that chronic disease, senility, syphilis, or exhaustion of the parent may be the cause. But Sir William Jenner rejects all these theories, and asserts that anæmia in the mother is the sole cause. In opposition to Wiltshire and Herring, he does not consider it to be directly hereditary; and, in conjunction with Lonsdale, he confirms Trousseau's hypothesis, that imperfect assimilation of food is the sole post-natal cause. Others consider that a want of sunlight plays a considerable part in its production; and altogether its causes are extremely obscure.[2]

In the same way convulsions in infancy cannot be set down to the influence of consanguineous marriage unless it can be shown that it has not been influenced by rickets, dentition, wrong-feeding, diarrhœa, etc. It may be caused by—

1. The advent of an acute illness.
2. The presence of a tumour, abscess, or syphilitic node, or other mental hurt.

---

[1] Reynolds' *System of Medicine*, vol. i. pp. 823, 825; vol. ii. p. 719; Thomsen's *Krankheiten, etc., auf Island*, pp. 11, 31; Campbell, *Deafness*, etc.

[2] Reynolds' *System of Medicine*, vol. i. pp. 805-808.

3. Chronic disease; such as rickets, emaciation, diarrhœa, and other exhausting types.

4. Temporarily, by an overloaded stomach, teething, worms, diarrhœa, etc.

5. Brain disease; such as epilepsy, etc.

Among its consequences are paralysis, amaurosis, defects of speech, strabismus, and mental hurts, such as idiocy, etc.[1]

The causes of phthisis are both general and local in their action. Hereditary predisposition has been placed in the first rank of general causes, and that it is often directly inherited is shown by the fact that infants are born with both large and miliary tubercles in their lungs. Nevertheless, the influence of inheritance has been much over-rated, since former observers have overlooked many of the now recognized causes, such as damp, inflammatory attacks, etc. On the whole, probably, the average is only about 12 per cent. for direct inheritance, and for family predisposition 48 per cent. Impure air, pythogenic fever, scarlatina, and measles, acting both directly and indirectly; the cessation of habitual discharges, miscarriages, bad confinements, over-lactation, and other weakening causes; mental oppression, through its effects on the habits of life; damp from the soil; trades or occupations in a dusty or gritty atmosphere, bronchitis, whooping-cough, pneumonia, pleurisy, and injuries to the lungs—all may lead to phthisis; in the parent, perhaps, first without his knowledge, and then be inherited by his child. In short, all causes which tend to a diminution of vital power are causes of phthisis; especially, perhaps, disease of the pancreas,

[1] Reynolds' *System of Medicine*, vol. ii. pp. 262, 263, 265.

by which the emulsion of fat is rendered difficult, while the assimilation of albuminous food, which constitutes a great part of the tuberculous matter, is specially favoured.[1] Unfortunately, very little is known concerning diseases of the pancreas, or how they are caused. It may, perhaps, be affected by disease in the neighbouring parts; by the abuse of mercury or tobacco; and, as the greatest authorities agree, by the abuse of alcohol.[2]

Hydrocephalus, when congenital, is generally, but not invariably, due to an arrest of development of the cerebral mass, assisted by slow inflammation of the arachnoid, especially of that part lining the ventricles. Syphilis is a frequent cause of this, both by the production of syphilitic nodes, and by the great tendency there is in these patients to serous inflammation. The inflammation may also attack a child after birth, and even begin late in life; but, in the last case, the dropsy is generally caused by the mechanical effect of a cancer, tubercle, or cyst, or of some hurt by which the veins are impeded, thus causing a watery exudation in the neighbouring part while it prevents the absorption of the serum naturally there. Hydrocephalus is often aroused when latent by whooping-cough; but in this case the whooping-cough is itself the result of general constitutional disease. Spina bifida, or hydrorachis, is another form of development of this tendency; a frequent result of hydrocephalus and the commonest congenital spinal affection. Here,

[1] Williams, J. B. and C. T., *Pulmonary Consumption*, etc., pp. 109, 110, 111 note, 115, 128-134; Reynolds' *System of Medicine*, vol. iii. pp. 546-549, 554.
[2] Reynolds' *System of Medicine*, vol. iii. pp. 410, 411.

again, is slow inflammation, but the result of a bifid spine is purely mechanical.[1]

The reader will, I hope, excuse me this long dissertation on the causes of these diseases. But it is necessary that we should distinctly see that, so far from their real origin being even tolerably clear, they are involved in great obscurity; and that we should know what other causes are generally assigned to these scourges of humanity, before we attempt to judge that alleged by the Parasyngeneiasts. We see what a variety of causes, totally independent of consanguinity, may produce scrofulous disease and degeneration; and how impossible it is—nay, how wicked, with only a few selected cases of consanguineous marriage for proof—to accuse these of the production of scrofula. M. Chazarain refers to Lugol for support to his statement that in ancient times before there were many roads, and when intercommunication was difficult, the country people as a rule intermarried among themselves; that they degenerated in consequence, and scrofula became more common than it is now.[2] Lugol, however, gives no support to this conclusion. Throughout the six or seven pages he devotes to the consideration of the effects of crosses on scrofula, he gives no authority for his wide and somewhat vague statements; while he only twice mentions even the names of any observers, to wit, Buffon and Alexandre Bodin. It is impossible to accept such an assertion on such vague authority.[3]

---

[1] Reynolds' *System of Medicine*, vol. i. pp. 58, 740, 823; vol. ii. pp. 410, 411, 719.

[2] Chazarain, *Du Mariage entre Cons.*, etc., p. 12.

[3] Lugol, *Recherches, etc., sur les Causes des Maladies Scrofuleuses*, p. 317.

Consanguineous marriages have been further accused of causing ichthyosis and leprosy; abnormally-sized heads and prognathous jaws; and dwarfing; and M. Aubé actually attributes hydatis in the liver to their effect! The same gentleman also attributes albinoïsm to these marriages;[1] but since he confines his proofs to the lower animals, we will defer the consideration of this last accusation to the next chapter, merely premising that we really know nothing as yet as to its production, but that Beigel shows some cause for the belief that a great shock to the nervous system of the mother, whether sudden, or by gradual depression, or by nervous disease, may cause albinoïsm in the offspring; while, as we shall see, the experiments of M. Legrain tend to corroborate this theory.[2]

[1] Devay, *Du Danger*, etc., pp. 52, 90, 104, 187; Balley, *Gaz. Méd. de Paris*, 1863, Dec. 5th, p. 804.
[2] Beigel, *Albinismus*, etc., Dresden, 1864.

CHAPTER VI.

THE RESULTS OF IN-AND-IN BREEDING ON THE LOWER ANIMALS.

IN the last chapter we have seen how many and how complicated are the causes which lead to the various diseases supposed to be produced by the intermarriage of near kin; and how impossible it is to lay our hands on any one case, and say, this is certainly due to the consanguinity alone, and can be due to no other cause. Nay, even if we be intimately acquainted with the whole history of a given case, and find that as far as our knowledge goes there is no apparent cause but the consanguinity of the parents, we are yet by no means sure that there may not be other causes as yet unrecognized, capable of producing that particular effect. All this applies more particularly to man, on whom we cannot experiment as we can on other animals, so as to exclude as far as possible all disturbing factors, and retain consanguinity alone. For this reason no amount of observations on isolated cases of consanguineous marriages, or on isolated communities who have continually intermarried among themselves, will enable us to determine, unless negatively, whether any observed disease in the offspring has been inherited, or whether it is owing to a morbid influence of consanguinity in marriage, whether, in other words, consanguinity pure and simple is a primary cause of disease.

A census would dispose of the practical question as to whether it is expedient that these marriages should be permitted or not. It matters little to the State whether consanguineous marriage is a primary or secondary cause of disease. But if it be found that there is so great a difference between the results of consanguineous and non-consanguineous marriages, that by permitting the former, a manifest increase of idiocy, deaf-mutism, or other disease is caused, by which the population is saddled with a heavy charge for the production and maintenance of individuals not only useless, but injurious to the community, in that case the State is both justified and bound to put what restraint it is able on individuals to prevent these marriages. On the other hand, if so marked a result is not discoverable, it would be both impolitic and dangerous for the Government to discourage them. No census, however, as I have pointed out, could determine whether consanguinity can be a primary cause of disease. For that we must interrogate Nature, as she has already been so successfully interrogated on other physiological questions. We must experiment on the lower animals, since we may not experiment on man.

It has been objected to this method of investigation of the subject of this work, by many authors who at the same time have not hesitated to avail themselves of it when it has suited their arguments, that observations on the lower animals are not applicable to man, that experiments on them would not result in the same way if applied to mankind. They remind us that opium does not produce tetanic spasms in man, while it does do so in some of the lower animals; that if a seton, or any piece of foreign matter, is passed under the skin of a rabbit, tuberculosis is frequently caused, while the same thing, though daily done to him, has not the same effect; that many diseases, such as *diabetes mellitus*, common to man, are not known to occur in the inferior animals; and that many other similar instances might be adduced.

Yet though I quite agree that there may be many cases such as these, as indeed we should expect from the inequality of the anatomical proportions in different animals, there is not a tissue in man which has not its fellow prototype in the lower animals; and many of our most valuable physiological facts have been obtained by experiments on these. Ask Brown-Sequard, Du Bois-Reymond, or any other of the world's leading physiologists what they think of the value of experiments of this kind. It was thus that Bell's discovery of the object of the double root of the spinal nerves was demonstrated to a hesitating world. It is by these means, and by these means alone at present, that physiology is keeping pace with the rest of the sciences. The more the diseases of the lower animals are studied, the more are they found to be identical with those of man, and to be brought about by the same causes in each case where the cause itself is not due to a special habit. We even find that animals are subject to nearly every brain disease, as well as to others which were formerly supposed to be found in man alone;[1] and the absolute identity of some of them, and by analogy of most of them, is established by such diseases as small-pox, hydrophobia, glanders, etc., which are intercommunicable among all the higher animals.[2] There are besides many more

[1] Swift, *Gulliver's Travels*, in the *Voyage to the Houyhnhnms*.

[2] Vaccination, indeed, is one of the most salient examples of this. Jenner found that cows got small-pox from horses which had been tended by the same persons who milked the cows, and its absolute identity is not only shown by the immunity guaranteed by the transference of small-pox from the cow to man; but also by the reverse experiment that it may be communicated by inoculation from men to cows, and that the matter from the vesicles thus produced acts in all respects like ordinary vaccine matter when again used on man (Sir Thomas Watson, *Lectures on the Principles and Practice of Physic*, vol. ii. pp. 940, 956, 957). Dr. Marsden thinks that inoculation from sheep-pox would be as efficient as that from the cow (Reynolds' *System of Medicine*, vol. i. p. 224), and Stephens asserts that inoculation on sheep has the same effect as on the human subject (Stephens, *The Book of the Farm*, vol. i. p. 487). Dogs also are known to suffer from small-pox ("Stonehenge," *On the Greyhound in* 1864, p. 81).

Hydrophobia may be communicated to man by dogs, wolves, foxes, cats, horned cattle, horses, pigs, goats, sheep, badgers, martens, deer, and other animals (J. and A. Gamgee, Reynolds' *System of Medicine*, vol. i. pp. 333, 334, 335, 337, 339).

Glanders is known to be communicable to man by the horse, ass, or mule; and may be produced either by inoculation or by a mere touch. The disease in man may be communicated to other men, or even to the ass (*Ibid.*, pp. 313, 314, 321, 324).

Not only did men catch the Pest (Black Death) of 1348-1360, but also animals which came in contact with pest-stricken persons, or with their clothes, were affected. "Maravigliosa cosa è ad udire," says Boccaccio, "quello che io debbo dire: il che so dagli occhi di molti e da'miei non fosse stato veduto, appena che io ardissi di crederlo, non che di scriverlo, quantunque da fededegno udito l'avessi. Dico che di tanta efficacia fu la qualità della pestilenzia narrata, nello appicarsi da uno ad altro, che non solamente l'uomo all'uomo, ma questo, che è molto più, assai volte visibilmente fece; cioè che la cosa dell'uomo infirmo stato, o morto di tale infermità, tocca da un altro animale fuori della spezie dell'uomo, non solamente della infermità il contaminasse, ma quello infra brevissimo spazio uccidesse. Di che gli occhi miei (siccome poco davanti è detto) presero trall'altre volte un dì così fatta esperienza: che essendo gli stracci d'un povero uomo da tale infermità morto, gittati nella via pubblica, e avvenendosi ad essi due porci, e quegli, secondo il lor costume,

diseases which occur both in man and in the lower animals, and which we know to be induced by the same causes in both cases, such as influenza, lead colic, ague, typhus fever, and others, both epidemic and chronic.[1] There are, of course, some differences

prima molto col grifo, e poi co'denti presigli, e scossiglisi alle guance ; in piccola ora appresso, dopo alcuno avvolgimento come se veleno avesser preso, amenduni sopra gli mal tirati stracci morti caddero in terra." Dogs, cats, fowls, and other animals were infected in crowds. In Gaza, 22,000 men and most of the animals died pest-stricken within six weeks. In England, domestic animals lay dead in every hedgerow ; and it is said that no beast or bird of prey would touch their carcasses (Hecker, *Volks-Krankheiten*, etc., pp. 26, 27, 45, 51; Boccaccio, *Decamerone*, Introduzione, pp. 8, 9.)

[1] Dogs, cats, horses, and possibly birds, have been affected simultaneously with influenza when that disease was rife (Dr. Parkes, in *Reynolds' System of Medicine*, vol. i. p. 37). And during the epidemic catarrh, which proved so fatal to horses in America, in 1872-73, it was noticed that pigs also were frequently affected when fed in the neighbourhood of horses suffering from that disease (See *The Times* for Dec. 21st, 1872).

Animals are certainly subject to ague, and they seem to desert aguish places. Bishop Heber states that the upper provinces of India are deserted by most animals during the ague months (Sir Thomas Watson, *Lectures on the Principles and Practice of Physic*, vol. i. pp. 769, 770, 781).

Rooks and gulls have been known to die in quantities when cholera was rife (*Ibid.*, vol. ii. p. 585).

Dogs, cats, and rats who inhabit houses or manufactories where lead is much used, are known to become affected with lead-colic (*Ibid.*, vol. ii. p. 558).

Brain diseases, oftenest dependent on some morbid growth in the brain, are common to cattle ; or they may suffer from phrenitis, and consequent coma and delirium. Horses, pigs, and dogs are also liable to inflammation of the brain ; and the latter are also subject to a disease known as "turnside" from injury to one side of the brain, and to chorea. Paralysis may occur in dogs, pigs (often induced by the *cystocercus*), horses, cattle, and fowls. Epilepsy is common to horses, pigs, and dogs. Tetanus to cattle, horses, and pigs. Apoplexy to cattle, horses, pigs, and fowls. The ophthalmias to cattle, horses, and dogs; who are also liable to amaurosis, cataract, otitis, and mutism. Fowls are subject to asthma. Cattle and pigs are liable to the foot-and-mouth disease. Coryza and catarrh are common to cattle, horses, sheep, dogs, and fowls. Laryngitis is known to occur in cattle, horses, and dogs. Bronchitis in cattle, horses, pigs, and dogs. Pleuro-pneumonia (sometimes differing from that in man in its contagious property) is common to

between the physiological actions of mankind and the
lower animals, but the general features are the same;
and though it might be easy to pick out one function
in one animal which does not exist in another, this
will be found to be either a special modification, or it

cattle, horses, pigs, and sheep. Tuberculosis is known to occur in cattle,
horses, pigs, sheep, and dogs; in rabbits, guinea-pigs, and fowls. In
animals, as in man, it is very hereditary. Quinsy is known in pigs.
Emphysema is common in horses, arising from the same cause as in
man. Diseases of the heart and large blood-vessels, the same as in man,
are common to cattle, horses, and pigs (though disease of the heart
is always on the left side). Peritonitis is common to horses, pigs, and
dogs. Dropsy, of various kinds, to cattle, horses, dogs, and fowls.
Diarrhœa, dysentery, constipation, and colic to sheep, cattle, horses,
pigs, dogs, and fowls. Hernia, umbilical and congenital, is known in
cattle and pigs. Cases of intus-susception of the bowel have been known
to occur in cattle; and protrusion of the rectum in pigs. Gonorrhœa is
known in cattle. Diseases of the kidney, such as nephritis, fatty degeneration, and hæmaturia, are common to cattle, horses, dogs, and cats.
Cystitis, as in man rarely idiopathic, is common to cattle and horses.
Calculi, both in the gall and urinal bladders, have been found in cattle,
horses, pigs, and dogs. Jaundice is common in horses, dogs, and sheep,
frequently caused by hepatitis; and, as in man, with well-marked sym-
pathetic pain in the shoulder, so that sheep thus affected frequently
seem to be lame. The true distemper in dogs is identified by Walsh with
typhus fever. Dogs are known to have bleeding and other cancers,
commonest in those parts where they are commonest in man; and to
suffer from anæmia, which is curable by the same tonics which cure a
like state in man. Rickets are common to dogs and deer. Congenital
hydrocephalus is known in cattle. Rheumatism occurs in cattle, pigs, dogs
(but only arthritic in old and worn-out animals), and in whales, in which
the bones are often found altered. Erysipelas is common to sheep and
pigs. Horses are said to be subject to nettle rash. Goitre occurs in
animals in goitrous districts. And albinoïsm is common to a great
many animals (Youatt, *On the Pig*, pp. 193-198, 206, 207, 210-213,
217, 222, 224-226, 230, 233, 235; *Ibid.*, *On the Horse*, pp. 189-199;
"Stonehenge," *On the Greyhound*, pp. 59-61, 78, 84, 89, 92-100, 104,
108, 109, 121-139; Stephens, *The Book of the Farm*, vol. i. pp. 162, 164,
212, 258, 268, 423, 424, 485-487, 536, 537; vol. ii. p. 471; Dobson,
On the Ox, pp. 6, 8, 67, 75-78, 80, 81, 89, 136, 153, 176, 181-186, 201-211,
216, 217, 248, 249; Sir Thomas Watson, *Lectures on the Principles and
Practice of Physic*, vol. ii. pp. 750, 682; Shirley, *Deer*, etc., p. 243;
Struthers, in *A Paper read before the British Association of* 1872, see the
Popular Science Review for Oct., 1872, p. 432; C. J. B. and C. T. Williams,

will be a minor function, or the experimenter has omitted to take into account something which is necessary for the success of the experiments. All this is more especially the case as regards generation, which is as unspecialized a function, indeed, a rather less specialized function, than that of the skin; and varies so little in its essential characteristics from the lowest organisms to the highest, that observations deduced from the breeding of domestic animals may very safely be applied to man.

Nevertheless, two kinds of objections have been made to this; the one, that domestic animals are usually from Nature's point of view nothing but monsters. By artificial means, it is urged, a certain part or quality of the animal is unnaturally developed, while at the same time another part or quality is as unnaturally diminished—withdraw man's care and attendance, and what becomes of the so-called improved animal then? It will languish and die; and in its place will appear the original stock from which it was developed. The other objection is, that any improvement which may occur is not really due to in-and-in breeding, but to the care of the breeders in feeding, tending, cherishing, and selecting their beasts; so that the weak are not allowed to breed, nor are they

Pulmonary Consumption, p. 113; St. Lager, *Etudes sur les Causes du Goître*, etc., pp. 58-61; Devay, *Du Danger*, etc., pp. 53-60; Boudin, *Mém. de la Soc. d'Anthrop. de Paris*, vol. i. 1863, p. 515; Legrain, *Bullet. de l'Acad. Royal de Méd. de Belgique*, vol. ix. 1866, pp. 284-308). There are doubtless many other animals liable to these and other diseases common to man; and the time will come when to doubt that animals may have the fevers would be as rare as doubts that snake bites are as poisonous to one as the other (or that entozoa are indifferent as to their habitation, this being determined really by the accident of their methods for multiplying), with about as many exceptions perhaps.

obliged to endure those evils and difficulties of existence which their wilder brethren are forced to undergo.[1]

But what an argument is this! Without denying one of these assertions, I yet maintain the animal itself is neither any worse nor any better than it was before. I say that it is better than a wild animal in the stall, and a wild animal is better than it in the Llanos. It is an argument which has already been refuted by wild Indians to the shame of white men. "We thank you," said the Iroquese to the Government of Virginia, in the year 1744, who had offered to educate some of their young men; "we thank you for your kind intentions. We have already had some experience of your education. Some of us whom you educated in all your sciences came back again bad runners, ignorant of all woodcraft, unable to bear heat or cold; they were ignorant how to trap deer, how to build a wigwam; they were ignorant even of our language—what good, then, either to themselves or to

[1] Gourdon, *Comptes Rendus*, vol. lv. 1862, pp. 269-273; Mitchell, *Edin. Med. Journal*, March, 1862, pp. 873, 874, 878; Mitchell, *Mem. of the Anthrop. Soc. of London*, vol. ii. 1866, pp. 451-452; Boudin, *Ann. d'Hygiène*, vol. xviii. 1862, pp. 67-70; Boudin, *Mém. de la Soc. d'Anthrop. de Paris*, vol. i. 1863, pp. 543-549; Geoffroy, cited by Perier, *Ibid.*, vol. iii. 1870, p. 257; Crossman, *Brit. Med. Journal*, vol. i. 1861, pp. 401, 402; Low, cited by Mitchell, *Mem. read before the Anthrop. Soc. of London*, vol. ii. 1866, pp. 450, 451; Devay, *Du Danger*, etc., p. 48; Flourens, *Comptes Rendus*, vol. lv. 1862, pp. 238, 239. M. Boudin "recommends to the notice of gastronomers" a fact noticed by "un voyageur étranger, qui a habité la France et l'Angleterre," who found himself obliged to eat twice as much beef in England as in France to satisfy his hunger; and which M. Boudin suggests is due to the fact that beef is more in-and-in bred in England than in France! (Boudin, *Ann. d'Hygiène*, vol. xviii. 1862, p. 69, note 5). The fact says a great deal for the taste of the intelligent foreigner, who probably went off singing "Oh, the Roast Beef of Old England!" and possibly, like the waters of the fountain of Trevi, it induced him to "come again."

us was their education? No. We cannot accept your offer, though we appreciate your goodwill. But send us, if you like, a dozen or so of your sons, whom we will educate as we do our own, and make *men* of them."[1] When animals are bred by man, he must select for every quality he wishes to develop, and not for one alone. If he selects for but one quality, the others will necessarily deteriorate, because they are neglected. It will not do, for instance, to select pigs only for rapid fattening qualities; or rabbits merely for long ears and a sleek coat; they must also be chosen from a large litter to ensure the hereditary transmission of fecundity. The fact is, that an animal properly bred in-and-in, and a wild animal, is each the most perfect according to its circumstances. Alter the circumstances, and the animal is at once unfit for its place. When the Kyloe breed of cattle was crossed, it was found only to become more delicate because the progeny were not so well acclimatized as the pure Kyloe; and the same was the case with the Galloways.[2] From the breeder's point of view in-and-in breeding improves the breed because it suppresses those qualities which under the circumstances are useless, and develops those which are useful, whether it be for racing, for wool, for the butcher, or for any other purpose; and without in-and-in breeding he cannot alter an animal to suit his purpose. He selects for breeding, it is true, but the one factor, the factor consanguinity, still remains; while by that means he excludes disturbing factors such as morbid inheritance, whereby he makes his experiments valuable for

---

[1] Quoted by Waitz, *Anthropologie*, vol. iii. pp. 170, 171.
[2] Macdonald, *Cattle, Sheep, and Deer*, pp. 268, 276.

our purpose. By in-and-in breeding he could as easily and as certainly breed animals with qualities which would elevate them above their savage brethren were they to rejoin them, as he can breed qualities which improve them for himself. Indeed, it may be said that those qualities in every state of existence which give an animal most chances of life and multiplication, are those qualities which improve it; for what else is improvement? It is equally undeniable that, in their present circumstances, our domestic animals are improved for themselves, as well as for the breeder, since they multiply and live, while the others are exterminated.

This very care which he bestows so unsparingly on the selection of his stock, is what would have to be done were we to experiment on the effects of blood-relationship in marriage undisturbed by other factors, the effects of which we are unable to calculate. An ideal experiment of this sort would be to take a fine and healthy set of people, divide them, and place them under the same conditions of occupation, food, and habitation. Make the smaller lot never marry any but their nearest relations, generation after generation, and the larger never marry any one related in the remotest degree, selecting in both cases only the most vigorous and healthy. Then note the result of these marriages and compare them. By such an experiment disturbing causes would, as far as is possible, be excluded, and if there were any morbid results, the sole factor which could cause it would be the consanguinity. It is precisely this, though not in so complete a form, that the breeders do; but with the advantage that, since the domestic animals breed faster in proportion to their lives than does man, a closer degree of consanguinity can be obtained from them than could be obtained from him. Nor is selection an unnatural process; for both selection and in-and-in breeding are common to wild animals. Most of those animals which do not pair for life, fight for possession of the females; the weakest thus necessarily is unable to breed, while the strongest not only has the advantage of propagating his strength, but in many cases, also, practises polygamy and incest, since he drives away all the young males and keeps the females.[1]

It is manifest, then, that if the question whether consanguineous marriages are harmful through their consanguinity, and not through inheritance (if they are harmful at all), is to be answered, that observations on in-and-in breeding in animals, and these alone at present, are able to do so. Naturally, persons with that preconceived notion which every one is bound to have on this subject who has not studied it, are apt to consider any evil result observed in the course of in-and-in breeding, as caused by that kind of breeding, without any previous examination whether there may not be other causes to account for it. It is so in the study of these unions among human beings; and the results obtained from in-and-in breeding in animals, though far more trustworthy than those obtained from consanguineous marriage cases, are not absolutely free from disturbing causes which affect their reliability. In the study of these cases, therefore, as in others, we must remember that one fact showing the harmlessness of in-and-in breeding is worth a hundred tending to show their harm-

---

[1] For those animals which are polygamous, see pp. 145-146 of this work.

fulness; since in the former the consanguinity is still a factor, but in the latter we are ignorant what other factors may have come into play. Let us now proceed to facts.

M. Allié, says M. Boudin, after a long experience, is of opinion that the system of in-and-in breeding is ruin to sheep. A flock at Petit-Bourg, he says, has diminished greatly in value since it passed into other hands, and this system has been practised.[1] The observations of Stephens lead him to the same conclusion; the progeny, he says, though improved in figure, firmness of bone, etc., are nevertheless delicate skinned, and therefore liable to the attacks of insects, and to inflammation; but this evil is only the result of long-continued in-and-in breeding, and by no means the immediate result.[2] M. Aubé asserts that sheep will produce a dark kind if bred in-and-in, which he explains as a step on the road to albinoïsm.[3] While " Mr. Giblett of Bond Street," quoted by Walker, asserted that sheep bred in-and-in on Bakewell's principle are fitter for the tallow-chandler than for the kitchen,[4] which is very likely true, but has nothing to do with consanguinity.[5]

On the other hand, M. Beaudouin gives the following account of a flock of 300 merinos bred in-and-in for a period of twenty-two years: the animals originally came

[1] Boudin, *Mém. de la Soc. d'Anthrop. de Paris*, vol. i. 1863, p. 520.
[2] Stephens, *The Book of the Farm*, vol. ii. p. 584.
[3] Devay, *Du Danger*, etc., p. 56.
[4] Walker, *On Intermarriage*, p. 335.
[5] Too much attention is given by most breeders to the fat-getting qualities of animals, in consequence of the prizes offered at cattle-shows being generally bestowed on the fattest beasts (See p. 249 of this work; and Macdonald, *Cattle, Sheep, and Deer*, pp. 258-260).

from Saxony, were renowned for the purity of their blood, and had only been a few years in the Côte d'Or when, in 1840, he commenced his observations. At that time, though suffering from no particular disease, the sheep were labouring under general debility seemingly attributable more to a want of acclimatization than to anything else. He began by a little judicious selection, eliminating about 15 per cent. yearly, and the flock soon became remarkably strong and healthy. There was no sensible sterility; altogether, perhaps, the cases of cryptorchis and monorchis were not more than 6 per cent., while in the females there were even fewer cases of barrenness. Cases of duplicate organs were about 5 per cent.; and in 1859, a year when these cases were unusually frequent in all the flocks about, there were as many as 7 per cent. in his. The sexes were produced in nearly equal numbers, and cases of miscarriage were not more numerous than among the neighbouring flocks. Far from degenerating, they became finer and far more to be depended upon to reproduce their proper type than is ordinary in flocks which are crossed. During the twenty-two years there was one instance of the birth of a Mauchamp sheep, which is a very rare occurrence. He concludes with the declaration that, in his belief, in-and-in breeding, combined with a moderate amount of selection, has no evil effect.[1] Close interbreeding, says Mr. Darwin, has perhaps been continued longer with sheep than with cattle, but perhaps the nearest relations have not been so frequently matched. Messrs. Brown, during fifty years, have never crossed their excellent flock of Leicesters, nor since the year 1810

---

[1] Beaudouin, *Comptes Rendus*, 1862, vol. lv. pp. 236-238.

has Mr. Barford crossed the Foscote flock. This gentleman asserts that when two nearly-related individuals are perfectly sound, no degeneracy is produced in their offspring by their union; or, in other words, that there is no danger in in-and-in breeding unless through morbid inheritance. But, on the other hand, he does not pride himself on breeding from the nearest relatives; and I may add that such is not a breeder's object: he does not choose a relative for its relationship, but for its qualities. In France, the Naz flock has been bred in-and-in for sixty years, without the introduction of any strange blood.[1] Ferdinand and Louis Fischer started a flock of 100 ewes of one family, and 4 rams of another; and these families have since been interbred without the admixture of a drop of fresh blood. Mr. Atwood's entire flock, which was so celebrated that it is now scattered by colonization into all the States of the North American Union, originated from a single impregnated ewe; and neither she nor any of her progeny or descendants while in his hands were interbred with any sheep not descended exclusively from Col. Humphrey's flock, from which she herself came. Mr. Hammond bought a small number of Atwood's flock in 1844, and he has since interbred solely between the descendants of these identical sheep. The Spaniards, in their sheep-breeding, guard against any admixture between the different cabanas, and they have been bred in-and-in for ages.[2] Hallam says that the fineness of Spanish wool is considered to be owing to an importation of English sheep about the year 1348, and again about 1465, in return for

---

[1] Darwin, *The Variation of Animals and Plants*, etc., vol. ii. pp. 119-120.
[2] Randall, *Fine Wool Sheep Husbandry*, pp. 108, 115.

which the Spaniards exported horses.[1] McCulloch says that the Spaniards themselves ascribe their superior breed of sheep to the introduction of a few from England by Catherine of Lancaster in 1394; while elsewhere he says the merino breed is said to have been introduced from Barbary.[2] These importations could not have been very great, and as it appears, the Spaniards have since bred them in very closely, with the result that they became so valuable, that up to the treaty of Basle their exportation was forbidden. By that treaty the French were allowed to buy 5,000 merino ewes, and as many rams; and from this stock the English sheep, which had also been carefully bred, were improved, while those of France and Germany were almost replaced by them.[3] These sheep, says M. Huzard, have been ever since bred in-and-in at Rambouillet, and have never been crossed except by a second importation under the First Consulate. The nearest relatives are generally put together, for the rams are usually put to their own progeny for several generations, and this without any sign of degeneration. The flocks of MM. Tessier, de Sylvestre, Perrault, Girod, and others testify to the same fact.[4] The merino, when introduced into Germany, was so immensely superior to all the native breeds, that it was everywhere accepted with enthusiasm. In Saxony, the greatest attention was paid to them, chiefly, however, as regards the *quality* of their wool, not as regards the quantity and quality, as well as the quality of the meat, as in England. To this

---

[1] Hallam, *Middle Ages*, vol. ii. p. 386, note.
[2] M'Culloch, *Geographical Dict.*, Arts. *Spain* and *Tunis*.
[3] *Ibid.*, Art. *Castile*.
[4] Bourgeois, *Quelle est l'Influence des Mar. Cons.*, etc., p. 13.

end they were kept in stables and fed on heating food, such as corn and hay, throughout the winter. The result is an unexampled quality of wool, but the animals have become a small and puny race. In England the breed of sheep was already so good that men were prejudiced in favour of their own breeds. Many merinos, therefore, fell into the hands of men who had no experience in breeding, and they were mismanaged; but, " in the hands of at least one practical breeder they were, however, eminently successful. He reports on them : soon after the king's flocks were imported, * * I purchased a considerable number of sheep from them, and selected from those of the Negrette blood, as being the largest sheep, and carrying the most and softest wool. These I continued to keep strictly pure, having no other sheep whatever, and I drew rams from the royal flock, so long as that was kept up; since which I have depended wholly on my own. By due attention in breeding, the wool, far from degenerating, has annually improved in softness and fineness, and these qualities have become much more uniformly even throughout the fleece; so that I now obtain for the whole a price beyond what any foreign wool brings in bulk in an unsorted state, whilst the fleeces of our own flock are full double the weight of those of the Saxon sheep. It is right, however, to state that the staple of my flocks having arrived at a length beyond that of other Merino sheep, has rendered it fit for combing, thus enhancing the value. The form of the sheep is also highly improved, whilst the disposition to fatten equals that of the Southdown. The mutton is of the first quality, and I can readily have for the fat wethers the highest price which any mutton brings in the London market."[1] The justly-celebrated New Leicester breed of sheep was entirely created as a distinct breed by this method. "Taking the native sheep," says Mr. Macdonald, talking of Bakewell, "he reduced his size, gave him small offals, induced him to lay on flesh and fat all along the breech, sides, shoulders, flank, and neck. He opened his wool, and also reduced it in weight, and a little in length. He increased the tendency to lay on fat in proportion to the food consumed, and made the animal take on fat at least, a year or two earlier; thus enabling two or three animals to be fed where one only was fed before. Nor was this change fitful or temporary, it was permanent and indelible; and, for nearly a century, the same breed of sheep has not only maintained its position, but has been used with more or less success to improve nearly every breed in the United Kingdom, and has, moreover, more or less, displaced almost every other breed."[2] A correspondent of Walker's says: "I have bred from rams from the same flock in Leicestershire, for fourteen years, which flock has not had a cross since the year 1799." Some of the new Leicester breed appear, however, to deserve the remark of the "Bond Street Butcher;" for Sir John Sebright said that Bakewell's principles were followed up too far; the propensity to get fat has increased so much that their stock has become small in

---

[1] Macdonald, *Cattle, Sheep, and Deer*, pp. 351-353.

[2] *Ibid.*, p. 446. At p. 415, however, Mr. Macdonald says, "The Leicester was notoriously a cross of various breeds in the first instance, although the success which supplied the cross is a secret buried in the 'tomb of the Capulets.'" At p. 447 he says, "It is not so clearly known what was the parent stock, though it is generally supposed to have been the Old Leicester."

size, delicate, and produces little wool. But another correspondent of Walker's points out that a propensity for fat-getting, and the production of the finest wool, are incompatible;[1] and it certainly appears from the fact that this breed has supplanted so many others that it cannot have degenerated. Too much fat is always a danger to a breed, for fat is a degeneration of tissue and a cause of sterility;[2] and, although by in-and-in breeding man is able to do a great deal in the way of alteration, he must still follow Nature, he cannot go contrary to physiological laws; he can increase the qualities which he wishes to get, chiefly only at the expense of qualities which he is content to do without; and can no more obtain an animal all fat with every other good quality, than he can teach his breed to live without food. We must remember that ill-directed breeding is as bad when there are frequent crosses as when there are none; that it is selection which is the great improver, when properly directed, and that breeding in-and-in is only advantageous because it fixes the breed, and obviates the necessity of crossing from an unimproved breed. Indeed a careless cross may diminish size—a charge against in-and-in breeding—just as in-and-in breeding under the same circumstances may do so. The Romney Marsh sheep were made smaller in this way; so were the Teeswater: and so are the mongrels of the Merino and Scotch, or the Southdown and Scotch breeds.[3] The sheep of Scotland, says Dr. Copland, are very small,[4] their fleeces fine and soft, their meat delicate

[1] Walker, *On Intermarriage*, pp. 236, 295, 351.
[2] See p. 249 of this work.
[3] Macdonald, *Cattle, Sheep, and Deer*, pp. 462, 478, 483.
[4] Animals living on islands are said to be frequently small breeds

and finely flavoured. In many parts they have much deteriorated since the introduction of Southdown breeds.[1] Indeed, the sheep themselves seem sometimes to have an antipathy to crosses, for on one of the Faroë isles it was observed that the half-wild native black sheep would not readily unite with imported white sheep.[2] The Shetlanders also tried to improve their native breed of sheep by crosses, and failed signally.[3] So bad are the effects of crossing an improved breed, which must necessarily comprise no very great number at first,[4] that some persons keep their animals in different families, and thus while they retain consanguinity, any tendency to disease peculiar to one family from the soil, habit, or what not, is obliterated. On the other hand, so valuable is in-and-in breeding to perpetuate any peculiarity either caused by selection, or by what is known as a "sport," that nearly all "created" breeds have been produced in this way, and valuable breeds, such as the Ancon

---

because they are bred in-and-in. But although the fact is true, the cause cannot be, since there are in many other and unisolated parts of the world also small breeds of the domestic and other animals. Thus, there are small breeds of horses in Ladakh, India; in Morbihan and Lorraine, France; in Servia; in the Szekler Mountains, Transsylvania; in Finland; in Galicia; and in Sweden: there are small breeds of cattle in Kumaon, India; in Manchooria; in Tunis; in the department of Isère; in Unterwalden; and in Sweden: and there are small breeds of sheep in Ladakh, in Manchooria, in Thibet, and in Hanover, etc. (M'Culloch, *Geographical Dict.*). Even had we not these facts, we should have no right to attribute any dwarfishness of island breeds to consanguineous unions among them, when the probability is that small stature, when not directly inherited, is due to hard work and poor food (Godron, *De l'Espèce*, etc., vol. i. pp. 17-18; vol. ii. pp. 286-294).

[1] Walker, *On Intermarriage*, pp. 349, 351.
[2] Darwin, *The Variation of Animals and Plants*, etc., vol. ii. p. 103.
[3] Macdonald, *Cattle, Sheep and Deer*, p. 529.
[4] Darwin, *ut sup.*, vol. ii. p. 414.

and Mauchamp, would have been entirely lost without it.

A majority of the most celebrated breeders and improvers of English cattle, says Mr. Randall, have bred closely in-and-in;[1] and this was necessary, as I have already pointed out, since an improvement cannot comprise a large number at first. Bakewell was one of these breeders, and his long-horns were for a considerable time closely interbred, though Mr. Youatt says that they at last became delicate, and the propagation of their kind uncertain,[2] a state which seems to have been due to bad management, for Bakewell himself was, as a rule, extremely successful. Knight once in the same season reared two young bulls of which the parents were nearly related; and both proved perfectly impotent, or at least failed to beget a single calf, yet the females bred well enough while young. But another correspondent of Walker's never found the generative power fail in consequence of in-and-in breeding of cattle; all that is necessary, he says, is to select carefully.[3] The half-wild cattle kept in British parks, at Cadzow Castle, Chillingham, and Chartly, are put forward as instances of long-continued in-and-in breeding without any evil result, by Culley, Dr. Brown, and Mr. Macdonald.[4] These cattle were parked 400 or 500 years ago, and are supposed to be the only remains of the ancient British cattle. Mr. Darwin, however, asserts that, compared to the wild cattle of South America these are bad breeders; and Dr. Smith says

[1] Randall, *Fine Wool Sheep Husbandry*, p. 117, note.
[2] Darwin, *The Variation of Animals and Plants*, etc., vol. ii. pp. 117, 118.
[3] Walker, *On Intermarriage*, pp. 228, 229, 238, 239.
[4] Darwin, *ut sup.*, vol. ii. p. 119; *Edin. Med. Journal*, March, 1862, p. 875; Macdonald, *Cattle, Sheep, and Deer*, p. 236.

that the Chillingham cattle now produce deviations from the original type of white, with black muzzles and red ears, which deviation he considers a degeneration.[1] It does not follow, however, that this is a degeneration in the ordinary sense of the word; while it must be allowed that that selection has not been practised with regard to their breeding which would prevent any selection on their own part sufficient to allow of the intensification of any particular colour, since, though the keepers may shoot these deviations from the original type, this will not prevent it in the first instance. The various colours are there, and it would be contrary to all the teachings of the evolution hypothesis if deviations did not occasionally occur, whether by sports, which would be rare in so in-bred a herd, or by selection among themselves, as explained by Mr. Darwin in his *Descent of Man*. The fact still remains, however, that these animals have been bred in-and-in for centuries, and still continue to breed without the help of crosses. The South American cattle are all descended from a few brought over from Spain and Portugal; the first by Garay, in 1580,[2] and they have since increased to such astonishing numbers that, even in 1587, there were 64,350 skins exported from New Spain. Vast herds of wild cattle are met with in all parts of the country, particularly in the plains of the southern provinces, where they exist in troops of 20,000 to 40,000; so that hides, jerked beef, horns, and bones have long formed, and still form, leading articles of export from Brazil. St.

---

[1] Darwin, *The Variation*, etc., vol. ii. p. 119; *Edin. Med. Journal*, March, 1862, p. 877.

[2] Or, according to M. Godron, in 1550 (*De l'Espèce et des Races*, vol. i. pp. 426, 427).

Domingo was also furnished with cattle by the Spaniards, and only 100 years after their first importation, the French found them in such quantities on the island, that the most reckless slaughter seemed to make no impression on their numbers.[1] On the Falkland Isles there are herds of magnificent cattle, all descended from a few brought over from La Plata about eighty or ninety years ago. They are now breaking up into separate herds of different colours, the white, on the highlands, breeding usually earlier than the others.[2] I wish to draw particular attention to this natural segregation, which is also common in horses and sheep,[3] and must be taken in connection with the tendency all polygamous animals seem to have to separate into families. Is this Nature's horror of in-and-in breeding? Is this her delight in crosses? Two bulls and five cows which escaped from the early colonists of Sydney in the year 1788, were found in 1795 to have multiplied to sixty individuals. In 1796, there were ninety-four. In 1797, *they divided into two troops*, one of sixty-seven, the other of one hundred and seventy individuals.[4]

Price, the most successful breeder of Hereford cattle on record, until twenty years ago, was a staunch advocate of in-and-in breeding; so were the Collings, Mason, Maynard, Wetherill, Bates, the Booths, Sir C. Knightly,[5] Bakewell, Culley, Ellman, etc.[6] The cow

[1] M'Culloch, *Geographical Dict.*, Waitz, *Anthropologie*, vol. iii. p. 494, note; Godron, *De l'Espèce*, etc., vol. i. p. 428.

[2] Darwin, *The Variation*, etc., vol. ii. p. 102.

[3] See pp. 278, 287 of this work; also an instance in dogs, p. 259.

[4] Godron, *De l'Espèce et des Races*, vol. i. pp. 428, 429, note 4; Darwin, *ut sup.*, pp. 102, 103.

[5] Randall, *Fine Wool Sheep Husbandry*, p. 117, note.

[6] Walker, *On Intermarriage*, p. 239.

Restless, almost an historical animal, was the result of in-and-in breeding to a degree which would not have been possible to obtain in man, owing to his long childhood. The bull Bolingbroke was put to his half-sister Phœnix, and produced the bull Favourite. Favourite was matched with his mother, and produced the cow Phœnix, a celebrated animal. Favourite was then matched with his daughter, and the produce was the famous bull Comet; then with his daughter's daughter; then with his daughter's daughter's daughter, he being the father in each case. The produce of this last union, a cow, had 93·75 per cent. of Favourite's blood in her, and was put to the bull Wellington, himself deeply interbred on both sides in the blood of Favourite, of which he had 62·5 per cent. in him. This union produced the cow Clarissa, an admirable animal. Clarissa was put to the bull Lancaster, who had 68·75 per cent. of Favourite's blood in his veins; and this union produced the celebrated cow Restless, a breeding-cow of Sir Charles Knightley's herd.[1] The rule of Mr. Bates was always to put the best animals together, regardless of consanguinity. His "Duchess" family, one of many families thus bred, ceased to breed; but he continued his former course of in-and-in breeding with triumphant success.[2] Mr. Darwin, however, points out that though Bates bred in-and-in for thirteen years, yet during the next seventeen years he thrice crossed his herd, not to improve them, but to increase their fertility; while Nathusius, after a careful study of pedigrees, finds that

---

[1] Darwin, *The Variation*, etc., vol. ii. p. 118; Randall, *Fine Wool*, etc., p. 117, note; Child, *Essays on Physiological Subjects*, pp. 24, 25.

[2] Randall, *Fine Wool*, etc., p. 117, note.

no breeder has continued in-and-in breeding all his life.[1] But, at all events many have bred in-and-in far more closely than would be possible in man, for a number of generations longer than the average of human families exist. Mr. Price, whose Herefords were the best in the world in his day, declared he had not gone beyond his own herd for a bull or a cow during forty years.[2] At Earl Ducie's sale in 1853, a white heifer, only five months old, sold for the enormous sum of four hundred guineas; she was the daughter of the bull Fourth Duke of York who was by Second Duke of York, and her dam was Duchess 59, also by Second Duke of York; consequently the sire and dam of the heifer were half-brother and sister. Many others which reach high prices are bred on this system.[3] Mr. Gardner gives a most successful case of breeding between son and dam.[4] M. Sanson points out that the Charolaise race of cattle has been greatly improved by in-and-in breeding; and that the small Bretton race of Morbihan, so celebrated for its milk and butter, are usually propagated in this way.[5] Our own small breed of Alderney, or rather Jersey, cattle, is so celebrated for its cream-giving qualities, that the importation of other breeds is forbidden by the local law, and in-and-in breeding is the rule.[6] At Rambouillet, in-and-in breeding was practised among the celebrated cattle of that place, a white hornless breed, with great success until they

[1] Darwin, *The Variation*, etc., vol. ii. p. 118.
[2] Randall, *Fine Wool*, etc., p. 118, note.
[3] Youatt, *On the Horse*, p. 121.
[4] Gardner, in the *Brit. Med. Journal*, vol. i. 1861, p. 290.
[5] Sanson, *Comptes Rendus*, vol. lv. 1862, pp. 123-124.
[6] M'Culloch, *Geographical Dict*.

were carried off by the cattle epidemic of 1815. M. Huzard also saw at Hohenheim and the royal farm of Holitzchen, herds of superior animals which were always bred in-and-in.¹ In this way, says Mr. Darwin, were in all probability bred the Niata cattle, from one individual sport.²

Earl Fitzwilliam keeps a herd of about a dozen Indian buffaloes, which *have* been crossed from Lord Derby's stock, but are also bred in-and-in very closely. These are exceedingly healthy.³

The eland, first acclimatized in England by the late Lord Derby between the years 1835–1851, at his menagerie at Knowsley, has been closely bred in-and-in. They were bequeathed to the Zoological Society in 1851, at which time there were two males and three females; and since then they have regularly reproduced without the loss of a single calf.⁴

At Fitzroy (Falkland Isles), near Mare and Island harbours, was, and is now probably, a herd of guanaco, numbering some twenty individuals, all sprung from a couple brought over as a present to the governor. For a long time they were kept in a paddock, but at last they were given to their present owner, Captain Packe, on account of a nasty habit they had of spitting at passers by. Captain Packe removed them to the neighbourhood of Fitzroy; where, though necessarily bred in-and-in, they have thriven and multiplied.⁵

Breeders are more unanimous on the evils of in-and-in breeding on pigs, says Mr. Darwin, than perhaps

---

[1] Bourgeois, *Quelle est l'Influence des Mar. Cons.*, etc., p. 13.
[2] Darwin, *The Variation*, etc., vol. ii. p. 414.
[3] Macdonald, *Cattle, Sheep, and Deer*, p. 231.
[4] *Ibid.*, p. 233.
[5] *The Field*, April 8th, 1871, p. 282.

on any other large animal. Mr. Druce says their
constitution cannot be preserved without a cross.
Lord Weston, the first importer of a Neapolitan boar
and sow, bred in-and-in till the breed was in danger
of dying out. Mr. J. Wright bred with the same
boar from its daughter, grand-daughter, great-grand-
daughter, and so on for seven generations; with the
result that the offspring in many cases failed to breed,
in others they produced few that lived, and of the
latter many were idiotic, without instinct to suck, and
unable to walk straight. The last two sows were put
to other boars, and produced several litters of healthy
pigs. The best in external appearance produced
during the whole seven generations, was one of the
last births, the sole one of the litter. She would not
breed with her sire, and yet bred at the first trial
with a stranger in blood. Nathusius imported a gravid
sow from England, and bred closely in-and-in from
the progeny for three generations, and with bad
results; yet he esteemed one of the latest sows a
good animal, and she bred well with a boar of different blood. On the whole Mr. Darwin thinks,
therefore, that in-and-in breeding does not affect the
external form, while it affects the general constitution,
the mental powers, and especially the reproductive
powers.[1] It must be remembered, however, that pigs
are precisely those animals which are cultivated most
for their fat, and that fat is very injurious to the
health of any animal, and especially to the reproductive powers. Crossing, on the other hand, gives
a tendency to reversion, and therefore a relief from

[1] Darwin, *The Variation of Animals and Plants*, etc., vol. ii. pp. 121, 122.

fat. Indeed, as I have already explained, facts against the harmlessness of in-and-in breeding have very little value compared to those in its favour; and this is too generally overlooked. Thus pigs with but little hair on their bodies have by correlation also very bad teeth, and this may be prevented by crossing with hairy breeds.[1] If a breeder, in beginning to breed in-and-in, chose an animal with rather less hair than usual, the progeny would have a tendency to bad teeth, bad digestion, and hence weakness; and he would naturally conclude on finding that this weakness was cured by a cross, that it was the in-and-in breeding itself which caused it, and not mere inheritance. Mr. Hobbs divided his stock into three families, and by this device, though he kept the consanguinity, he avoided any chance inheritance of a morbid tendency, and obtained more latitude for selection. Mr. Coate, who won the prize for the best pen of pigs at Smithfield Club Show five times, says: " Crosses answer well for profit to the farmer, as you get more constitution and quicker growth; but for me, who sell a greater number of pigs for breeding purposes, I find it will not do, as it requires many years to get anything like purity of blood again."[2] So Mr. Youatt says: " A useful pig in these days may easily be bred; but if you want fixity of type, or, as it is well called, 'character,' you must adopt pure blood." Red pigs are "invaluable for giving vigour and constitution to black breeds, when demoralized by over coddling, over feeding, and injudicious in-and-in breeding."[3]

[1] Darwin, *The Variation of Animals and Plants*, etc., vol. ii. p. 327.
[2] *Ibid.*, pp. 121, 122.   [3] Youatt, *On the Pig*, pp. 6, 7.

In Circassia there are six sub-breeds of horses, three of which are asserted by a native proprietor of rank, almost always to refuse to mingle and cross whilst living a free life, and will even attack each other.[1] It is a crime punishable by death to forge the mark of pedigree on an animal.[2] The Arabs are equally particular as to their breeds, and their horses are better able to stand a change of climate than are European horses.[3] Mr. N. H. Smith, long a resident among the Arabs, is of opinion that colts bred in-and-in show more blood in their heads, are of better form, and are fit to start with fewer sweats than are others; but when the breed is continued incestuously for three or four generations, the animal degenerates.[4] It is difficult to know what is meant by "breeding incestuously." Mr. Meynell, it appears, did not think breeding from even sire and daughter or son and dam, was close in-and-in breeding;[5] and Mr. Bowly says the term in-and-in breeding "ought to be applied only to animals having precisely the same blood, as own brother and sister. Now, breeding from such relationship as this, seeing that the male has only half the blood of the dam, and the female only half the blood of the sire, can scarcely be called pure 'in-and-in' breeding, but may on the contrary, if carried out with caution, be done with advantage."[6] Our race-horses are derived from a mixture of Persian, Barbary, Arab, and native horses;

---

[1] Compare Aristotle, *Hist. Anim.*, vi. xvii. 7.
[2] Darwin, *The Variation*, etc., vol. ii. p. 102; M'Culloch, *Geographical Dict.*, Art. *Circassia*.
[3] Perier, *Mém. de la Soc. d'Anthrop. de Paris*, vol. i. 1860, p. 82.
Walker, *On Intermarriage*, p. 294. [5] *Ibid.*, pp. 234-235.
Macdonald, *Cattle, Sheep, and Deer*, p. 413.

but from the first they have been bred closely in-and-in.[1] Rachel, the dam of Highflyer, was the daughter of Blank and grand-daughter of Regulus; yet both Blank and Regulus were sons of Godolphin. Fox was born under similar conditions of relationship. The dam of Goldfinder was the daughter of Blank and grand-daughter of Regulus. The grand-dam of Buckhunter was a daughter of Bald-Galloway, who was also the sire of Buckhunter. The great-grand-dam of Flying-Childers, one of the most famous race-horses, was a daughter of Spanker, while his dam was also the dam of this last. The sire of the Knight-of-St.-George, a winner of the St. Leger, was also his grand-sire and great-grand-sire.[2] Smith, in his work on breeding for the turf, gives "once in and once out" as the rule for breeding; but "twice in" and once out, says Mr. Walsh, is more in accordance with the practice of our most successful breeders.[3] The breeder can have no hesitation, continues Mr. Walsh, in coming to the conclusion that in-and-in breeding carried out once or twice is not only not a bad practice, but is likely to be attended with good results. The evidence of repeated success in resorting to the practice of in-and-in breeding is too strong to be gainsaid.[4] "For the racecourse," says Dr. Elam, "the pure south-eastern breed is adhered to; but different *stocks*

---

[1] M'Culloch, *Geographical Dict.*, Art. *England*, p. 266.

[2] Sanson, *Comptes Rendus*, vol. lv. 1862, pp. 122, 123.

[3] Mr. Youatt, however, says that the most talented breeders have avoided in-and-in breeding during the present century, because they are convinced that it does harm (*The Horse*, p. 121). But Cecil, the editor of this edition, says in his own work (*The Stud-Farm*, etc., p. 79) that though he thinks in-and-in breeding undesirable, yet he would allow it "after three or four generations."

Walsh, *The Horse*, etc., pp. 139-141.

MARY
daughter of Charles
Bold, Duke of
Burgundy.
m. 1477.

PHILIP
Archduke
d. 1506

ELEANOR
her uncle by
Emmanuel of
2nd, Fran-
ance.

JOHAN
=John of
her first-co
both sides.

MATTHEW
=Ann of
Tyrol.

of the same breed, and those brought up in different localities are selected."[1] However, by "crosses," breeders by no means understand the introduction of fresh blood. There are scarcely two thorough-bred horses in the stud-book, says Mr. Walsh, that cannot be traced back to the same stock in one or more lines. An absolute freedom from relationship is not to be found, or if so, very rarely. Yet continued in-and-in breeding in the closest relationship he does not think advisable—it is apt to develop weak points in the constitution. "The cautious breeder, therefore, will do well to avoid running this risk, and will strive to obtain what he wants without having recourse to the practice, though, at the same time, he will make up his mind that it is unwise to sacrifice a single point with this view."[2] Mr. Darwin says that statistics show nearly one-third of our race-horses have proved barren, or have slipped their foals; a fact which he ascribes to their high nurture and close inter-breeding.[3] This is very probably the case, since a racing horse or mare, however delicate it may be, is too valuable not to breed from. Indeed, it is generally a disabled animal, one that has gone lame, and is therefore deprived of exercise and, with this, much of its natural health, which is set apart for breeding. Nor are they chosen for their fertility, but solely for their running powers.

In-and-in breeding in horses is carried on at any rate to a very great extent, and with decidedly beneficial effects on the race. "Nimrod" concludes

---

[1] Elam, *A Physician's Problems*, p. 68.
[2] Walsh, *The Horse*, etc., pp. 141, 142.
[3] Darwin, *Descent of Man*, vol. i. p. 303, note 40.

a comparison between the thorough-bred and half-bred hunter in these words: "As for his powers of endurance under equal sufferings, they doubtless would exceed those of the 'cocktail,' and being by his nature what is termed a better doer in the stable, he is sooner at his work again than the other. Indeed, there is scarcely a limit to the work of full-bred hunters of good form and constitution and temper."[1] Napoleon's celebrated State horses were directly derived, says M. Huzard, from the Arab breed of Count Huniady, who had bred continually from the same two stallions.[2] Indeed, it is the natural state of horses to breed in-and-in.

Donkeys, says M. Godron, are bred most carefully in the East, as carefully as the Arab horse, and their genealogy is as rigidly preserved to prevent any defilement of their blood. They are large, lively, rapid, and much esteemed.[3] In Egypt they fetch some thirty or forty pounds sterling.

Some goats turned out on the Island of Juan Fernandez in the year 1660, multiplied so prodigiously that it became necessary to turn out dogs to keep them down.[4] M. Aubé states that the Angora goats have occasionally to be crossed by a coarser sort; and that they are very subject to pleuro-pneumonia in consequence of in-and-in breeding.[5]

Sir J. Sebright declares that by breeding between brother and sister he has seen strong spaniels dwindle into diminutive lap-dogs; and Mr. Darwin, in citing

[1] *Westminster Review*, July, 1863, *cit.* p. 100.
[2] Bourgeois, *Quelle est l'Influence des Mar. Cons.*, etc., pp. 13, 14.
[3] Godron, *De l'Espèce*, etc., vol. i. pp. 390-391.
[4] *Ibid.*, pp. 407, 408.
[5] Devay, *Du Danger*, etc., pp. 60, 61.

this, adds a case in which some blood-hounds on being bred in-and-in were afflicted with a bony enlargement of the tail, a malformation which was obliterated by a single cross; while Mr. Scope attributes the rarity of Scotch deer-hounds to close interbreeding, the few specimens yet existing being all related.[1] But why should they diminish for this reason, when sheep, cattle, horses, and other domestic animals have been bred in-and-in for long periods without dying out? M. Aubé gives an instance of a farmer who received as a gift two magnificent dogs, brother and sister, and who wished much to breed from them. The offspring, however, proved very inferior; their bodies were smaller in proportion to those of their parents, with their heads and tails relatively large, and their backs weak and bowed. Though they retained the hunting instincts of their parents, they had none of their endurance, and the whole family died out at the third generation.[2] M. Boudin mentions a case where the offspring of a brother and sister were mute.[3] In these cases there was doubtless morbid inheritance, for Sir J. Sebright himself created many races of animals by in-and-in breeding, and was, as we shall see, a staunch advocate of the system. Dr. Child instances a case in which a dog was put to his daughter, and in succession to her puppies, and also to one of the puppies of the latter, yet the dogs were all good dogs.[4] Mr. Meynell's famous hounds were bred most closely in-and-in from sire and daughter, and dam and son,

[1] Darwin, *The Variation of Animals and Plants*, etc., vol. ii. p. 121.
[2] Devay, *Du Danger*, etc., p. 62.
[3] Boudin, *Mém. de la Soc. d'Anthrop. de Paris*, vol. i. 1863, p. 508, note 1.
[4] Child, *Essays on Physiological Subjects*, Note on Essay I., p. 54.

and sometimes brother and sister.[1] Turnspits and pugs were probably derived from sports, and must necessarily have been much bred in-and-in.[2]

M. Bertrand, after forty years' experience in the breeding of sporting and other kinds of dogs, considers in-and-in breeding to be injurious. He used never to cross his breed if he could possibly avoid it, and then noticed that after several generations they became more delicate, and, though better than their ancestors, less robust, and more subject to canine diseases, which were the more inveterate the longer the dog had been bred in-and-in. The males often became impotent, the females sometimes ceased to be fruitful at an early age, but every cross gave them back their vigour. M. Chipault coupled two magnificent dogs, son and dam, with the result of four puppies, two of which died after sixty days of paralysis.[3] But if M. Bertrand has been unsuccessful after forty years of experience, others have been successful. From an examination of their pedigrees, says Mr. Walsh, talking of greyhounds, " it will be obvious that ' Jason' and ' Rosebud' were first-cousins, once removed as regards ' Butterfly,' and twice removed as regards ' Majesty.' Mr. Randell's celebrated bitch ' Rival,' as stout a greyhound as ever ran, was by a grandson of ' King Cob,' out of a daughter of that dog; and Mr. Sharpe's ' Maid of Islay ' * * * * is another illustration of success attending this mode of breeding, she being by ' Jason,' a grandson of ' Monarch,' out of ' Molly Malone,' a grand-daughter of the same dog, and she has also plenty of bone. Mr. Long's

[1] Darwin, *The Variation*, etc., vol. ii. pp. 120, 121; Walker, *On Intermarriage*, pp. 234, 235.
[2] Darwin, *ut sup.*, p. 414.
[3] Chipault, *Etudes sur les Mars. Cons.*, etc., pp. 112, 113.

THE MARRIAGE OF NEAR KIN.

'Lizzie' again may be quoted, being a large and bony bitch, yet out of an aunt by her nephew. 'Motley,' and his sisters 'Kitty Brown,' 'Miss Hannah,' and 'Money-taker,' were also in-bred, and have been of great use at the stud, especially the two first named, to which we are indebted for 'David' and his numerous winning progeny, and for 'Chloe,' the winner of the Waterloo Cup last year. All the litter were, however, small, but as 'Tollwife,' their dam, was a diminutive bitch, and had a strain of the Italian greyhound, no conclusion can be arrived at on that score. 'Mustard,' sire of 'Monarch' and other good greyhounds, was three times in-bred to 'King Cob,' yet he possessed great size and enormous bone, and his son 'Monarch' resembles him in both these particulars. As one of the strongest modern instances of close in-breeding without loss of constitution, size, or bone, I append his pedigree, which is more remarkable from the fact that 'Mathilda Gillespie,' and 'Vraye Foy' were also much in-bred."

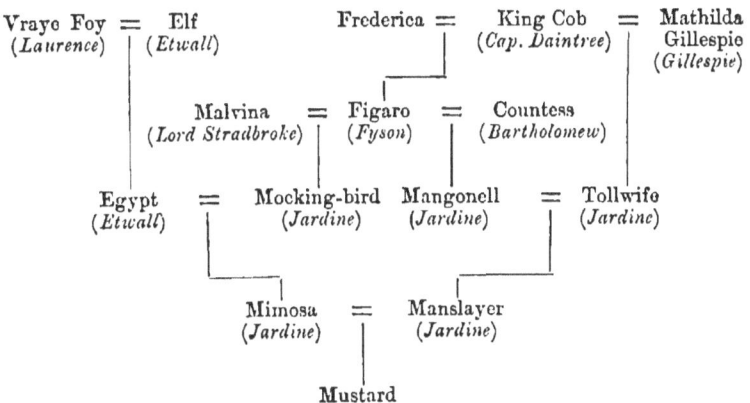

"The argument," continues Mr. Walsh, "that

in-breeding, when not carried too far, is advantageous, is now so generally admitted that it is a loss of time to support it by facts. Mr. A. Grayham's rule of 'once in and twice out' is perhaps the most prudent course to pursue; but 'twice in and once out,' is, in my opinion, not carrying in-breeding too far when the out-crosses are decided."[1] Mr. Laverack, a most successful breeder, and the creator of the "Blue Belton" race of setters, is a most determined advocate of in-and-in breeding. His system is to select strong parents of the quality he wants. "My dogs," he says, "are more *inter-crossed*, and *inter-bred* than directly bred in-and-in. There are several secrets connected with my system of intercrossing that I do not think advisable to give to the public at present. I can only say better constitutions, better feeders, and hardier animals than I have, do not exist. It must not be supposed I am prejudiced and obstinate in my system of breeding. I have *tried* crossing, or letting my blood loose ten or a dozen times, but the result has always been unsatisfactory; therefore I stick to intercrossing with my own strain, as I have ever found it answer best." He considers that the setter has generally degenerated in consequence of crosses, and because their blood has not been kept pure.[2] Whatever may be the difference between *intercrossing* and *in-and-in breeding*, it certainly appears from the subjoined genealogical table of the dog Dash, that Mr. Laverack in practice breeds habitually between brother and sister, and in the direct ascending and descending line :—

[1] "Stonehenge," *The Greyhound in* 1864, pp. 259, 260.
[2] Laverack, *The Setter*, etc., pp. 2, 3, 29, 30.

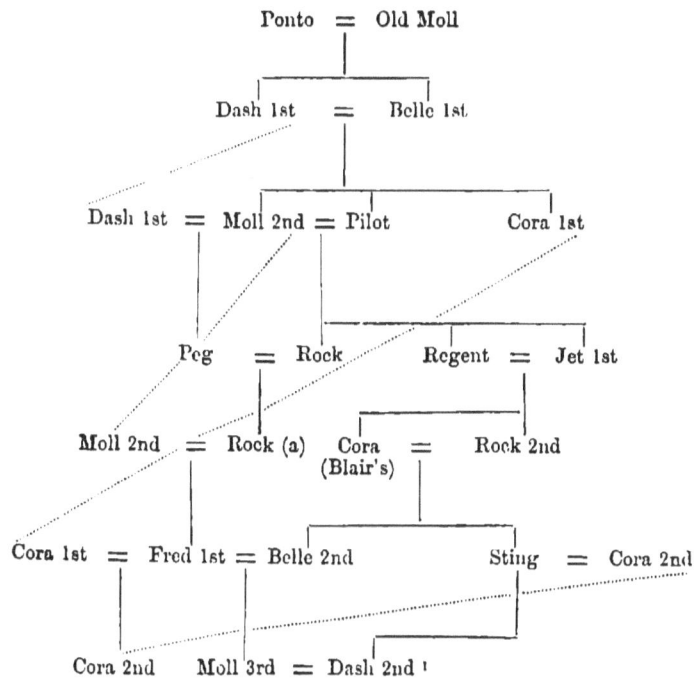

He says himself " a dog named Trimmer (blue Belton) was one of the best I ever possessed. * * this extraordinary animal was bred between brother and sister." [2]

The dogs of Cairo form distinct tribes, each of which confines itself to its own division or quarter of the town. Should a dog of one tribe intrude upon the territory of another, whether urged by some tempting piece of carrion, or by the mere spirit of adventure, he is immediately set upon by the others, and if

[1] The dotted line calls attention to closer cases of interbreeding. This table is re-arranged from that given by Mr. Laverack (*The Setter*, etc., p. 21). It may possibly be slightly inaccurate, but I do not think it is wrong in any important particular.

[2] *Ibid.*, pp. 21-23.

caught severely punished; should he be lucky enough to escape back to his own territory, he boldly faces his pursuers, and they in their turn fly.[1]

In many of the British deer-parks the deer have been allowed to breed uncrossed for long periods, without any degeneration showing itself, or loss of general health. The dark herds of deer in the Forest of Dean, in High Meadow Woods, and in the New Forest, supposed to have been brought by James I. from Norway, have never been known to mingle with the pale-coloured herds, although kept together with them.[2] Another case showing the rarity of crosses when animals are left to themselves. Dr. Davy mentions the case of a pair of red deer, who about the year 1850 were taken from the herd and put into a paddock of twenty or thirty acres adjoining Stornoway Castle, Isle of Lewis; these have multiplied yearly, and numbered ten years after, twenty-three, not including several which were killed, all descendants of the original pair, and all very much improved in comparison with the deer of the forest.[3] Nevertheless it is the practice, says Mr. Darwin, to infuse new blood into the fallow deer of the British parks, and this, he says, proves of the greatest benefit in removing the taint of *rickback*, and improving their size and appearance.[4] Rickbacked deer are too generally found in many parks, says Mr. Shirley, supposed to be due to weakness, brought on both by breeding in-and-in too much, and also by insufficient

---

[1] Lane, *Modern Egyptians*, vol. i. pp. 360, 361.
[2] Darwin, *The Variation of Animals and Plants*, etc., vol. ii. p. 103.
[3] Child, *Essays on Phys. Subjects*, pp. 53, 54.
[4] Darwin, *The Variation*, etc., vol. ii. p. 120; Shirley, *Deer and Deer Parks*, pp. 241, 242.

food.[1] In other words, we may say that the cause is unknown. The Scotch deer, however, breed naturally in-and-in, and the red deer generally breed between brother and sister for generation after generation,[2] and yet they are, as a rule, perfectly healthy.

M. Aubé, by means of a most extraordinary series of arguments, attempts to substantiate an hypothesis that in-and-in breeding is a cause of albinoïsm. Besides a few experiments which he made on breeding rabbits in-and-in, of which we shall see more anon, he gravely states that he saw three partridges, two sparrows, and a whole single jackdaw perfectly white; and hence he argues they became albinoes because, in all probability, their parents were related. Nor was this all; M. Aubé actually saw two deer, albinoes from the park of Raincy, and in this case, also, had no doubt that they had become albinoes through in-and-in breeding! But let us leave these childish arguments, gravely repeated by M. Devay, and test his experiments. M. Aubé obtained guinea-fowl with an admixture of white plumage after the third generation of in-and-in breeding, and further asserts that if rabbits are bred in-and-in, the first generation will assume a grey colour spotted with white, or a pale russet colour; the next generation will be black, or black and white; the next a slate colour; and the sixth generation will be albinoes.[3] No particulars are given as to the conditions of his experiment, at least none are given in M. Devay's quotation. M. Legrain,

---

[1] Shirley, *Deer and Deer Parks*, p. 243.
[2] See p. 145 of this work.
[3] Devay, *Du Danger*, etc., pp. 54, 56.

according to his own account, made a series of experiments on rabbits, with a view to disproving these assertions of M. Aubé. He says that for his first experiment he took four male rabbits and four females, marked with black and white patches in proportions almost exactly equal. These he coupled, and kept in a dark place, and under otherwise unfavourable conditions of health, as to food, cleanliness, etc., with the result of twenty young in the four litters. Six months after birth a brother and sister from each litter were coupled, and then kept under the same circumstances as their parents had been, with the result of sixteen young. He again waited six months, to allow the young to grow up, and repeated the experiment, and so on until in the third generation, when, as he relates, the results proved disastrous, the progeny being feeble, and only living a short time, and since the whitest pairs were always chosen for breeding, in the third generation were already red-eyed albinoes.

Experiment No. 2, he conducted in the same way as the first, with the exception that the rabbits had abundance of light and air, besides good food, etc. As before, he chose rabbits marked with black and white in equal proportions to start from, always paired brother and sister, and always selected the whitest individuals to breed from; with the result, he says, that in the fifth generation he obtained perfect albinoes. There was not, however, as in the former experiment, the slightest diminution of prolificness, or the least degeneration in health.

In experiment No. 3, M. Legrain says he sought to preserve the colour of the original parents, while at the same time he wished to breed in-and-in as closely as before. To this end he started and continued this experiment, as in experiment No. 2; that is, the original parents were black and white; he bred generation after generation from brother and sister, and then from brother and sister the issue of brother and sister, and so on; and he kept them under all conditions of health; but in this experiment, in place of always selecting the whitest couples for pairing, he says he chose those couples which most nearly resembled the parents. The fifth generation, according to M. Legrain, was quite healthy, nor was there any individual either quite white or quite black.

Experiment No. 4 was to show that albinoïsm might be as easily produced by crosses as by breeding in-and-in. To this end, he says he took from four different sources rabbits the same as before; and by always selecting the whitest for pairing in each generation, though he always crossed, he got perfect albinoes in the sixth generation.

Since M. Aubé states that black is merely a stage on the road to albinoïsm, M. Legrain recalled to his mind that he had the good fortune to know a gentleman who, for a long time, had kept a breed of perfectly black rabbits, and had rigorously excluded every other colour, so that for a long time none but black rabbits had been produced by them; from these he says he obtained two males and two females, all belonging to the same family, and coupled brother and sister in each instance. This he continued to do generation after generation; but, even in the sixth, none but black rabbits were born. Though in every instance a brother and sister were paired, yet the finest pair from the litter was chosen in each case, so as to exclude every cause but consanguinity for possible degeneration. But they degenerated in nothing, says M. Legrain, neither in size, in health, nor in prolificness.

Experiment No. 6 was to show how a tendency to disease might be intensified by in-and-in breeding. Hence, M. Legrain says he coupled a pair of rabbits whom he afterwards proved by dissection to have been tuberculous, as in the former experiments on breeding in-and-in, always pairing brother and sister. The results he gives as bad, and adds that no particular care was taken to ensure healthy conditions of life.

He then tried experiment No. 7, to show that the bad results of the last one were not due to the in-and-in breeding, but to carelessness of sanitary conditions. He therefore, as he says, repeated the experiment, with the exception that the rabbits were kept under the best possible conditions of health. Finally, he says, he tried an eighth experiment, in which he started from one tuberculous and one healthy parent, taking the utmost care to ensure perfect sanitary arrangements.[1]

Such are the details of the experiments which M. Legrain pretends to have carried out, but which have been conclusively proved never really to have taken place;[2] not because they are in themselves impossible, but since, in his account of them, M. Legrain inci-

---

[1] Legrain, *Bullet. de l'Acad. Royal de Méd. de Belgique*, vol. ix. 1866, pp. 287-295, 305-307.

[2] When this work was first published, these pretended experiments were given as having actually occurred. Since then, Mr. George Darwin has informed me that they have been proved never to have taken place. It appears that soon after their publication in the *Bullet. de l'Acad. Roy. de Méd. de Belgique* several members questioned their authenticity, and a commission was appointed to inquire into them, with the results as detailed by Dr. Crocq, that—(1) From internal evidence the alleged experiments were made to take place consecutively, and *not* simultaneously. (2) That since they took place consecutively, the time they took must have been:—

dentally shows that they took place consecutively, and not simultaneously, the time they must have taken would point to the fact that they were begun when their author was yet at school, and about ten years

For the first series, twenty-three months and eight days.
,, second ,, thirty ,, sixteen ,,
,, third ,, thirty ,, sixteen ,,
,, fourth ,, thirty-seven ,, twenty-four days
,, fifth ,, forty-two ,, eighteen ,,

or a total of thirteen years, eight months, twenty-two days; or if one of them, which M. Legrain does not show must have taken place consecutively, be deducted, ten years, six months, and twenty-eight days. The second lot of experiments must have taken a further time of—

For the first series (four generations), twenty-three months, eight days.
,, second ,, (six ,,), thirty-seven ,, twenty-four days
,, third ,, (six ,,), thirty-seven ,, twenty-four ,,

or eight years, two months, and twenty-six days. The least time necessary for these experiments must, altogether, have been sixteen years and ten months. (3) Now, M. Legrain became a veterinary surgeon in the year 1858; his paper was sent to the Academy of Brussels in the year 1865; hence, M. Legrain must have begun his experiments nine years and ten months before he took his degree, and while yet at school. (4) Further, the experiences of M. Aubé, which M. Legrain sought to controvert, were published in 1859; but M. Legrain must have begun his experiments in 1849. (5) M. Legrain was, besides, shown to have an exceedingly bad character for literary honesty.

In consequence of this report, the Academy, without allowing M. Legrain to be examined in his defence, voted that the memoir of M. Legrain should be held as "non-avenu."

On April 5th, or 9th, M. Legrain wrote a letter to the Academy concerning their hostile vote, and a committee was appointed to consider it. This committee reported on May 21st, as follows:—"Après discussion, la Commission, considérant qu'il n'y a rien dans la lettre de M. Legrain qui puisse engager l'Académie à revenir sur sa décision, est d'avis de proposer à la Compagnie de passer à l'ordre du jour sur cette communication." This conclusion was adopted.

I greatly regret that, not having seen this latter discussion (published in another volume of the bulletin), I published M. Legrain's experiments as true, and have therefore substituted this sheet for the one originally published in my work—the only reparation in my power. At the same time, I certainly think that it would be a great advantage were the Academy to adopt the excellent suggestion of M. Gluge, "qu'il conviendrait d'entendre d'abord une commission *avant* d'imprimer," and not shut the stable door after the steed is stolen. (See the *Bulletin de l'Académie Royale de Médicine de Belgique*, 1867. Third Series. Vol. i., No. 1, pp. 26-49; No. 4, p. 388; No. 5, p. 534.)

before M. Aubé wrote—that, in fact, he began to refute M. Aubé ten years before that gentleman had written anything to refute.

I have quoted so much on these experiments because they are exceedingly well thought out, and would serve as a model to any one who might wish to experiment on breeding in-and-in. Had they been ever really undertaken, it would have been impossible to exaggerate their value in the present state of our knowledge; for the in-and-in breeding of our domestic animals, the nearest approach to direct experiment which we have, is subject to have its value lessened by the breeder's desire to obtain as result, a condition which is incompatible with healthy reproduction—that is, they do not always exclude every cause of degeneration but that possible cause—consanguinity; since they breed pigs for fat, and do the same in some cases with sheep and cattle.

The few other cases mentioned by M. Boudin and by M. Chipault,[1] are under the circumstances unworthy of mention, since there is no account of their method of procedure in obtaining albinoes by in-and-in breeding. In the same way, M. Aubé's assertion that fish become barren, and gold-fish become albinoes, if shut up in basins together so that they must breed in-and-in, cannot yet be accepted, since we do not know what are the causes of albinoïsm, or how the experiments were conducted.[2]

[1] Boudin, *Mém. de la Soc. d'Anthrop. de Paris*, vol. i. 1863, p. 515; Chipault, *Etudes sur les Mar. Cons.*, etc., p. 111.

[2] Devay, *Du Danger*, etc., pp. 57, 58. M. Aubé also asserts that from his own observations and from those of other naturalists, breeding in-and-in from rare moths results in barrenness (*Ibid.*, p. 59). I may suggest that this, considering their state of captivity, is by no means extraordinary, and considering also that they were rare species, that is that they were unsuited even to a free life.

In the year 1418 or 1419, J. Gonzales Zarco turned out one female rabbit with her litter on Porto Santo, near Madeira. There they became such a nuisance that the island had to be abandoned, and Cada Mosto, only thirty-seven years afterwards, describes them as innumerable. It is true they are smaller than the English wild rabbit,[1] but, as I have already had occasion to show, this is probably due to over-crowding, and consequent scarcity of food.[2]

Sir J. Sebright asserts that his fowls got long in the legs, small in body, and bad breeders, from too close in-and-in breeding. Mr. Clark continued to breed in-and-in from his own kind of fighting cocks, till they became under the weight required for the best prizes, and lost all their pluck. On one cross from Mr. Leighton's they again resumed their former courage and weight. This breeder found that breeding from father and daughter produced a greater loss of weight in the offspring than breeding from mother and son. Mr. Eyton, of that ilk, says his Dorkings became smaller and less prolific if not occasionally crossed. Mr. Hewitt says the same of Malays, as to size at least. But fanciers with large stocks can breed from their own stock without this danger, because they keep various families separate for crossing purposes. Mr. Ballance, who breeds in this way, says that breeding in-and-in does not necessarily cause deterioration, "but all depends upon how this is managed." "My plan has been to keep five or six distinct runs * * * and select the best birds from each run for crossing. I thus secure sufficient crossing to prevent

---
[1] Darwin, *The Variation*, etc., vol i. pp. 112, 113.
[2] See pp. 277-278, note 4, of this work.

deterioration."[1] Jumper and frizzled fowls were in all probability derived from monstrosities of their kind,[2] and must, therefore, have been much in-and-in bred.

Mr. Darwin crossed some half-bred Penguin and Labrador ducks with Penguin, and afterwards bred them in-and-in, "and they were extremely fertile." The hook-billed duck must have been bred in-and-in sometime, as it must have been derived from a "sport."[3]

There are a great many sorts of fancy pigeons, says Sir J. Sebright, and each variety has some particular property which constitutes its supposed value, and which amateurs increase as much as possible, both by breeding in-and-in and by selection, until the particular property is made to predominate in such a degree, in some of the most refined sorts, that they cannot exist without the greatest care, and are incapable of rearing their young without the assistance of other pigeons kept for the purpose.[4] He continued closely interbreeding some owl-pigeons, till they arrived at such a degree of sterility that he nearly lost the breed. Mr. Brent crossed a trumpeter pigeon with a common pigeon, and afterwards bred him with his female descendants to the fourth generation, but failed to get a fifth. All fanciers cross their breeds, says Mr. Darwin, but he admits that their excessive delicacy cannot be accounted for by the amount of in-and-in breeding which does take place. He himself

---

[1] Darwin, *The Variation*, etc., vol. ii. pp. 124, 125.
[2] *Ibid.*, vol. ii. p. 414.
[3] *Ibid.*, vol. i. p. 279, note; vol. ii. p. 414.
[4] Walker, *On Intermarriage*, p. 227.

paired a brother and sister of a hybrid breed, and found them perfectly fertile. The short-faced and tumbler breeds are probably derived from an individual of their respective kinds.[1]

It is evident from the facts narrated in this chapter that if consanguineous marriage is injurious otherwise than through inheritance, this cannot in any way affect mankind, for wherever any experiment has shown that after continuous in-and-in breeding, barrenness is caused, this in-and-in breeding has always been so very close that it could not possibly have been done in experiments on man. On the other hand, many other experiments, or rather instances, have been detailed, in which, though there has been close in-and-in breeding, rather beneficial than injurious results have been obtained. The experiments of M. Legrain especially, show what an immense part inheritance plays in in-and-in breeding, and how dangerous it is to argue that because in-and-in breeding has apparently proved injurious in the hands of one breeder, therefore it is always injurious; and they further show that the same results may be obtained for good or for evil by means of crosses, as by means of in-and-in breeding, the only difference being that to improve by crosses is almost impossible, since many improvements have been obtained from the in-and-in breeding of monstrosities or sports which could not have been perpetuated by crossing, while crosses tend to produce mediocrity, or even reversion to a primitive and unimproved type, and therefore to degeneration.

[1] Darwin, *The Variation of Animals*, etc., vol. i. p. 192, note; vol. ii. pp. 126, 414.

At the same time, the evidence is abundantly sufficient to show the great value of crosses when judiciously carried out in removing an inherited disease, and, as Mr. Darwin shows, in the increase of size it often gives to the offspring; but the fact still remains that consanguinity, as far as our evidence has gone, cannot be said to originate malformations, disease, or sterility.

CHAPTER VII.

On the Alleged Benefits of Change of Blood in Mankind.

IN the last chapter I attempted to show that when in-and-in breeding proves injurious to the offspring, and this evil is removed by a cross, it does not prove the in-and-in breeding is itself injurious, but only shows that carelessness may make it so; and however doubtful the above evidence may leave us, whether in-and-in breeding generally leads to good or to bad results, there can be no rational doubt, after studying the effects of crosses, that in-and-in breeding cannot create disease, but at most is able to intensify an hereditary taint. For if in-and-in breeding is in itself bad, then crosses must be beneficial. If crosses act by virtue of being a cross and not by virtue of removing an hereditary taint, then the greater the difference between the two animals crossed, the more beneficial will that act be. But it is an undoubted fact, that the wider the difference the less good is the result as a rule. Hence crosses do not and cannot act by virtue of their difference, but simply act for good when the general health of the family crossed is not so good as that of the individual introduced into it.[1] In mankind, at least, a cross is always a dangerous thing, since if he marries into a family not related to him, he knows, as a rule, nothing whatever of the pathological history of its members, while he can avoid intermarriage in his own family if there is the least suspicion of hereditary disease. If, again, the marriage is between two persons of different races, such as a white with a Hindoo or a negro, the marriage will prove barren, or the progeny will prove barren, and besides have so unsettled a temperament, that these beings are known all over the world as the worst class of mankind.

There have lately appeared several articles in various periodicals on the moral state of the *Eurasians*, as they are called, or half-breeds between British, Portuguese, Dutch, or French Europeans, and Hindoos. "The dangerous character of this element in the society of our Eastern Empire," says the *Times*,[2] "has long been recognized. The Eurasians, with a few remarkable exceptions, have shown all the vices of both the parent races, and hardly a sign of the virtues of either Europe or Asia. Truth, honour, and honesty have been almost as rare among the men as chastity among the women. The sincerity and endurance of the Anglo-Saxon, the gravity, dignity, and temperance

---

[1] That crosses are beneficial in very often effecting an increase of size in the progeny, exceeding that of either parent, is established beyond reasonable doubt. But at the same time this cannot be continued without certain degeneration and the ultimate extinction of the organisms on which it is practised, and even in the first generation the offspring are only improved in one way, at the cost of an ill-balanced growth, and the loss, more or less, of their generative power. I shall treat more fully on this in the next chapter.

[2] *The Times*, Sept. 3rd, 1874.

of the Hindoo, are lost in the Eurasian character." "The lower classes of half-castes in India," says the *Pall Mall Gazette*, "lead the life of pariah dogs, skulking on the outskirts between the native and the European communities, and branded as noxious animals by both. In a higher class the lads pick up a living as menial servants, or on the river or wharfs, but constantly lose their places from drunkenness, and are reduced to starvation and the gaol."[1] The life of the women is even more degraded, according to the Archdeacon of Calcutta,[2] and even their apologists cannot put them on a level with either of the parent races, and only succeed in showing that many of them are capable of cultivation. "Undoubtedly," says Mr. Sendall, late Director of Public Instruction at Ceylon, "there are characteristic failings and vices incident to a mixed pedigree, interesting to the scientific observer, and requiring due recognition on the part of administrators and statesmen. * * * * If not destined for high achievements, they are still capable of attaining a fair average level of moral and intellectual development."[3] And Sir Alexander Arbuthnot says of them, that "there is nothing in the Eurasian nature which precludes the expectation that, in their case, education will develop many of the qualities which people in India are accustomed to regard as belonging exclusively to the ruling race."[4] De Warren says they are well gifted, but he adds, they inherit more often the vices of both parents than the virtues of either;[5] and

---

[1] *Pall Mall Gazette*, Sept. 1st, 1874. [2] *Ibid.*
[3] Letter to *The Times*, Sept. 9th, 1874.
[4] *Macmillan's Magazine*, Oct., 1874, p. 558.
[5] Perier, *Mém. de la Soc. d'Anthrop. de Paris*, vol. ii. 1865, p. 293.

Görtz says that these half-breeds are nearly all of relaxed temperament, and their mental power is poor."[1] But whatever may be the cause of their moral degradation, it is certain that they are neither fertile among themselves, nor are unions of the two parent races fruitful. It is the exception when a Hindoo female has children by a European husband;[2] and the half-castes themselves rarely rear their children.[3]

The Topas, a set of half-castes of Pondicherry between Hindoo women and French or Portuguese men, are, according to an anonymous author, debased, ignorant, superstitious, lazy, and debauched. Négrin confirms this; and Dr. Collas asserts that they are far more subject to phthisis than either parent.[4] According to the *Revue Coloniale*, says M. Broca, the Topas of Pondicherry show a mortality even greater than that of Europeans, though these are considerably shorter lived in India than in Europe.[5]

The Dutch-Singhalese half-castes are no more favoured by Nature than others of their kind. Though some of the women are delicately pretty, the men are so slight and so badly made, that they are at once distinguishable from the parent races. They are lazy, effeminate, without a virtue, but with all the vices of both parents.[6]

---

[1] Waitz, *Anthropologie*, vol. i. p. 201.
[2] I give this on the authority of several gentlemen who are well acquainted with India.
[3] The *Pall Mall Gazette*, Sept. 1st, 1874.
[4] Perier, *Mém. de la Soc. d'Anthrop. de Paris*, 1865, vol. ii. pp. 294, 296.
[5] Broca, *Hybridity*, etc., Blake's Trans., p. 39.
[6] Perier, *ut sup.*, p. 295.

The Dutch-Malay half-breeds of Java, who go by the name of Lipplapens, are stupid and nerveless, says Görtz. M. Boudin says they can be employed in nothing by the Dutch on account of their stupidity. As to their fertility, Dr. Yvan says of them, that they may produce among themselves up to the third generation without any necessity of crossing with one of the parent races, but they then produce only daughters, who are generally sterile.[1] The Portuguese-Malays, according to Dr. Yvan, are also indolent, small, meagre, degraded, without either any of the energy of their savage parent, or anything in common with their conquering fathers.[2] It appears that the Portuguese can bear the climate far better than the Dutch; for the latter are generally sterile even among themselves at Batavia, since Steen Bille asserts that their children are weak and poorly, and frequently sterile in the second generation, while the Portuguese are everywhere fertile, and Dr. Yvan believes that in other Dutch colonies the half-breeds are continuously fertile. It is suggested therefore by M. Broca, that the sterility of the Lipplapens may be due to the effect of the climate. Yet he himself considers this to be but a partial cause, and attributes their sterility to the fact that they are proportionally numerous to the parent races, since the Dutch are numerous compared to the natives, and hence a large colony of Lipplapens is produced, who marry among themselves instead of with one of the

---

[1] Waitz, *Anthropologie*, vol. i. pp. 201, 207; Perier, *Mém. de la Soc. d'Anthrop. de Paris*, 1865, vol. ii. p. 304; Broca, *Hybridity*, etc., Blake's Trans., pp. 39-41.

[2] Perier, *ut sup.*, pp. 300-301.

parent races.[1] The Spanish-Malay half-breeds, says Souven, are tolerably prudent and industrious. On Mindanao, where the half-breeds are in the majority, they are very proud of their Spanish blood, and speak Castilian with greater purity than it is spoken in many parts of Spain.[2] In Amboyna the mixed breeds between Portuguese and natives are far darker than are even the natives.[3] Everywhere in the South Seas crossing has debased the native races, fully as much perhaps from their fellowship with the scum of Europe as from the hybridity. The fact remains, however, that as soon as the Ladrone Isles fell under the yoke of foreigners, death, pestilence, and depopulation began. At present these unfortunate islands are inhabited by a mixed race of Spaniards, Tagals, Caroliners, Polynesians, and Chinese—a good-natured and soft-hearted race, but exceedingly ugly, indolent, and listless; while the original pure race was handsome, well formed, and stronger than most Europeans.[4] Neither do the unions of Europeans and Polynesians seem to be particularly fertile. Jacquinot saw only two half-breeds, Anglo-Tahitians, a perfect welding of both races. Alfonsi saw not even one half-breed at Tahiti, during a residence of eighteen months on the island, and this, notwithstanding that there was a French garrison of about 400 men stationed there, besides about 200 other Europeans and Anglo-Americans. During a resi-

---

[1] Waitz, *Anthropologie*, vol. i. p. 207; Broca, *Hybridity*, Blake's Trans., pp. 40-43; Perier, *Mém. de la Soc. d'Anthrop. de Paris*, vol. iii. 1870, p. 230.

[2] Perier, *ut sup.*, vol. ii. 1865, p. 303.

[3] Waitz and Gerland, *Anthropologie*, vol. v. part i. p. 107.

[4] *Ibid.*, part ii. pp. 47, 161, 162.

dence of four years on Nukuhiva, the same gentleman saw only one half-breed, while there were 100 French soldiers in the place. In the years 1862–1863, there were twenty-six marriages between Frenchmen and native women on Tahiti and Morea, which produced "at least double the number of children," or about two per marriage. Davis, however, states that Europeans and Sandwich islanders are very fertile together, and that it often happens that a woman who is barren with her native husband becomes a fertile wife to a European! On the Island of Owahu, to which all his personal observations relate, he says the European-native half-breeds form a large class, and are very handsome, but nearly barren among themselves.[1] In New Caledonia, de Rochas saw but two half-breeds. In Fiji, these mixed marriages seem proportionally far more fertile. In 1851, Binner found on Ovalau a good school with about eighty half-caste children; and after he had been on the island some time he got double the number to attend. There is a story too of an Irish convict who escaped from New South Wales to Mbau or Rewa, where he became the father of about fifty children. Du Bouzet talks too of the numerous progeny of a few English and American sailors at Lefuka; but Pritchard says that when these half-breeds marry among themselves, they are far more sterile than when they intermarry with one of the parent races.[2]

Cruise, after a long residence in New Zealand, says he saw but two European-Maori half-breeds, and

[1] Perier, *Mém. de la Soc. d'Anthrop. de Paris*, vol. ii. 1865, p. 307; vol. iii. 1870, pp. 234, 235, 238.
[2] Perier, *Ibid.*, vol. iii. 1870, pp. 237-238.

thinks that they must practise abortion, which is very probable, since M. Perier says that there are now 500 there. According to Polack, these half-breeds are a fine, healthy, and strong race; but Savage says they are not either mentally or bodily superior to either parent.[1] Gliddon says that before the extermination of the Tasmanians, only two European-Tasmanian half-breeds were known to have been born. Quoy and Gaymard saw only one; Jacquinot saw none; but according to the account already given of a colony in Bass' Straits,[2] these unions must be occasionally fertile. Jeffreys, indeed, saw one half-breed, a girl, who, he says, had a very agreeable figure; and asserts that the half-breed infants were burnt by the Tasmanians, and their mothers maltreated—which would certainly account for the death of them among the natives, but not for their absence when born under the protection of resident Europeans.[3] In Australia, as in Tasmania, the absence of half-breeds is so remarkable that Waitz is driven to account for it by an accusation of infanticide. They are born, according to Macgillivray, but do not seem to live—at least not at Port Essington—and in New South Wales every half-breed *girl* is said to be killed.[4] But, at all events, the chief Bongarri kept his *son* alive, a half-breed, the offspring of the chief's wife and a convict.[5] If these unions were even moderately fertile there

---

[1] Perier, *Mém. de la Soc. d'Anthrop. de Paris*, vol. iii. 1870, pp. 236, 243; Waitz, *Anthropologie*, vol. i. p. 201.

[2] See p. 192 of this work.

[3] Perier, *ut sup.*, pp. 241-243; Broca, *Hybridity*, etc., Blake's Trans. p. 47.

[4] Waitz and Gerland, *Anthropologie*, vol. i. p. 203, vol. vi. p. 779.

[5] Broca, *ut sup.*, p. 52.

should be a number of half-castes, even if the natives habitually killed them when they had the chance; since there must be permanent unions of white settlers and native women in a country where the male population has always been in excess.[1] Cunningham and Lesson speak only of one, and that is the same one, the son of Bongarri's wife. Mone says that in 1842 only three half-breeds were known in all Australia. Indeed the very word *cross-breed* was used in Australia in a totally different signification than what it is elsewhere.[2]

The Portuguese-Chinese half-breeds at Macao are given up to Asiatic vice, says Castano. The men are robust and big, with regular features and animated eyes. The women are yellow-skinned, with the nose flat, the mouth enormous, and a massive frame.[3]

The half-breeds between the Dutch and the Hottentots are distinguished, says Moodie, for uniting in their persons the vices of both people. They are superior to the Hottentots in understanding, and perhaps under other circumstances many of them would also be superior to the Dutch. As it is, however, they are a drunken and depraved lot, unwilling to work or lead a regular life. They consider themselves to be superior to the Hottentots with whom they live, and hate the white population, from whose society they are altogether shut out. Physically they are taller and stouter than the Hottentots, well made, with a

[1] Broca, *Hybridity*, etc., pp. 47-52, 58. I can confirm both the general argument here, and the sterility of mixed unions in Australia from conversation with old Australians.

[2] Broca, *ut sup.*, pp. 47-48; Perier, *Mém. de la Soc. d'Anthrop. de Paris*, vol. iii. 1870, pp. 238, 239.

[3] Perier, *ut sup.*, vol. ii. 1865, pp. 301, 302.

skin of a disagreeable sallowness, their hair long and not so woolly as that of the Hottentot, and they share in some degree the tendency of the Dutch to fat.[1] Le Vaillant and Barrow both say that Hottentot women are more fertile with whites than with Hottentots,[2] but it is possible that the one has merely copied from the other. In the year 1801 a horde of these half-breeds, or Bastaards, as they were called, established themselves near the Orange or Gariep River, together with a lot of Bosjesmen, Namaquas, Kaffirs, and pure Hottentots, under the leadership of a chief named Kok. In the same year a missionary began to organize them into a settled community in the village of Klaarwater. In 1805, when visited by Lichtenstein, there were about thirty families, one-half of which belonged to the Bastaard race, the rest were Namaquas, or Hottentots, and the village began to grow apace by the arrival of refugees, by intermarriages with Bosjesmen, and with Koranas of the neighbourhood. The missionaries seemingly had not much influence over the colony, for the people practised polygamy, they lived by pillage and hunting, smeared their bodies with paint, covered their hair with grease, and were ignorant and uncivilized generally. About the year 1810, however, the missionaries made another earnest endeavour to civilize them; they tried to make them take to agriculture, changed their old name of Bastaards to that of Griquas, an old Hottentot name, while Klaarwater was re-named Griqua-town. Griqua-town was now joined by a

[1] Waitz, *Anthropologie*, vol. ii. p. 305; Walker, *On Intermarriage*, p. 362; Perier, *Mém. de la Soc. d'Anthrop. de Paris*, vol. ii. 1865, pp. 320, 321.

[2] Perier, *ut sup.*, p. 321; Reich, *Ehe*, etc., p. 326.

large body of Koranas, of which there were 1,341 in the year 1813, on a total population of 2,607, or more than half were Koranas. In the year 1814, owing to an attempt of the Governor of Cape Town to make them join the local army, Griqua-town was nearly revolutionized; a part of its people went off to the mountains of the neighbourhood, turned robbers, and, under the name of Bergmaars, pillaged the country round about. Joined with bands of Koranas, they robbed and murdered the Betchouanas and Bosjesmen, carried off their women and children, and thus further mixed their race. At last, in 1825, the Bergmaars were persuaded to return to Griqua-town and live peacefully; but in the meanwhile, those who remained at Griqua-town quarrelled among themselves because the Governor of the Cape had sent a political agent to Waterboer, a Bosjesman by origin, whereas Kok had always been in possession of supreme authority. Kok, therefore, with his party went off and formed another village; a chief named Berend formed a third; and Waterboer remained head of Griqua-town.[1] It follows, therefore, from this history that the Griquas are by no means an instance of any continuous fertility among half-breeds. They have mixed with Hottentots, Bosjesmen, Kaffirs, Koranas, Namaquas, and Betchouanas, with probably additions of real Bastaards or Dutch-Hottentot half-breeds, from time to time. They have consequently become almost a pure African race, and though they have increased very fast, this increase is due in a great measure to immigrants.

[1] Broca, *Hybridity*, etc., Blake's Trans., pp. 3, 4, note; compare also Godron, *De l'Espèce*, etc., vol. ii. pp. 354, 355; Waitz, *Anthropologie*, vol. i. p. 206; Perier, *Mém. de la Soc. d'Anthrop. de Paris*, vol. ii. 1865, pp. 320-324.

The European-Negro half-breeds, known as Mulattoes, are said to be preferred for house work in America on account, as Sir Charles Lyell suggests, of their better education, since they have associated so much during childhood with the whites. In Guadelope nearly all the manufactures are in the hands of Mulattoes, and some at least, says Cassagnac, are rich and industrious. In Peru many study theology, and most of the medical practitioners in Lima are Mulattoes. This, however, is saying but little for their capabilities, since both professions there are compatible with great ignorance. A. de St. Hilaire considers the Brazilian Mulattoes to be even superior, if anything, to the Creoles or native whites, except that they have not one spark of chivalry, and are as changeable of mood as their African parents. Nott asserts that Mulattoes cannot endure physical exertion as Europeans do; and that the women especially are delicate, have many chronic diseases and miscarriages, and only produce weakly children. They are, he says, of lower viability than any other race of human beings in South Carolina; but at Mobile, New Orleans, and Pensacola (Gulf of Mexico), such was not so evidently the case, since their European parents were here generally of the Latin race, and not of the Anglo-Saxon. He further made the curious observation that the nerves of their limbs were a good third less in thickness than in any pure race, light or dark. Labat, who travelled over the Antilles more than a century and a half ago, says that the hair of the Mulattoes is much less woolly than that of the Negroes, and of a light chestnut colour, and he saw one at Cadix with red hair; they are generally well made, vigorous, strong, skilful, and industrious; courageous, and impudent beyond imagination, are proud, false, devoted to pleasure and to thievery, and are capable of the worst crimes. The Spaniards themselves, he says, are not better soldiers or worse men. Demersay says that in Paraguay the Negroes are preferred to Mulattoes, who are proud and treacherous. The Spaniards have an old proverb, "No se fie de mula y mulata." Freycinet, too, says that in the Isle of France (île Maurice), they are proud and lazy, devoted to gambling and debauchery, soon squander any money their wives have brought as a dowry, make them jealous, and take no thought for the morrow. According to Dr. Hancock, the Mulattoes of South America are perhaps not naturally inferior to their fathers, either mentally or physically, but certainly far superior to the primitive African race.[1] From all which it appears that the Mulattoes are far better morally in South America than elsewhere, and here they have also the best position. Mulattoes of the same degree are seldom mutually fertile, according to Van Amringe, Knox, Hamilton Smith, Day, Nott, Etwick, and Long. Lewis, however, asserts that though Mulattoes are weak and delicate, and their children are of low viability, yet they are as fertile together as Europeans and Negroes, which is not saying very much, for, according to Hombron, Negroes are more fertile with Indians than with Europeans, and Indians are more fertile with Europeans than with Negroes. Seemann also says that

---

[1] Waitz, *Anthropologie*, vol. i. pp. 198, 199; Perier, *Mém. de la Soc. d'Anthrop. de Paris*, vol. ii. 1865, pp. 357-359; vol. iii. 1870, pp. 277, 279; Walker, *On Intermarriage*, p. 363; Broca, *Hybridity*, etc., Blake's Trans., pp. 33, 34.

the half-breeds are fertile among themselves at Panama, though the children are weakly; and Bachman gives one instance of a Mulattoe family at New York who have never crossed with either parent race, and have perpetuated themselves for five generations.[1] But the balance of evidence seems to show that Mulattoes— the result of a first cross between a white and a negro—though not absolutely sterile together, are yet very nearly so; and even where an union of this sort does turn out fertile, there seems to be an irresistible tendency towards atavism, the child will resemble more nearly one of the pure races from which it is descended than its parents. Thus Burdach says that Mulattoes return to one of the parent stocks about the third generation; Knox says every Mulattoe returns to one of the races from which he has sprung; Livingstone said the same of the Portuguese Mulattoes of Angola; and M. Perier gives many other citations of authors who are not quite so explicit, but seem to say the same thing.[2] These observations, moreover, are very difficult to collect, since Mulattoes only seldom intermarry among their own class, but form by their complicated intermarriages terceroons, quadroons, quinteroons, griffes, negritoes, etc., and thus can very well continue their race aided by constant additions of Mulattoes.

The Mestisos, or European-American-Indian half-breeds, and the Cholos, are in Peru short, sallow, coarse-haired, with a low forehead, lazy, treacherous, and careless; in short, more like their Indian than

---

[1] Waitz, *Anthropologie*, vol. i. pp. 205-207.
[2] Perier, *Mém. de la Soc. d'Anthrop. de Paris*, vol. iii. 1870, pp. 270-273.

their European parents according to Raynal, Pöppig, Stevenson, and Botmiliau. In Chili they are often bigger than the Indians, but in their small feet and hands, in their hair, and flattish large-nostrilled nose, says Pöppig, they more nearly resemble their Indian parents. The Spanish-Guarani half-breeds are perhaps the best specimens of the Mestiso. Azara considers them somewhat superior even to the Spaniards of Europe in height, build, and whiteness of skin. D'Orbigny also considers that the European type prevails in these half-breeds, who " almost rival the whites " in looks, while they are their equals in general intellectual power. The Spanish-Paraguay half-breeds are, according to Brackenridge, also superior to the Spanish Creoles. Ulloa asserts that the Mestisos of Concepcion are scarcely distinguishable from the Spaniards. Some of the Mestisos of La Plata, says De Moussy, are magnificent men, but so, according to D'Orbigny, are their parents, yet only some of these half-breeds are " magnificent." According to Spix and Martius, Tschudi, and Humboldt, the Mestisos are in character superior to the Mulattoes, less changeable, and more industrious, both in Brazil and Peru. Koster says they have more self-respect, and are braver than the Mulattoes, but they are nevertheless also weak, indolent, soft, and pitiful beings. Both the Mestisos and Mulattoes, says Pöppig, in the stock but probably truthful phrase, inherit the vices of both parents with the virtues of neither. With the pride of the whites exaggerated to a mania, he combines the laziness and apathy, the changeability and improvidence of the Indian. He is as licentious as the Mulatto, and as tyrannical to the Indians as the Mulatto is to the

Negro ; but in mechanical things he possesses a great talent for imitation. Raynal says that in Mexico they are very intelligent, witty, and imaginative, but without the bravery of the natives, or their skill, or their love of liberty. They have no aptitude for work, nor application to accomplish any great work, neither have they the patriotism of the Spaniards. Gambling and revolution are their chief delights.

They seem, however, to be tolerably fertile. Jourdanet estimates the half-breeds at one-third of the population, while the whites scarcely form a tenth in Mexico. In Nicaragua, says Waitz, there are 145,000 Mestisos, 80,000 Indians, 10,000 Whites, and 15,000 Negroes. In Paraguay the Mestisos marry mostly among themselves, and their offspring form the greater part of the so-called Spanish population. We have seen that Hombron considers marriages between Europeans and Indians as more fertile than those between either and negroes.

The European-North-American-Indian half-breeds are not at all numerous, for the various nationalities in North America do not mix so much as in the South. Kohl believes the unions between French settlers and Indians to be very fertile, and their daughters especially are more like their French parents. If these half-breeds, however, marry among themselves they usually have only daughters, and their children are often stunted and of low viability. The Scotch-Indian half-breeds, he continues, form a powerful race. Barnard Davis also says that at the Red River settlement the half-breeds are handsome and fertile. Dr. Landry, of Quebec, however, says the half-breeds of that neighbourhood are peculiarly subject to phthisis, and the greater number die early.[1]

The Paulistas, like the Griquas, are a half-breed race, who have bred so much with the natives of the country that they have in reality become a pure race. They are thought to have been originally half-breeds between the Portuguese and people of Guiana, and are said to be brave, fierce, strong, and cruel. Muratori says that from the unions between the Portuguese and Indians were born children having all the faults of both parents and the virtues of neither; and, though of Portuguese origin, they were considered to be unworthy to bear that name, which they dishonoured by their infamous conduct. He adds that their country was the asylum of all Portuguese, Spanish, English, Dutch, and Italian criminals; that they had depopulated the country round about them, and that during the space of 130 years they had carried more than two millions of Indians into slavery. Froger, who wrote before Muratori, in the year 1695, says they took their origin from criminals of all nations. They go, he says, in bands of forty or fifty together, travel over all Brazil, and come back after four or five months with sometimes over three hundred slaves. Their courage certainly is indisputable, and their character no less so. The Jesuits of Paraguay represent them as a robber race of the worst type. Lacordaire says that for a century and a half the Paulistas were on land what the Buccaneers were on the sea. They continually attacked the Jesuit missions, robbed their churches, and carried off all their Indians.

[1] Waitz, *Anthropologie*, vol. i. pp. 193, 260, 202, 207, 209; vol. iv. pp. 195, 196; Perier, *Mém. de la Soc d'Anthrop. de Paris*, vol. ii. 1865, pp. 342-346, 369, 370.

Dom Vaissette, Charlevoix, De Surgy, Raynal, D'Orbigny, Page, and De Moussy, all agree in these accusations; but Gaspar, writing in 1797, says that the elevation of their character, the delicacy of their sentiments, their probity, their industry, their public spirit, cannot be a heritage from a gang of ruffians and vagabonds; and indeed it appears they have now become better. The population of St. Paul is estimated at 30,000, including individuals of all colours; some of them who have kept themselves from intermarriage with the Indians, says D'Orbigny, are whiter even than the Creoles; others are of all shades from light yellow to coffee colour. They still show traces of their Indian origin in their high cheekbones, small black eyes, and hesitating look.[1]

The Gauchos are another half-breed race of South America, of the same description as the Paulistas, and thought to be descendants of Spaniards and Indians. Azara describes them as robust and healthy, hospitable, but dreadfully addicted to gambling, and utterly careless of human life.[2]

The Zamboes, or offspring of Negroes and American-Indians, are, according to Dr. Hancock, remarkable for their physical superiority over their progenitors on either side, and this, he says, "is a well-known fact."[3] Lavayssé also describes him as superior in mental power to either parent and to the Mestiso; yet, he says, that in Caracas they are considered the worst class of the population; and in spite of A. de St. Hilaire's praise, especially of those of the Parana-

---

[1] Perier, *Mém. de la Soc. d'Anthrop. de Paris*, vol. ii. 1865, pp. 346-351.
[2] See also Page and Demersay. Perier, *ut sup.*, pp. 351-355.
[3] Walker, *On Intermarriage*, p. 363.

hyba district, Waitz points out that nearly everywhere they have the reputation of being very bad characters. The Zamboes of St. Vincent are accused of having on one occasion murdered all the colonists on the isle, and this without provocation. In Lima, says Tschudi, four-fifths of the worst crimes are committed by Zamboes.[1] St. Venant describes them as a debased race, whose chronic state of misery, laziness, and of apathy, among all the means to wealth it is impossible to paint. In Peru, says Lacroix, they form the worst part of the population; they are cruel, vindictive, implacable, stupid, and quarrelsome. In stature they are rather tall, and, like the Indians, have scarcely any beard; their hair is distinctive—indeed, it has been described to me in conversation as that of the Indian and Negro commingled, but each hair like that of one of the parent races. Radiguet describes the Zambo as often vigorous and tall; his woolly hair comes down over his low forehead, his eyes are bright and intelligent, his thick lips (always half open) disclose a set of brilliantly white teeth, his face is not pleasing, though expressive and animated, it is often hard and scoffing. In Columbia, Famin says the Zamboes are fierce, thievish, and little susceptible to civilization; they are robust, and their skin is a dark copper colour. Squier observed that most of the criminals of Nicaragua were Zamboes, and he describes them as bigger and better made than their parents, without possessing any of their good qualities.[2]

The Cafusos, a half-breed race like the Paulistas,

---

[1] Waitz, *Anthropologie*, vol. i. pp. 200, 201.
[2] Perier, *Mém. de la Soc. d'Anthrop. de Paris*, vol. ii. 1865, pp. 363-365.

but originating from Negroes and Brazilian Indians, according to Spix and Martius, live isolated in the forests of Tarama. They are slender, muscular, weak legged, and of a copper colour; their eyes are slightly oblique, and more frank looking than those of the Indians; their hair not so woolly as that of the Negro, and less lank than that of the Indian, stands up like that of the Mop-headed Papuans of New Guinea. D'Orbigny says that the Cafusos do not live perfectly isolated, and Broca points out that though the Cafusos are said to marry among themselves, we know very little about them, or how much they may have intermarried with the parent stock.[1]

The Negro-Caraïb half-breeds were formed by the intermarriages of runaway Negroes and Caraïbs, after they were settled together on the islands of Dominica and St. Vincent in 1660. After many struggles the half-breeds obtained the mastery over the true Caraïbs, who were most of them compelled to fly, so that in the year 1763 there were 3,000 black Caraïbs and only 100 red or true Caraïbs. These half-breeds also found time to massacre all the white colonists, and were all transported by the English to the island of Roattan, whence, with the help of the Spaniards, they escaped to Honduras. Here, says Waitz, they always intermarried among themselves, and are very able and active labourers.[2]

In the Cherokee Republic, it is noteworthy that the half-breeds between the North American Indians and

---

[1] Perier, *Mém. de la Soc. d'Anthrop. de Paris*, vol. ii. 1865, p. 366; Broca, *Hybridity*, etc., Blake's Trans., p. 3.

[2] Waitz, *Anthropologie*, vol. iii. pp. 353-354.

the Negroes, do not possess the franchise.[1] Of all half-breeds in Peru, says Stevenson, those between Negroes and Mestisos, or Negroes and Mulattoes, are by far the most abandoned. Yet Brackenridge asserts that the half-breeds in Peru between the Indians and Mestisos are superior in everything to either parent.[2] Lacroix endorses the opinion of Stevenson on the Negro and Mulatto half-breeds, and adds that these are more robust than the Mulatto.[3]

The Negro-Hottentot half-breeds, according to Le Vaillant, have an agreeable figure, and their skin is of a much less disagreeable colour than that of the Bastaards. They are much sought after as servants, for, with great activity and docility, they combine a fidelity which is not to be found in any other half-breed.[4]

The Darfur women and Arabs produce offspring which are feeble and of low viability, according to Mohammed ibn-Omar el-Tounsy; while in the same place these pure races are among themselves perfectly fertile.[5] The people of Fezzan are a mixed race of Negroes and Arabs, tolerably well formed, but ugly, with small eyes and protuberant lips. They are selfish, insincere, inhospitable, and wholly destitute of either physical or mental energy or enterprise.[6]

The Maroons of Surinam are also a hybrid race between Negro runaways from the Dutch estates and native tribes. They are a vigorous and athletic

[1] Waitz, *Anthropologie*, vol. iii. p. 295.
[2] *Ibid.*, vol. i. pp. 201, 202.
[3] Perier, *Mém. de la Soc. d'Anthrop. de Paris*, vol. ii. 1865, p. 364.
[4] Waitz, *ut sup.*, vol. ii. p. 305; Perier, *Mém. de la Soc. d'Anthrop. de Paris*, vol. ii. 1865, p. 322.
[5] Waitz, *ut sup.*, vol. i. p. 203; Perier, *ut sup.*, vol. iii. 1870, p. 281.
[6] M'Culloch, *Geographical Dict.*

set of men, active and enterprising, and superior to either parent stock.[1]

The mixed breed between Persians and Arabs, inhabiting the Bahrein, or Aval Islands, possess "more of the indolence and cunning of the former than the bold frankness of the latter."[2]

The Arab-Abyssinian half-breeds are said by Dr. Pruner to be a fine race,[3] yet the Memlouks never succeeded in propagating their race in Egypt. This may have been due to the effect of the climate, for Lane says that European women in Egypt are often barren, and the children of foreigners from comparatively cool countries seldom live, even though the mother be a native.[4]

The Kuruglis, or half-breeds between Turks and Moors, are superior even to the Turks, says Wagner, in strength and beauty.[5] Indeed, the Turks themselves and the Persians are often cited as examples of the advantage of crosses.[6] It is necessary, however, first to show that the assumption that the Turks were originally an ugly Mongol race is a true one, and De Gobineau points out that the only evidence in favour of this theory is the affinity of language, which, though valuable enough when corroborated, is not of much value alone. But the traditions point to an origin from Central Asia; they always speak of the Turks

[1] Walker, *On Intermarriage*, pp. 363, 364.
[2] M'Culloch, *Geographical Dict.*
[3] Waitz, *Anthropologie*, vol. i. p. 202.
[4] Lane, *Modern Egyptians*, vol. i. p. 199; Perier, *Mém. de la Soc. d'Anthrop. de Paris*, vol. iii. 1870, p. 229.
[5] Waitz, *Anthropologie*, vol. i. p. 202.
[6] Walker, *On Intermarriage*, p. 240; also Bory de St. Vincent, Morel, Chardin, and Bomare, cited by Perier, *Mém. de la Soc. d'Anthrop. de Paris*, vol. i. 1860, pp. 73, 74.

as a handsome race, and, indeed, the people of Central Asia, and between Central Asia and Europe are nearly all handsome races.[1] Even supposing it to be an unquestionable fact that the Turks were originally an ugly Mongol race, the supposition that their beauty is the result of crosses with Greek, Georgian, and Circassian women, is clearly untenable, since only the upper classes can afford these wives, and the small fecundity of the polygamous classes is too notorious[2] to allow of the belief that these few rich people could have influenced the whole race.[3] Even Waitz abandoned the theory that their beauty was due to crosses, and attributes it to the influence of civilization.[4]

There are yet a few more half-breeds, an account of whom I have met with. The Chinese-Cambojias are said by Gutzlaff to be productive at the first generation when they intermarry among themselves, but grow gradually more and more sterile, until at the fifth generation they are perfectly barren.[5] The Chinese-Tagal half-breeds are the sole capitalists of the Philippines; they are economical, prudent, patient, and manage to do a good business where the Spaniards ruin themselves.[6] The same, however, might be said of their Chinese parents.

---

[1] De Gobineau, *De l'Inégalité des Races Humaines*, vol. i. pp. 216, et seq.

[2] See *The Trans. of the Brit. Association*, 1872, *Stat. Section*, a paper read by Mr. H. Clarke.

[3] Perier, *Mém. de la Soc. d'Anthrop. de Paris*, vol. i. 1860, pp. 73, 74, who cites Pritchard, De Salles, Grellois, and Pouchet.

[4] Waitz, *Anthropologie*, vol. i. pp. 83, 84.

[5] Broca, *Hybridity*, etc., Blake's Trans., p. 43, note 1.

[6] Perier, *ut sup*, vol. ii. 1865, p. 305.

The European-Greenland Eskimo half-breeds resemble more nearly the European parent in feature, and the Eskimo in moral qualities.[1] At present, says Waitz, 14 per cent. of the population are half-breeds.[2]

These observations are sufficient to show that crosses are not so beneficial as they are generally supposed to be. In them we see none of that wonderful beauty, fertility, and moral and physical strength, surpassing the races on both sides from which they are derived, nor do we see the mixed races elbowing away the pure in the great struggle for existence. We see on the other hand the persistence of the great law of inheritance, as exemplified in that atavism which leads back a hybrid race to the parent stock from which it is sprung. We see how simple is Nature's method of reproduction in the fact that differing races are not absolutely barren together, though the one race has not been accustomed to produce offspring of another kind. We see further, that if Nature is led and not dragged she will sometimes allow of the production of new types of men. New races may be formed by the more gentle fusion of two or more others, by the intermarriage of the half-breeds sometimes with the one parent stock, and sometimes with the other, or by a cross with another race not extremely different, and subsequent breeding in-and-in. Though the evidence detailed above leads us to believe that the generality of half-breeds are worse than their parents and incapable of continuous

---

[1] Perier, *Mém de la Soc. d'Anthrop. de Paris*, vol. ii. 1865, p. 370.

[2] Waitz, *Anthropologie*, vol. iii. p. 301.

reproduction among themselves, some individuals may even be better than their parents, and even have that power of indefinite procreation among themselves. The instance given by Burdach of a Mulatto family of New York, which was still fertile after five generations of marriages with mulattoes, the family of Souza, and the half-breeds of Norfolk Isle,[1] are sufficient to show that indefinite procreation of half-breeds is not impossible. Yet that evidence emphatically shows that a cross in itself is not a good thing. As in other realms of the organic world, the body of the half-breed is least often injuriously affected, though it may be weedy and inharmonious; but the brain and reproductive system are too often seriously hurt. Waitz attempts to excuse these unfortunate beings on the score of their exceptional position, rejected by the superior race and themselves rejecting the inferior;[2] yet we do not find them any better where they are actually the most numerous class, as in South America, and might therefore be reasonably expected to find sufficient consolation in their own society. Indeed, half-breed races dwelling apart from the superior race, such as the Paulistas and Griquas, have been notorious for their depravity as long as they could reasonably be held to be half-breed races. Mr. Darwin attributes this ferocity and depravity of character to atavism, which causes a relapse to the type of their savage ancestors[3]—a very probable cause, but hardly sufficient, it appears to me, to account

---

[1] See pp. 320, 161, 159, of this work.
[2] Waitz, *Anthropologie*, vol. ii. p. 305; and Viroy, cited by Perier, *Mém. de la Soc. d'Anthrop. de Paris*, vol. ii. 1865, p. 359.
[3] Darwin, *The Variation of Animals and Plants*, etc., vol. ii. p. 46.

for it alone, without the supposition that the mind is as much affected by the mixture of the characters of the two races as the hair of the Mestiso, a mixture without a blending. Attempts have also been made to account for their sterility by a supposition that this is due to the injurious effects of the climate;[1] but in most places both pure breeds are perfectly fertile among themselves, as the English and East Indians, or the Arabs and Dafurians, while the crosses are as injurious as ever. There is an absence of homogeneity, an inharmoniousness, an unfittingness the more obvious as the races are more distinct, just as the welding of two different metals is more difficult than the welding of two pieces of the same. Thus the Saxon and Latin races of Europe differ so little that the produce of their unions is as good, perhaps, as the pure race; the European races are more similar to the American Indians than to the Negro, hence we find that the Mestiso is as a rule superior to the Mulatto, who is, perhaps, the worst of all. In short, the greater the difference between the two parent races, the worse will the produce be; and we must fully agree with the apothegm of an observant Portuguese gentleman quoted by Livingstone, that " God made white men, and God made black men,— but the devil made half-castes."[2]

[1] Perier, *Mém. de la Soc. d'Anthrop. de Paris*, vol. iii. 1870, pp. 225, 227-230.

[2] Livingstone, *Zambesi*, p. 50.

CHAPTER VIII.

Why are there Two Sexes?

THE most powerful deductive argument of those who affirm the harmfulness of the marriage of near kin is, that if these marriages are not harmful, and if crosses are unnecessary,

> "Oh, why did God,
> Creator wise, that peopled highest heaven
> With spirits masculine, create at last
> This novelty on earth, this fair defect
> Of Nature, and not fill the world at once
> With men as angels without feminine;
> Or find some other way to generate
> Mankind?"

Indeed, if the ultimate object of the two sexes was not to ensure crosses, what was the ultimate object?

The two greatest living thinkers on this very obscure biological question are Mr. Darwin and Mr. Herbert Spencer. Their views are discordant with each other on the precise nature of reproduction, but they both agree that the whole arrangement of dual sex is chiefly in order that crosses may occur. My own knowledge on this subject is, of course, immeasurably inferior to that of these gentlemen, and it is therefore with extreme diffidence that I venture to examine their theories, to the end that we may see whether there may not be some other explanation of this arrangement of Nature sufficient to explain it without resorting to the hypothesis that crosses are necessary; an examination to which we are encouraged by the fact that they also found part of their argument on those observations which have already been discussed in the above pages.

The theory of Mr. Herbert Spencer is the most decidedly adverse to my argument of the two, for the whole essence of his theory turns on the supposition that the germ and sperm cells have arrived at a state of molecular stability which is upset by their union, because they are different from each other, and would not be upset if they were not different; just as the mixture of bismuth, tin, and lead, to use his own illustration, forms a compound of looser molecular stability than the mixture bismuth with bismuth, or tin with tin.[1] He argues that the reproductive cells must have arrived at this state of stability, because the rest of the organism in adult age has arrived at this state; but his reasoning on this point cannot be summarized, and I must refer the reader who, perchance, has not read his very deep and suggestive works to the *First Principles* and the *Principles of Biology*. Hence, if his theory is true, it follows that the more alike two individuals are, the less likely are they to be fertile; and the more unlike two individuals are, the more likely are they to be fertile.

Mr. Darwin, on the other hand, in his celebrated

[1] Spencer, *Biology*, vol. i. pp. 274-281.

theory of *pangenesis*, considers that the higher organisms are merely aggregations of simple organisms or cells; and that each cell of the compound organism reproduces itself, just as they are known to reproduce themselves when simple and alone, by means of germs which they throw off, and which grow again into facsimiles of the parent cell, when placed under favourable circumstances. These germs he supposes to be constantly thrown off by every cell in the body throughout the life of each; they are collected in the spermatic fluid and ova of these compound organisms, and therefore a fertilized ovum contains the germs from the cells of two organisms, a double set of germs, some of which being more hardy or more favourably circumstanced as to nutrition, are developed, while others remain dormant.[1]

It will be observed that this theory differs essentially from Mr. Herbert Spencer's, inasmuch as it supposes every organism to be competent to reproduce itself without the aid of any other organism. It would therefore be favourable to my argument as it stands, did not Mr. Darwin further go on to say, "that it is a general law of nature that the individuals of the same species occasionally intercross, and that some great advantage is derived from this act."[2] "No two individuals," he further explains, "and still less no two varieties, are absolutely alike in constitution and structure; and when the germ of one is fertilized by the male element of another, we may believe that it is acted on in a somewhat similar manner as an individual when exposed to slightly

---

[1] Darwin, *The Variation*, etc., vol. ii. pp. 377-388.
[2] *Ibid.*, p. 91.

changed conditions."[1] We might allow the benefit of occasional change, but Mr. Darwin expressly excludes the only rational explanation of the way in which it benefits, as far as our understanding yet goes; for he says, talking of the advantage of change of soil to seeds: "Considering the small size of most seeds, it seems hardly credible that the advantage thus derived can be due to the seeds obtaining in one soil some chemical deficient in the other soil;" and he then partially adopts Mr. Herbert Spencer's view, that change tends to upset the equilibrium at which the seed may have arrived.[2] Combining this with the doctrine of pangenesis, it appears that the theory held by Mr. Darwin as to genesis, is that the union of the sperm and germ cells is not necessary to upset the equilibrium of both, since each can develop separately without this; but it is necessary to afford a change of circumstances to each cell which develops, a change which is not a chemical or physical change, but which acts in a way of which at present we can form no conception. Hence Mr. Darwin considers that the object of dual sex is to secure change, which is generally beneficial, just as change of air is generally beneficial; the only question we have to ask ourselves therefore, is whether this kind of change is such a vital necessity that it obliged the creation of two sexes instead of one, or whether there are not other reasons for the creation of two sexes which supply a more probable explanation. Mr. Herbert Spencer, as we have seen, gives an explanation which, if true, is amply sufficient to account for the sexes.

[1] Darwin, *The Variation*, etc., vol. ii. p. 145.
[2] *Ibid.*, pp. 147, 148.

He lays great stress on the fact, that even in the lowest organisms in which there is hardly any distinction of sex, a cell from one part impregnates a cell from another, and hence argues that this can only be accounted for by the theory that these two cells are somewhat differentiated, and that the one upsets the stability of the other. But is it necessary to believe that these cells are differentiated? Is it not more probable that they are identical in constitution since they are both derived from very undifferentiated parts of the organism, at least in the lower organisms? Mr. Herbert Spencer points this out himself.[1] He draws this conclusion from the fact that in some animals and plants a small piece of tissue taken from any part can produce a new organism; and, moreover, since both sperm and germ cells are like unspecialized cells in general, it follows that they are not very unlike each other. Thus, in the common polype, sperm and germ cells are produced in the same layer of tissue; and in *Tethya*, one of the sponges, they are mingled together in the general parenchyma; the pollen-grains and embryo-cells of plants rise from the same part, the cambium or medium layer, and therefore a part which is little specialized; and sometimes one kind of cell graduates into another.[2] It is strange, to say the least of it, that if the sexes are divided, in order that the sperm and germ cells should be sufficiently differentiated to react on each other, they should differ so little; and yet this difference remains little even in the more specialized higher organisms. Mr. Meehan obtained

---

[1] Spencer, *Biology*, vol. i. p. 220.
[2] *Ibid.*, pp. 220-222.

cuttings from *Cuphea leiantha*, a diœcious plant—some of the cuttings produced males, others female plants; unisexual trees will also in some seasons produce a sex different to that they usually do.[1] Even in the higher animals there is not an organ in the one sex, perhaps not a part that has not its strictly analogous part in the other sex; we even find that the higher unisexual organisms partially return sometimes to a lower stage, and have both sexes present in the same individual.[2]

[1] Hardwick's *Pop. Science Review*, vol. xi. 1872, p. 250.

[2] There appears to have been an ancient belief that man was formerly hermaphrodite. Thus the Jewish Talmudists taking the Hebrew noun in the Pentateuch in its individual and not collective sense, considered that our original parent was hermaphrodite. Plato also in his *Symposion*, introduces Aristophanes as holding the same opinion: "The ancient nature of man was not as it now is, but very different, for then he was androgynous both in form and name" (Simpson, in *Todd's Cyclopædia of Anatomy and Physiology*, Art. *Hermaphroditism*, vol. ii. p. 686, note). In one part of Australia the natives had a legend that Tarrotarro first separated the sexes (Waitz and Gerland, *Anthropologie*, etc., vol. vi. p. 800). Mirabeau, taking up the Talmudist line of argument, says, "'Dieu créa *l'homme* à son image, il *les* créa mâle & femelle.' Il est bien clair, il est bien ovident que Dieu a créé Adam androgyne; car au verset suivant, (verset 28), il dit à Adam: 'Croisset & multipliez-vouz; remplissez la terre.'" He ingeniously points out that this happened on the sixth day, Eve was not created till the seventh, and in the meanwhile, since we cannot look upon them as literal days, a great deal of business was got through. "Donc Adam ayant été créé hermaphrodite le sixieme jour, & la femme n'ayant été produite qu'*à la fin du septieme*, Adam a pu procréer en lui-même, & par lui-même tout le tems qu'il a plu a Dieu de placer entre ces deux époques" (Mirabeau, *Errotica Biblion.*, pp. 30-32). So Butler says (Butler, *Hudibras*, Part III. Canto i. ll. 761-770):—

> "Man was not man in Paradise,
> Until he was created twice,
> And had his better half, his bride,
> Carv'd from th' original, his side,
> T' amend his natural defects,
> And perfect his recruited sex;
> Enlarge his breed, at once, and lessen
> The pains and labour of increasing,
> By changing them for other cares,
> As by his dried-up paps appears."

I say this partial return to hermaphroditism shows the little real differentiation of the sexes one from another. Nor is it so very uncommon. Lateral hermaphroditism has been observed among the crustacea in the lobster; among fish, in the genera *Salmo*, *Gadus*, *Cyprinus*, and the *Merlangus vulgaris*, *Accipenser huso*, and *Esox lucius*; among birds, in the *Gallinaceæ*; and among quadrupeds, in the calf. Cases of this kind of hermaphroditism, that is, where there is a genuine ovary on one side and a testicle on the other, have also been observed in mankind.[1] Transverse hermaphroditism, with the external organs female and the internal male, has often been observed among our domestic quadrupeds, and is particularly frequent in black cattle. When of these latter twin calves are born one of which is male and the other apparently female, the female is generally in reality an hermaphrodite of this type. Hermaphroditism of this sort has also been observed in goats, deer, dogs, sheep, and in mankind; but the reverse case, of external male organs and internal female, has only been recorded in mankind. Cases of double hermaphroditism, in which the proper organs of one sex co-exist with one or more of the other, have been observed in cattle, goats, dogs, asses, sheep, and pigs.[2]

[1] Of recorded cases perhaps that of Marie Derrier, otherwise known as Charles Doerge, is the most remarkable. In this individual there was an uterus, vagina, *two* Fallopian tubes, and an ovary; a testicle, prostate gland, and penis. The right Fallopian tube had in place of the ovary at its end, the one testicle. It is alleged that during the twentieth year of his or her life there were three catamenial discharges; and during the last few years before death, he or she was subject to epistaxis and hæmorrhoids.

Simpson, in Todd's *Encyclopædia of Anat. and Phys.*, Art. *On Hermaphroditism*, vol. ii. pp. 696-698, 700-703, 705, 706.

Dr. Knox has propounded the theory that the earliest state of the fœtus is hermaphrodite; but Simpson does not accept this theory since, first, it is directly opposed to the views of Rathke, Meckel, Müller, Valentine, and other anatomists; and, secondly, such a theory does not explain the presence of multiple organs of the same sex,[1] and these double forms of the reproductive organs are far commoner than double forms of any other parts or organs; from which he concludes that hermaphroditism is not a return to a lower type of the organism, but due to local inheritance,[2] and Mr. Darwin points out that it is the same as the inheritance of spurs, combs, and other male adjuncts by females.[3] But if what physiologists say, that the male and female special organs are merely different developments of the same, be true,[4] then the occurrence of double organs of the same sex in one individual is just as remarkable as the occurrence of organs of both sexes in the same individual, and looks like extremely remote atavism modified by local inheritance. Indeed, Simpson himself says: "We should look upon this possibility with a less degree of scepticism when we consider that a double hermaphroditism exists as a normal sexual condition of some of the lower tribes of animated beings."[5]

The most powerful objection, however, to Mr.

[1] Cases have been recorded of three or five mammæ on the same individual, two penes, two clitores, double vesiculæ seminales on each side, double uterus with four Fallopian tubes, etc.

[2] Simpson, in Todd's *Encyclopædia of Anat. and Phys.*, Art. *On Hermaphroditism*, vol. ii. pp. 727, 728, 733.

[3] Darwin, *Descent of Man*, vol. i. p. 280.

[4] Carpenter's *Human Physiology*, pp. 890-891.

[5] Simpson, *ut sup.*, p. 733.

THE MARRIAGE OF NEAR KIN. 341

Herbert Spencer's theory of the object of dual sex, as he himself points out,[1] is that very high organisms are able to produce assexually, or by parthenogenesis;[2] and it is this also on which Mr. Darwin relies when he considers that the union of the sperm and germ cells gives a superfluity of reproductive cells.[3] It is far too general to be a monstrous method of reproduction, since it is the rule in all budding plants; it occurs in many insects, such as the honey-bee, *Psyche helix*, *Solenobia clathrella* and *lichenella*, the silkworm moth, and probably in many others of which only females are known; the *Nematoda*, and some *Rotifera*, produce by parthenogenesis; and many other organisms are at some period of their life entirely assexual, and reproduce themselves by means of buds. We even see it, or rather attempts at it, in some of the higher organisms, which seem to indicate a redundance of fertility; I allude more especially to the so-called ovarian dermoid cysts, which pathologists begin now to discriminate from the true dermoid cysts. They both contain chiefly hair, skin, teeth, and other products of the cutis, but the latter contain other things which do not appear in the former. The true dermoid cysts are besides always congenital, are nearly always situated on the upper eyelid and root of the nose, and are rarely larger than a hazel nut. Ovarian cysts, on the other hand, in all probability proceed from a Graafian vesicle, and are often found in the Fallopian tubes or uterus; they occur at the period generally in which

---

[1] Spencer, *Biology*, vol. i. p. 233.
[2] I use this word in the sense of its derivation.
[3] Darwin, *The Variation*, etc., vol. ii. pp. 385-386.

other ovarian cysts are commonest, namely, the age of fecundity, and though they may occur at any age, yet this is not extraordinary since ova are found in the infant at birth. Further, they are commoner than true dermoid cysts, and are far more commonly multiple. Out of 188 cases of dermoid cysts, Lebert found 129 in the ovaries. Lastly, they contain substances which are certainly not productions of the skin, since in the ovarian dermoid cysts are found besides hair, skin, epithelium, teeth, glands, etc., also bone, fragments of brain and nerve matter, and striated muscle. Muscular fibres were observed by Virchow "having the same form and general characters as those of the embryo." Brain matter has been described by Gray, who found a tumour consisting of five cysts, of which three contained fat and hair, one of the three contained also bone and a tooth; the fourth, about the size of a walnut, enclosed a brain-like mass in which the elements of the grey-substance and nerve fibres were clearly discernible in a sort of meshwork which resembled the pia mater; the fifth and smallest had similar contents. Rokitansky "found an independent nervous apparatus, arising from a ganglion, in a cylindrical osseous new formation, covered with true cutis, growing into an ovarian cyst. The mass was also vascular. The reddish ganglionic substance was enveloped in a capsule formed by two layers of the cell wall. A nervous cord issued from the ganglion, and sent ramifications into the osseous body, which were ultimately distributed in the same way as the nerve fibrils of the cutis." Chalice and Friedreichs also found nervous matter in these cysts. Teeth often grow in regular alveoli, the greater part of the teeth are rudimentary, but the milk teeth are often seen atrophied by the pressure of permanent teeth beneath. Bones, too, are often found with a periosteum. From these facts Dr. Ritchie concludes that every dermoid cyst of the ovary is an ovum which has undergone a certain amount of development; that it is a perverted attempt of parthenogenesis. And though Mr. Wells does not seem to hold this belief, he advances no valid argument against it;—for surely these cysts cannot be compared to other morbid growths of the body![1] Of course actual parthenogenesis after ages of sexual reproduction is a natural impossibility; but the fact that ova will develop sometimes, even so far as this, is very significant. Indeed, agamogenesis seems to be the normal manner of reproduction for the organic world, and only modified by sexual reproduction. For budding is as much a method of reproduction as seeding, and seeds, according to Mr. Herbert Spencer, are nothing but latent buds modified, when food is not abundant, to a form more likely to reach a suitable place for development, and which will not develop until it does arrive at such a place. Hence in some organisms reproduction by means of seeds is only occasional, but in the higher organisms it is constant, since in locomotive animals buds would be an intolerable burden, and therefore natural selection prevents the process of budding from going any further than the lower organisms.

[1] Ritchie, *Contributions to Assist the Study of Ovarian Physiology and Pathology*, pp. 169-175, 198, 199; and Wells, *Diseases of the Ovaries*, etc., pp. 65, 67-70.

But if we reject Mr. Herbert Spencer's theory accounting for dual sex, how shall we account for it? There is obviously a great advantage in this division of the sexes; what is this advantage? In the higher animals it is obvious enough: power of attack by the male while the female protects her offspring, a gain in locomotive power, and reduction of bulk, are in themselves sufficient to account for a change from unisex to dual sex; it is a division of labour whereby the functions are better performed than were the sex undivided. In the lower animals, and even in hermaphrodites it is also a division of labour; for do we not see that the higher an organism is developed, the more specialized are its various parts; that the simple functions of the lowest organisms are divided in the higher? It would be strange indeed if the reproductive organs alone remained unchanged and undivided. If there is any physiological gain in the differentiation of the various cells which make up the digestive canal of the higher organisms; or if there is any gain in the division of function between the skin, lungs, and kidney; surely there must be a gain in the division of the organs of reproduction. We see this more clearly if we compare the organism with a hive of bees. There we see one set of bees elaborate the wax and fetch the honey, another set builds the comb and feeds the young. The drones are analogous to the sperm cells of an hermaphrodite, the queen to the ova, the rest to the various accessory organs. Each chooses for itself the most suitable food, just as the cells of our organism select each a different food; and who can doubt that by thus working together they accomplish infinitely more, and do that infinitely better, than were each to work for himself alone? In hermaphrodites there is then a perfection and saving of labour; in separate sex, a greater perfection and greater saving of labour; and, doubtless, were the oxidizing and digestive functions not essential to the life of the individual and not in constant operation, they also would be divided as the function of reproduction.

Here we see the advantage in the ripening of one set of organs before another, possessed by some hermaphrodites, for the whole system of nutrition may thus be devoted to the development of one at a time. It is therefore a division of labour which is very apt from its nature to be hereditary; and in this way an hermaphrodite might become unisexual. Should a plant exhaust itself in the production of anthers, it will not develop the pistil at all. Or, the plant may from change of circumstances have its constitution so upset that it will be unable to produce more than one sex, or become irregular in the production of the sexes. We see this change particularly common when the plant is obliged to produce great quantities of pollen to neutralize the waste of its distribution by the wind. " The most successful cultivators in Ohio," says Mr. Darwin, " plant for every seven rows of ' pistillata,' or female plants, one row of hermaphrodites, which afford pollen for both kinds; but the hermaphrodites, owing to their expenditure in the production of pollen, bear less fruit than the female plants."[1] Now, such an hermaphrodite is apt to produce male plants in the course of generations, for the self-fertilized seeds

[1] Darwin, *The Variation*, etc., vol. i. p. 353.

are few in proportion to the others. We see in many plants abortive anthers inherited just as the stump of a tail in tailless breeds of animals, certain useless muscles in man, or vestige of ear, is inherited. Kölreuter found that by crossing male flowers that possessed the rudiment of a pistil with an hermaphrodite species, the offspring possessed a much larger and more developed pistil, which shows that these rudiments are really the same as the perfect organ.[1]

But how can crosses, when both sexes are produced at the same time in the same flower, be an economy? Most of the orchids, as Mr. Darwin points out, have beautiful contrivances for crossing, while they are totally unable to fertilize themselves, and their pollen is sometimes barren, and sometimes even injurious when applied to their own stigma. "Considering how precious the pollen of Orchids evidently is, and what care has been bestowed upon its organization, and on the accessory parts;—considering that the anther always stands close behind or above the stigma, self-fertilization would have been an incomparably safer process than the transportal of the pollen from flower to flower. It is an astonishing fact that self-fertilization should not have been an habitual occurrence. It apparently demonstrates that there must be something injurious in the process. Nature thus tells us, in the most emphatic manner, that she abhors perpetual self-fertilization."[2] The reason that these orchids habitually cross by means of insects cannot be that there is any saving of

---

[1] Darwin, cited by Spencer, *Biology*, vol. i. p. 386.
[2] Darwin, *The Fertilization of Orchids*, p. 359.

physiological power, for there is probably a greater expenditure in affixing the anthers to the proboscis of a moth than would be necessary in contracting the anther on to the stigma. Neither can it be due to any saving in generative power, since both sexes are produced at the same time, and yet they do not fertilize each other on the same flower. There must evidently be some advantage in crossing; but it by no means follows that because we cannot just now state precisely wherein this advantage consists, it must necessarily consist in the cross, *as* a cross. In that case the necessity of crosses would not be the exception that it is; instead of nearly all hermaphrodites being merely *able* to cross, they would be *obliged* to do so, as are most orchids. It seems to me rather a redundancy of fertility superadded to unisex, just as parthenogenesis seems to be a later stage than dual sex. If orchids were originally hermaphrodites fertilizing themselves as well as their neighbours, and if from any cause, such as poorness of soil, greater expenditure on reproduction or some other cause, certain among them became unisexual, producing only undeveloped pistils or anthers, might they not resume their double sex under more favourable circumstances, and yet from long disuse be unable to fertilize themselves? If in their unisexual state the females had been fertilized by the agency of insects, would not the apparatus for this means of propagation in all probability still continue, for what cause would it have to change? It seems to me that the way in which the pollen is distributed is particularly safe and economical compared to the broadcast distribution by the wind, as occurs in other plants, or the scattering of it all over the flower by the movements of an insect. In the orchis, the pollen is mathematically applied to the stigma, and little or nothing is ever lost; perhaps, indeed, it was the perfection of this plan which proved so great a saving in pollen that the plant was induced to develop again its dormant sex. To me it is far more inexplicable that pollen should be wasted as it is in the *coniferæ*, grasses, and other plants. The only explanation of this waste seems to be that these organisms, as they are below the orchids in the vegetable world, have not yet attained to their perfection of economy.

It might be urged, that if the sexes were divided for the reason given by Mr. Herbert Spencer, their object is fully attained by dual sex; but if they were divided on economical grounds, then why are they not further divided, for the greater the division of labour the greater the economy? But this division of sex also has its drawbacks. The chances against the meeting of these sexes increase in a ratio, greater as the sexes are more numerous, entirely out of proportion to the gain by subdivision. Where the sexes are only dual, reproduction is only twice as perfect as when there is no sex, but where there is no sex no meeting is requisite between two individuals, while the chances against the meeting of a male and female individual when two individuals do meet, is two to one, since two females may meet, or two males. In the same way the chances against meeting, as compared to the gain in economy of reproduction, is greater in an increasing ratio as the division of the sexes.

The theory that two sexes are necessary to secure crosses, is largely dependent on the supposed observation that breeding in-and-in is in itself and of itself injurious, exclusive of inheritance. But I hope I have already shown these observations to be fallacious, as also the complement that a cross must necessarily be an improvement. Gärtner, Herbert, and Lecoq, all bear witness to the advisability and advantage of crossing plants, though not from a different variety; and Mr. Darwin, dissatisfied with these general statements, himself experimented, and found the offspring of crossed plants nearly always more vigorous than the parent kinds.[1] Now, the effects of crosses are, first, variability, which depends, according to Mr. Darwin, "on the reproductive organs being injuriously affected by changed conditions;"[2] and, secondly, on reversion, which is generally a change for the worse, as the organism thus reverts to its former unimproved state, and the good effects of natural or artificial selection are thus lost.[3] In cultivated plants crosses are particularly disastrous, since the crossed offspring of plants which have had their organization more quickly altered by cultivation, are more liable to reversion than when crossed in their natural state.[4] It follows, that if animals or plants are at all improved by selection and consequent in-and-in breeding, a cross simply as a cross is positively injurious to them. It is, moreover, most difficult and generally impossible to raise a hybrid from distinct species; a difficulty which Mr.

---

[1] Darwin, *The Variation*, etc., vol. ii. pp. 127-129.
[2] *Ibid.*, p. 394.   [3] *Ibid.*, pp. 43-46.
[4] *Ibid.*, p. 83.

Darwin thinks, apparently, due to difficult development,[1] by which he means there is a mechanical difficulty, that the reproductive organs of the two sexes are not sufficiently adapted to each other. It may be rude unfitness of the external organs, or it may be a more delicate unfitness of the semen to the ovule, which prevents the penetration of the latter by the former. It is true there are many cases where near allied kinds will not interbreed, while quite distinct species will readily do so; but this proves nothing beyond the fact that the difficulty of crossing is less sometimes in a different species altogether than in a different variety of the same kind. Plants widely differing in every part of the flower, in the pollen, in the fruit, and in the cotyledons, can be crossed, while with others belonging to the same genus this has been found impossible. It is the same, however, with reciprocal crosses. *Mirabilis jalappa* can be easily fertilized by the pollen of *Mirabilis longiflora*, but Kölreuter tried two hundred times the reverse process, and failed.[2] When a cross is successful even, the superiority over the parent is only purchased by a loss of fertility. Gärtner, indeed, shows that certain hybrids may combine unusual vigour with great fertility,[3] which may be the case in some instances, but it certainly is not the rule. Nor does Mr. Darwin take that view. "Seeing," he says, "that hybrid plants, which from their nature are more or less sterile, thus tend to produce double flowers; that they have the parts including the seed, that is the fruit,

---

[1] Darwin, *The Variation*, etc., vol. ii. pp. 179, 180; *Origin of Species*, pp. 263, 264.

[2] *Ibid.*, *Origin of Species*, pp. 257, 258.

[3] *Ibid.*, *The Variation*, etc., vol. ii. p. 131.

perfectly developed, even when containing no seed; that they sometimes yield gigantic roots; that they almost invariably tend to increase largely by suckers and other such means;—seeing this, and knowing * * * * that almost all organic beings when exposed to unnatural conditions tend to become more or less sterile, it seems much the most probable view that with cultivated plants sterility is the exciting cause, and double flowers, rich seedless fruit, and in some cases largely-developed organs of vegetation, &c., are the indirect results—these results having been in most cases largely increased through continual selection by man." While the fact that they are sometimes fertile as well as vigorous " is probably in part due to the saving of nutriment and vital force through the sexual organs not acting, or acting imperfectly, but more especially to the general law of good being derived from a cross."[1] But I have already given reasons for dissenting from the view expressed in the last sentence, the more so, indeed, since the first explanation is amply sufficient. For who, at all acquainted with physiology or pathology, can for a moment doubt the immense influence of reproduction on the strength, vigour, size, and general health. Mr. Darwin also gives some explanation of the sterility of crosses and of hybrids: "A hybrid partakes of only half of the nature and constitution of its mother, * * * it may be exposed to conditions in some degree unsuitable, and consequently be liable to perish at an early period; more especially as all very young beings seem eminently sensitive to injurious or unnatural conditions of life."[2] Now, he

---

[1] Darwin, *The Variation*, etc., vol. ii. pp. 131, 172.
[2] *Ibid.*, *Origin of Species*, p. 264.

observes that when animals are placed under changed conditions, when, for instance, they are confined or tamed, or when plants are cultivated, the sexual organs are chiefly, and the first to be injuriously affected;[1] and he argues that hybrids being necessarily under changed and unnatural conditions "the reproductive system, independently of the general state of health, is affected by sterility in a very similar manner."[2]

Seeing, then, that crosses lead to many imperfections; that if organisms are not nearly allied they can rarely be made to interbreed; and that the result of such crosses is an offspring of weedy growth, ill-balanced intellect, often as susceptible to unfavourable circumstances as an unacclimatized animal, and generally sterile; it is impossible that crossing can be considered in any way beneficial except inasmuch as it may relieve a possible hereditary tendency to disease.

[1] Darwin, *The Variation*, etc., vol. ii. p. 148, *et seq.*
[2] *Ibid.*, *Origin of Species*, p. 265.

CONCLUSION.

WE have seen in the preceding pages that not one of the many reasons which have been advanced why marriage between near kin should be prohibited by the State can stand inquiry. We have seen that there is no natural horror of incest, and that many peoples have practised and habitually do practise it; while, on the other hand, we have seen that whatever may be the reason of certain prohibitions which exist, they are certainly not due to any conscious or unconscious experience of any evil results. We have seen that the statistics on which so much reliance has been placed, as a proof of the harmfulness of consanguineous marriage, are, when not absolutely false, miserably misleading and defective. And, finally, we have seen that the great argument of biologists that crosses must be beneficial or there would not be two sexes is by no means proved, and the presence of dual sex can be accounted for in another way.

On the other hand we have seen many cases of in-and-in breeding in isolated communities, and more especially among domestic animals, in which no evil effects have been observed; and I must here again call attention to the very superior value of this argument in comparison to the converse one, that in many cases evil effects have been observed in the offspring, since in the latter we have no proof that these effects are not the result of morbid inheritance. We have seen also the failure of arguments against marriages between near kin, and cannot but see in that failure a further series of arguments in their favour. It remains for our consideration whether we should rest satisfied with this result, and whether it is advisable that the laws concerning the prohibited degrees should be altered in any way.

The first question scarcely needs consideration, for no question ever really remains stationary. Yet at the same time it would be very desirable that a little more study should be expended before such positive assertions on either side were made, and that those who believe in the baneful effects of consanguineous marriages would rather devote their energies to the only way the practical question can be answered, and, like Sir John Lubbock, assay to obtain a census on a scale sufficient to set at rest for ever the doubts of legislators. Several times, indeed, within the last few years have attempts been made in this direction by scientific men: in Belgium by M. Uytterhoeven, in Italy by M. Mantegazza, in London by Sir John Lubbock, and various abortive, because half-hearted, attempts have been made in the United States of America, and by the Ministry of the Interior of France. That there should hitherto have been so great a difficulty is perhaps not extraordinary, seeing that the population at large, despite their traditions, continue to contract these marriages with a recklessness which plainly shows how little they really fear them; while at the same time they do not feel inclined to submit to what they ignorantly consider a useless and inquisitorial inquiry into their private affairs. It is difficult for an individual to picture to himself his utter insignificance in the eyes of the Registrar-General. But surely if he is content to make known his age, the number of his children, and their mental state, it will not hurt him to state the relationship, if any, which exists between him and his wife. Mr. Lock and Mr. Hardy, in their discussion of Sir John Lubbock's motion to ask this question in the Census of 1871, seemed to think that "speculative philosophers" were meddling animals, whose curiosity it was dangerous to satisfy; give them an inch and they will take an ell, let the measure pass this time and they will apply by degrees every kind of mental torture, so that the Census Bill, instead of being a blessing, will become a curse. But where is the use of the census, unless it lies in the deductions made from it by "speculative philosophers?" Or what does it matter to us if we have to answer once in ten years concerning the health and number of our families? Far better would it be to accustom the people gradually to inquiries of this sort, and show them at once how entirely all individuality is lost in the mass, and how valuable these answers are both to the community and the individual himself. If desirable, let the papers be returned in sealed envelopes, and let them be opened in a town different to the one in which they were collected; or adopt any other plan that for the moment may keep family secrets from the knowledge of any acquaintance. The census, properly worked, is a mine of wealth; and if made secret, who can object to answer even the most "inquisitorial" question?

In the meanwhile, however, let us not have any more such rodomontade as that of Reich, who "imperatively demands" that governments should forbid all marriages within the first four degrees;[1] or of M. le Comte de Maistre quoted by M. Devay, who says the time has now come when the Pope should refuse his consent to consanguineous marriages.[2] Governments and the Pope know better what they are about, and as they are in a measure forced to march with the times, we see if anything a loosening of the prohibitions, and a wise tendency to greater freedom in marriage. When they find it vain to forbid marriage between brothers and sisters-in-law, and even find some difficulty in the prevention of marriages between uncles and their nieces, is it likely that it is in their power to forbid those between first-cousins?[3]

[1] "Ich verlange von der Gesetzgebung auf das Bestimmteste, an gar kein Kirchen-Gesetz sich zu kehren, sondern alle Ehen in den ersten vier Graden der Verwandtschaft auf das Strengste zu verbieten" (Reich, *Ehe*, etc., p. 531).

[2] Devay, *Du Danger*, etc., p. 216.

[3] I would refer any one who may be doubtful as to the inability of a government to prevent marriage within certain degrees in which the people themselves see no harm, to the *Report of the Royal Commission on Marriage Law of* 1848. It will be seen that despite the laws and that amount of public opinion which upholds them, more than one per cent. of all marriages contracted in England are between widowers and their deceased wife's sister; while the proportion of these marriages to all marriages of widowers, is one in thirty-three. The proportion must therefore be enormously great compared to those cases where a widower's deceased wife has left an eligible unmarried sister. The recognition of these marriages has been forced upon the governments of France, Den-

It is of the greatest importance that the marriage law should be as little subject to change as possible, to the end that it may be certain in its operation, and not induce people by any foolish provisions, either to break it, or entirely do without it. Marriage should therefore be only prohibited in those degrees which by general consent are considered incestuous; that is, in the direct ascending and descending line, between brother and sister, and uncle and niece; or those degrees which as a rule imply an unsuitable difference of age between the parties. It is true that uncle and niece are often at a suitable age for marriage, and brother and sister generally are; but in the one case an uncle has an undue advantage over other suitors, while as a rule he is proportionally too old, and if some uncles were permitted to marry their nieces, all must be allowed to do so; in the other case it is probable that if brothers and sisters were allowed to marry, they would do so while yet too young, while as they generally do not desire it, and the chances of morbid inheritance are thereby greatly increased since selection cannot be exercised, it is as well to forbid it. With cousins it is very different. A man may have fifty cousins and only one or two sisters; he is much more likely therefore to fall in love with a cousin; moreover, the chances of morbid

mark, and all the Protestant States of the Continent. The Australians have squeezed a recognition of such marriages out of their reluctant mother-country. Even in England there is a powerful society organized in their favour; indeed in this country the feeling against them is entirely due to the Statute of Henry VIII., created to gratify his own passions. For instances of immorality to which this prohibition has led in England, see p. 2, q. 8, p. 4, q. 27, p. 6, q. 46, p. 9, d., p. 12, b., p. 14, b., p. 16, q. 175, of the *Report of* 1848.

inheritance are practically reduced to nothing, and the fact that they do occur is sufficient to warn any government that a prohibition would be useless. On the other hand, it is for many reasons commendable to marry a relative, for here one can exercise some selection, since a man generally knows the state of health and the disposition of members of his own family. With prohibitions against the intermarriage of connections we must apply the same rule. Marriage between step-mothers and step-sons, or father and daughter-in-law, should be prohibited, since they must usually be of an unsuitable age for marriage; but brothers and sisters-in-law should be allowed to marry, as they are usually of suitable age, and more nearly represent the deceased party than any other person could.

The conclusions we arrive at, then, after due study of the foregoing matter, are these:—

I. That any deterioration through the marriage of near kin *per se*, even if there is such a thing in the lower animals, is impossible in man, owing to the slow propagation of the species.

II. That any deterioration through the chance accumulation of an idiosyncrasy, though more likely to occur in families where the marriages of blood-relations was habitual, practically does not occur oftener than in other marriages, or it would be more easily demonstrated.

III. That seeing the doubt, to say the least of it, which exists concerning the effects for harm of marriages between near kin, and on the other hand the certainty that whenever and wherever marriage is impeded a direct and proportionate impulse is given to the practice of immorality, it is advisable not to extend the prohibitions against marriage beyond the third collateral degree, and to permit all marriages of affinity excepting those in the direct ascending or descending line.

THE END.

APPENDIX.

I MAY, perhaps, be blamed, for giving all the following cases in full, worthless as they are, from a statistical point of view, but they are nevertheless useful to show on what foundation the fear of consanguineous marriage stands; and as they are probably facts, and certainly widely scattered through various periodical publications, their collection may prove serviceable in a way I cannot foresee, just as one man may treasure the shrub which another pulled away to get more easily at some fossil, while a third might centre his whole endeavours in trying to catch some rare moth they may have disturbed, so each may draw something valuable from these cases, or skip them as he pleases.

It is a mystery to me, indeed, why they have been collected at all; since most impartial authors agree that they are always, though not purposely, mere selected cases. "Memory searches for instances of unions of kinship," says Dr. Mitchell, " from the history of which the answer is to be framed. Now, it is certain that all those which have been marked by misfortune will be first called up, while many of those which have exhibited no evil effect or no peculiarity of any sort will be passed over or forgotten. The attention, in all likelihood, has been frequently drawn to the first, while nothing may have occurred in the progress of the last to keep alive the recollection of relationship in the union. I need scarcely say that facts collected in this manner are almost sure to lead to inferences beyond the truth, yet it is from such data that conclusions on this subject have frequently, if not usually, been drawn."[1] M. Mantegazza also says—"Questi fatti * * * * non sono la fotografia della società umana in una vasta regione di paese : ma son fatti scelti qua e là secondo l'opportunità di molti osservatori sparsi e divisi. È naturali poi che in ogni paese si sia raccolto un numero maggiore di fatti contrarj alle unioni fra parenti ; perchè erano i primi a cadere sotto gli occhi, a fermar quasi l'attenzione dell' osservatore. I più fra i medici e legislatori sono persuasi

[1] Mitchell, *Mem. read before the Anthrop. Soc. of London*, vol. ii. 1866, pp. 402, 403.

che questi matrimonj sono nocivi alla prole; per cui quando hanno sotto gli occhi figli robusti e senza mende, non si curano di domandare se siano il frutto di due cugini o di uno zio o di una nipote."[1]

The collection of cases of consanguineous marriage, where the results as regards the offspring are good, is on the other hand very valuable; since if it is true that the intermarriage of near kin will of itself without any previous taint or hereditary tendency whatever produce offspring who suffer from some disease of the nervous system or none at all, then why do not all marriages of this sort produce these effects? The cause is there—a very powerful cause if we are to believe these writers—but it only acts in a very uncertain way, indeed, so uncertain is it, that we have good cause totally to deny its action in any case. If, forsooth, we are to believe with Morris, that over 64 per cent. of the children of these marriages are affected in some harmful way,[2] how could it fail to manifest itself in a way that could not be ignored even by the greatest advocate of consanguineous marriage? If, again, we are told to believe that every consanguineous marriage must lead in time to disastrous results, how much more convincing to the contrary is a set of observations, like that of M. Seguin or M. Bourgeois, where after continued consanguineous marriages in the family no harmful effect is there, than isolated cases of which nothing is known either of family idiosyncrasies or family habits!

Though I must confess strongly tempted, I have made no remarks on the following cases, except where some explanation has been necessary I have added it in an occasional foot-note. Any one bearing in mind the manifold causes of the diseases said to result from consanguineous marriage, will readily detect the weak points in these observations for himself, and will at once see that these cases do not and cannot represent the whole population, any more than a set of cases taken from a hospital would do so; or rather, they are even less faithful than these would be, since the latter taken in sufficient quantity might afford some data for a certain class of the population, while the former, since they are selected cases, and by no means even with the best intention impartially chosen, would not be reliable for any class of the population, however great the quantity.

1. Two first-cousins, perfectly healthy, married, and, as far as M. Balley is aware, had only two children, a boy and a girl. The boy was an albino, and the intelligence of the girl was but poorly developed.

[1] Mantegazza, *Studj sui Matrim. Cons.*, pp. 27, 28.
[2] Boudin, *Mém. de la Soc. d'Anthrop. de Paris*, vol. i. 1863, p. 511.

2. A chemist of Bourbonne-les-Bains married a German lady, who was the daughter of first-cousins. Both were healthy, and they had four children. Of these, the eldest was malformed, but intelligent; the second, a deaf mute; the third, well formed and healthy; the fourth, a girl, was imbecile.

3. A captain at Rome married his first-cousin. They had several still-born children, besides some who lived a short time, and were malformed; and one still living who besides being malformed, is rickety, and has been almost from his birth upwards affected with chorea.[1]

4. A doctor married his first-cousin, and had two children, seemingly of poor intelligence.

5. Two first-cousins married, and had eight children; but every alternate child was a deaf-mute.[2]

6. The three daughters of a lady married the three sons of her sister. Of these, one had a boy and two girls, who were all perfectly healthy.

7. The second had three boys and two girls; the eldest boy had a slight impediment in his speech; the second was a deaf-mute, and married a person not related to him, by whom he had two children who could speak; the third boy was also deaf-mute, and remained unmarried; the two girls were perfectly healthy, but one of them stammered slightly on certain letters.

8. The third marriage produced two boys and one girl still living, and a malformed individual who died. Both boys were congenital deaf-mutes. The elder married a person not related to him, and has a child who can speak. The girl first began to speak when aged six years.[3]

9. A marriage between first-cousins, which produced a puny boy, who was an albino, and died in his thirteenth year.[4]

10. In a family of six children, two of them boys, two girls and a boy married their first-cousins, and the remaining three married strangers. The first consanguineous marriage produced eleven children,

---

[1] Cases Nos. 1 to 4, reported by Balley, *Gaz. Méd. de Paris*, vol. xviii. 1863, p. 111, and *Comptes Rendus*, vol. lvi. 1863, pp. 135, 136.

[2] Case No. 5, reported by Brochard, *Comptes Rendus*, vol. lv. 1862, pp. 43, 44.

[3] Cases Nos. 6 to 8, reported by Q. de Ranse, *Comptes Rendus*, vol. lv. 1862, pp. 405, 406.

[4] Case No. 9, reported by Aubé, who asserts that albinoism as a result of consanguineous marriage is very common. He accordingly instances three albinoes, of which two were brothers, and *their father was unknown;* the third is Case No. 9 (See Devay, *Du Danger*, etc., pp. 53, 54).

ten of whom died of hydrocephalus in their babyhood, and one died of the same disease in his fourteenth year.

11. The second marriage produced eight children, of whom six died young, and the remaining two are delicate.

12. The third marriage produced five children. Of these, one died on the fifteenth day of its life; one lingered three years; and one died of meningitis, aged twelve years. The three non-consanguineous marriages produced six, seven, and six children respectively. Of these, two of one family died young, and one of another.

13. A still-born infant was born after a protracted labour of ten hours. It was anencephalous, club-footed, hare-lipped, and had a fleshy appendage fifteen centimetres long on the right shoulder, only four toes on the left foot, and only three on the right. The parents were first-cousins; the mother aged thirty-three, of nervous temperament, and subject to hysteria, but otherwise of good constitution; the husband was aged thirty-nine, and had a good constitution.

14. A child of first-cousins had an extra toe on each foot. His father was a distinguished chemist.

15. An illegitimate child of second-cousins, born in the hospital of St. Charité, at Lyons, was destitute of a cranial arch, and the brain was but feebly developed.

16. A boy, twelve years of age, had at the crown of his head a patch of black and white hair. His parents were first-cousins.

17. In a village in the district of Yverdon, two brothers married two sisters, their first-cousins. They were all healthy well-to-do peasants without any hereditary tendency to disease of any kind. The one had two children by his marriage, both albinoes, and one of which died young.

18. The other had five children, and all of them albinoes. Three of these died, one from a fall, and the other two from some disease not reported by the *Echo Médicale Suisse*, whence M. Devay extracts this case. The husband in this last case lost his wife and married again a person not related to him, by whom he had four healthy children.

19. A boy of twelve years of age, who was an idiot, was descended from a noble family in which consanguineous marriages were customary. He had an elder brother who was perfectly healthy and successful in his studies, but subject to a most violent temper. Epilepsy had appeared occasionally in the family.

20. A woman married her uncle, and has as yet only two children. Both of these are deaf-mutes. The parents are healthy, and there is no hereditary tendency to disease in the family.

21. In a family of fourteen children, the parents of whom were related, all died early excepting one. Those who died young were attacked by convulsions, the others by consumption.[1]

22. Two first-cousins, of good constitution, married, and had eight children. The first at eighteen months of age got an acute fever attended with delirium, but no convulsions; on convalescence, his lower members atrophied, he became a cripple, and besides lost his hearing, though he retained his intelligence up to the date of his death, aged fifty-one. The second died of meningitis, aged five years. The third is very intelligent and healthy, but has become completely deaf. The fourth, a congenital deaf-mute, is very intelligent. The fifth is healthy and intelligent, but his hearing is slightly defective. The sixth, a girl, was congenitally deaf, though intelligent; at the age of thirty altered for the worse, and she became subject to hallucinations. The seventh was perfectly healthy at birth, but afterwards became idiotic. The eighth was a congenital deaf-mute, and became a professor in an asylum for the deaf and dumb.[2]

23. An uncle married his niece, and had eight children. Of these, seven died before they reached their fourth year from convulsions, hydrocephalus, etc. The eighth is now thirty-three years old, delicate, and affected with psoriasis diffusa, almost congenitally. Both parents were healthy.[3]

24. A marriage between first-cousins, which produced ten children, of whom five died young. The three remaining girls were puny and bent. One of the two boys is healthy, but slightly bent; the other was very small, and got articular rheumatism at twelve years of age, from which he is rarely free. The father and mother were both healthy, and the father's two brothers who married persons not related to them have healthy children.

25. A marriage between first-cousins, which produced seven children, of which one died young. The eldest, a boy, is healthy, but has a scrofulous tendency. The younger brother, aged fourteen, was scrofulous, and had when aged six years a white tumour first on his left elbow, and then on his right, which took a long time healing, and came again on his knee, so that at last his leg had to be amputated. The youngest is vigorous, but articulates badly and is malformed. The eldest girl was small and malformed, and had to be

---

[1] Cases Nos. 10-21, reported by Devay, *Du Danger*, etc., pp. 99-103, 111, 112, 139, 143, 144.

[2] Case No. 22, reported by Forestier, see Devay, *ut sup.*, pp. 120, 121.

[3] Case No. 23, reported by Potton, see Devay, *ut sup.*, p. 145.

treated as a child up to the age of twenty-seven. She is now forty-one. The second was always delicate. The third, aged thirty-five, has now been married several years, and has no children. The mother of these children has two sisters who married out of the family, and both have healthy children.[1]

26. A boy, whose parents were the children of two brothers, had his toes malformed and undivided. He has a sister who is perfectly well formed.

27. A girl, the only daughter of an uncle and his niece, was epileptic. There was no family tendency to this disease.[2]

28. Two relatives married and had six children, all affected with scrofula or rickets. One who was rather better than the rest was not suckled by the mother, but had five wet-nurses. Both parents were healthy.[3]

29. A marriage between uncle and niece, which produced only one child, a puny girl, who while yet in the cradle had croup, convulsions, etc. She lingered several years, though troubled with excessive nervousness, and finally died from general hæmorrhage from the skin and mucous membranes. Both parents were healthy.[4]

30. In a family where the parents were first-cousins, and where for generations consanguineous marriages have been the rule, were born three girls, all congenital deaf-mutes, squat in figure, puny, and scrofulous. One is epileptic. They have a brother who hears and speaks.

31. A marriage between second-cousins, which produced two children, a boy and a girl, the latter a congenital deaf-mute. The parents were healthy, one was aged thirty-three, the mother twenty-six. Their habitation was healthy, but they had three nieces, who were also deaf-mutes.

32. Two children, boy and girl, congenital deaf-mutes, were born from parents who were first-cousins. The boy died at the age of four. The mother's mother had died of tuberculosis.

33. Three children were born to parents distantly related, but perfectly healthy. One, a boy, hears and speaks; but the other two who are girls, are congenital deaf-mutes.

34. A marriage between uncle and niece which produced but one child, a girl. At the age of eighteen months, during dentition, she

---

[1] Cases Nos. 24, 25, reported by Viennois, see Devay, *Du Danger*, etc., pp. 237, 238.
[2] Cases Nos. 26, 27, reported by Bondet, see Devay, *ut sup.*, pp. 101, 146.
[3] Case No. 28, reported by Teissier, see Devay, *ut sup.*, p. 145.
[4] Case No. 29, reported by Doyon, see Devay, *ut sup.*, p. 146.

was taken with convulsions and became perfectly deaf. The father was slightly deaf. He was, besides, twenty-one years older than her mother.

35. A marriage between first-cousins which produced seven children, three girls and four boys. Two girls are deaf-mutes; the rest of the children are all dead. The father is two years younger than his wife, and their dwelling is damp. One of the girls recently died of phthisis.

36. A marriage between first-cousins which produced four children, two boys and two girls. One girl is a congenital deaf-mute, subject to congestion of the cervical ganglia, and suffers from chronic conjunctivitis. Her sister is too young to determine whether she hears. Her brothers both speak and hear. The parents are healthy, but live in a damp house.

37. Two girls, cousins of case No. 31, the daughters of first-cousins, are congenital deaf-mutes. Their habitation is perfectly healthy. Two cousins of the father married their own cousins, and have deaf-mutes in the family. Deaf-mutism was previously unknown in the family.

38. Two first-cousins, aged respectively thirty-four and twenty-eight, had two sons, of whom one was a congenital deaf-mute, and the other could hear and speak. Their dwelling was healthy.

39. A marriage between first-cousins which produced eight children, seven boys and one girl. Four of the boys were congenital deaf-mutes. The father was fourteen years older than his wife; and when the eldest deaf-mute was born, the parents were aged respectively forty and twenty-six. Deaf-mutism was previously unknown in the family, and their dwelling was perfectly healthy.

40. A marriage between second-cousins, which produced two children, boys, of which one was a congenital deaf-mute. The father was aged thirty-six when the mother was twenty-two, but both were perfectly healthy.

41. A marriage between second-cousins, which produced three children, all boys and congenital deaf-mutes. The parents were healthy, their dwelling dry, and deaf-mutism was previously unknown in the family.

42. A marriage between second-cousins, which produced two children, one of whom was a congenital deaf-mute. The father was aged thirty-five at the time of inquiry, and the mother thirty-four. Their dwelling was healthy.

43. A marriage between first-cousins, which produced two children,

a boy and girl. They were both deaf-mutes. The mother was younger than her husband.

44. A marriage between first-cousins, which produced two children, a boy and girl, both congenital deaf-mutes. The boy was also lame from birth, puny, and of bad health. The girl died at the age of four years. Their parents were well-to-do people.

45. A marriage between first-cousins which produced four children, two boys and two girls. The boys were both of them congenital deaf-mutes. Their dwelling was healthy, and there was no hereditary tendency to deaf-mutism in the family.

46. A marriage between first-cousins, which produced five children. Of these three were deaf-mutes. There was no hereditary tendency to this disease, nor was the age of the parents disproportionate, nor was their dwelling unhealthy.

47. A marriage between third-cousins, which produced four children, three girls and one boy, all deaf-mutes. Their parents were healthy, and their grand-parents were not deaf-mutes.

48. The Mayor of C—— (Dordogne) married the daughter of his first-cousin, and had by her a son and daughter not only exempt from all disease, but endowed like their parents with the best health. The daughter when she was twenty, married a person who was not related to her, and a few years older than herself, and gave birth to a congenital deaf-mute. They lived in a healthy place, and there was no deaf-mutism previously in her family, nor was it common where they lived.[1]

49. A healthy man married his first-cousin, by whom he had a daughter, who stammered, and is now aged twenty-three. She had a child which died of hydrocephalus at the age of three.

50. The same man married again, and again he married a first-cousin. Three children were born to this second marriage, of whom two died young from convulsions, and one is healthy.[2]

51. A marriage between first-cousins, which produced one epileptic child, now aged twenty-five. The parents at his birth were aged respectively thirty-five and thirty. There was no hereditary tendency to the disease in the family.

52. A marriage between first-cousins, which produced two children,

---

[1] Cases Nos. 30-48, reported by Chazarain, *Du Mariage entre Cons.*, etc., pp. 35-40, 43, 44.

[2] Cases Nos. 49-50, reported by Chipault, *Etudes sur les Mar. Cons.*, p. 44. They are given somewhat differently in the *Comptes Rendus*, vol. lvi. 1863, p. 1001.

a girl who was epileptic, and a boy who was scrofulous. The husband was two years his wife's senior, and there was no hereditary tendency in the family to these diseases.

53. A marriage between first-cousins, which produced four children, three of whom died young of convulsions, and one is healthy. The mother is now ill, but up to the time of the birth of these children, and till six years after the birth of the last, she was quite well.

54. A marriage between healthy first-cousins, which produced only one child, and that died young.

55. A marriage between persons not related, which produced two children, who died of typhoid fever between the ages of five and of twenty. On the death of their mother the man married again, this time his first-cousin. Two children were born to this marriage, both of them hunchbacked.

56. Two illegitimate children of first-cousins were born, the one with two extra little fingers and one extra big toe; the other with two extra fingers, both great toes double, and one extra little toe. These children died at the respective ages of fifteen days, and six weeks. The mother afterwards married a stranger, and had three well-formed children.[1]

57. A marriage between first-cousins, which produced two children, both deaf-mutes. There was nothing of the sort in the family before.[2]

58. Three children, whose parents were first-cousins, were deaf-mutes.

59. A Neapolitan, who married his niece, had four children by her, of whom one, a girl, was very eccentric; the second, a boy, was epileptic; the third, a boy, was very intelligent; the fourth, also a boy, was an idiot and epileptic. Both parents were healthy.[3]

60. A woman who remained barren during her marriage with her first-cousin, had several children on her second marriage with a person not related to her.[4]

61. An intelligent officer married his niece. His first child was a boy, healthy in all respects. Three years after he had another boy, afflicted with hare-lip and cleft-palate, and who only lived a few weeks.[5]

62. A marriage between persons related in the second degree. Seven children were born, of which one with hare-lip died six days after birth, and the remaining six died before they reached the third year.

---

[1] Cases Nos. 51 to 56, reported by Chipault, *Etudes sur les Mar. Cons.*, etc. pp. 43, 44, 47, 48.
[2] Case No. 57, reported by Duteval, see Chipault, *ut sup.*, p. 34.
[3] Cases Nos. 58, 59, reported by Trousseau, see Chipault, *ut sup.*, p. 43.
[4] Case No. 60, reported by Lisle, see Boudin, *Mém. de la Soc. d'Anthrop. de Paris*, vol. i. 1863, p. 518.
[5] Case No. 61, reported by Potier-Duplessy, see Boudin, *ut sup.*, p. 516.

63. It is to be presumed the husband in Case No. 62 married again, for he had a son who married his first-cousin, and had three children, of which one had hare-lip and died young.

64. A son, one of the two remaining children from Case No. 63, married his second-cousin, and had five children. Of these four died young, and one of the four was a deaf-mute.

65. The remaining son, in Case No. 64, when aged thirty-three, married his first-cousin, who was aged twenty-one. They had four children, the eldest of whom, a girl, had a cleft-palate and hare-lip, and died when twenty-nine days old. The second, a boy, was well formed, and is now five years old. The third, a girl, is also well formed, and now aged three years. But the fourth, a girl, had a cleft-palate and hare-lip, and died when seventeen days old.[1]

66. A marriage between first-cousins, which produced three children. The eldest, a girl, is much bowed; the second was born blind; and the third was hunch-backed. The mother in this case was of remarkably good constitution, and her husband was a fine-looking man, aged twenty-seven at marriage.[2]

67. In a Protestant family on the Isle de Ré, three brothers married three sisters, their first-cousins. To the first marriage five children were born, of which the eldest died from convulsions when ten months old; the second, a girl, is scrofulous; the third, a girl, died from convulsions at the age of eight months; the fourth, a girl, has an impediment in her speech; the fifth, a boy, is scrofulous, and slightly idiotic.

68. In the second marriage also five children were born. The eldest, a boy, is insane, scrofulous, and has an impediment in his speech. The second, a girl, articulates slowly. The third, a boy, is scrofulous and a deaf-mute; he married a person not related to him, and has two healthy children able to speak. The fourth, a boy, is a deaf-mute. The fifth, a girl, is healthy.

69. To the third marriage eight children were born. Of these, the first was stillborn. The second, a boy, was a deaf-mute; he married a person not related to him, and has a child able to speak. The third, a boy, was scrofulous, and died of hydrocephalus, aged three years. The fourth, a girl, is scrofulous, and could not speak before she was four years old. The fifth, a boy, who died in convulsions at the age of twelve months, was thought to have been deaf. The sixth, a boy, is a deaf-mute. The seventh, a boy, was scrofulous, and died at the

---

[1] Cases Nos. 62 to 65, reported by Rizet, see Boudin, *Mém. de la Soc. d'Anthrop. de Paris*, vol. i. 1863, p. 516.

[2] Case No. 66, reported by Robillard, see Boudin, *ut sup.*, p. 521.

age of five years. The eighth, a boy, died at the age of ten months from convulsions, and was thought to have been deaf.[1]

70. A marriage between uncle and niece; both were of lymphatic temperament, and she much younger than he. They have two daughters, now aged respectively fifteen and twenty-five; like their parents both lymphatic, but both rejoicing in perfect health.

71. A marriage between second-cousins, both healthy, produced one child, a healthy boy, and now aged thirty-five years.

72. The boy mentioned in Case No. 71 married, at the age of twenty-five, a cousin related to him in the sixth degree, and eleven years older than himself. Two boys were born to them, and both are perfectly healthy.

73. The brother of one parent in Case No. 71, and sister of the other, married, and had three perfectly healthy sons; and one daughter, who became delicate when about twenty years old. Both parents were healthy.

74. The daughter, in Case No. 73, a few months after her twentieth year, married her second-cousin, who was to all appearance of poor constitution. They had a stillborn child, and the mother died in childbirth consequent on the grief, anxiety, and fatigue she had gone through in nursing her parents, who were both carried off by cholera a few weeks before her confinement.

75. A marriage between first-cousins, which produced up to the present two children, who are both still young, and healthy.

76. A marriage between first-cousins, who have as yet one child, four years old, and perfectly healthy.

77. A marriage between cousins related in the sixth degree. The husband was unhealthy and died young, leaving a child of feeble constitution, but without any pronounced disease.

78. A marriage between cousins related in the fourth degree. The parents are perfectly healthy, and have a child scarcely two years old, and also perfectly healthy.

79. A marriage between first-cousins, not in good health. They have a puny child, now five years old. The husband became consumptive, and a second child died at the age of a few months, having been constantly ill.

80. A marriage between cousins of the sixth degree. They have a son now aged thirteen years, and perfectly healthy.

81. A marriage between first-cousins of good constitutions. They

[1] Cases Nos. 67 to 69, reported by Ponsin, see Boudin, *Mém. de la Soc. d'Anthrop. de Paris*, vol. i. 1863, pp. 522, 523.

have no children, but the wife's sister, also perfectly healthy, is likewise barren.

82. A marriage between cousins related in the sixth degree, in a family where carcinoma was hereditary. The wife died when she had nearly attained her thirtieth year, having given birth to four children, two of whom died young, and the remaining two are poorly. The father is healthy.

83. The sister of the husband in Case No. 82, and the brother of his wife, married, and had two children poor in health. One of these died at the age of ten years, and the mother died about the same time of a tuberculous disease.

84. A physician married his niece, and had two children, whose physical and intellectual powers enabled them to follow with distinction their father's profession. Both parents were healthy.

85. Another physician, aged thirty-five, married his niece aged twenty, who was of lymphatic temperament. In the course of four years he had two children, aged at present respectively three and fifteen months. Both are healthy.

86. A third physician in good health, married his first-cousin, who was of poor constitution. He has one child, now aged ten years, and in poor health.

87. A marriage between first-cousins, which produced two daughters, now aged respectively twenty-three and twenty-five years. They are not very healthy, but more so than their mother, who, however, is still living.

88. A marriage between cousins related in the sixth degree. Two children were born to them, now aged respectively twenty-one and twenty-three years. Although one of them at the age of three years had an attack of acute meningo-encephalitis, which destroyed his hearing, they are both very healthy.

89. A marriage between second-cousins, which produced three children. Of these, one died at the age of forty-one of an acute disease. The other two are now aged respectively fifty-four and fifty-six, and are in excellent health. The parents died at the respective ages of seventy and ninety-two.

90. One of the children in Case No. 89 married a first-cousin, and had seven children, now of ages ranging between thirteen and thirty years. They are all very healthy.

91. The son mentioned in Case No. 89, who died of an acute disease at the age of forty-one, had married his first-cousin, and had seven children. Of these four died of epidemic diseases, all under the

ago of eleven years; but the three others, whose ages range at present between twenty and twenty-eight, are all perfectly healthy.

92. One of the children in Case No. 91 married her first-cousin, who was one of the children mentioned in Case No. 90 and had within four years three children, all healthy. The eldest died of croup, aged eighteen months.[1]

93. A marriage between first-cousins, which produced three daughters, who all subsequently became the mothers of large families. All were perfectly healthy and well formed.

94. A nephew of the wife in Case No. 93 married his first-cousin, and had seven children, of which five are still living, and in perfect health. Of the two who died, one death was from croup, and the other child only lived eight days after birth.

95. One of the children in Case No. 94 married her first-cousin, and has two children, whose command of voice is only too manifest.

96. M. G——, Dr. at C—— (Aveyron) married his first-cousin, and has three pretty children, all perfectly healthy and well formed.[2]

97. Sp. Ligustinus married his first-cousin, and had eight children, six boys and two girls. At the time that Livy makes him narrate this case the two girls were married, and four of the boys grown up.[3]

98. A marriage between persons in no way related, which produced two children, who both died, was followed by another marriage on the part of the husband, with his first-cousin. This last marriage has resulted in one healthy girl.

99. A marriage between first-cousins. The young wife had three miscarriages one after the other, each at six weeks. She was from puberty subject to dysmenorrhœa; and at every return of the catamenia suffered from uterine congestion, ending in hæmorrhage.

100. Two Russians, first-cousins, married, and have five children. The husband was scrofulous, and his wife, though naturally of good constitution, is far from healthy. Yet all the children enjoy the best health, though one was delicate during infancy.[4]

---

[1] Cases Nos. 70 to 92, reported by Bourgeois, *Sur l'Influence*, etc., pp. 39-41.

[2] Cases Nos. 93 to 96, reported by Devic, *Gaz. Méd. de Paris*, vol. xviii. 1863, p. 158.

[3] This case is interesting as bearing on the Roman law. It is related by Livy, *Hist.*, book xlii. 34: "Quum primum in ætatem veni, pater mihi uxorem fratris sui filiam dedit; quæ secum nihil attulit, præter libertatem pudicitiamque, et cum his fecunditatem, quanta vel in diti domo satis esset. Sex filii nobis, duæ filiæ sunt, utræque jam nuptæ. Filii quatuor togas viriles habent, duo prætextati sunt."

[4] Cases Nos. 98 to 100, reported by Dally, *Anthrop. Review*, pp. 94-95, May, 1864.

101. A marriage between first-cousins, which proved barren.

102. A marriage between first-cousins, which produced five children, three boys and two girls. The eldest son is now married;[1] another is as yet unmarried, and aged twenty-five; the third died of epilepsy, aged twenty. The eldest girl married a person not related to her; and during the three years of married life she has passed, has had one child The second girl married a little before her elder sister.[2]

103. The eldest son mentioned in Case No. 102, also married a first-cousin, and already has two healthy children.

104. The second daughter mentioned in Case No. 102, also married her first-cousin, and has as yet three children, all perfectly healthy.[3]

105. A marriage between first-cousins, which produced four children, three girls and one boy. The girls are all healthy, but the boy is an idiot. He was the last birth.

106. A marriage between first-cousins, which produced twins, but Dr. Down does not say whether any more were born. The second twin, a girl, is small, deaf, an idiot, and had fits for many years. The father is healthy, but his brothers and sisters have a tendency to consumption. The mother's parents were also distantly related, and she died from a tumour in the brain. Several of her relatives died of consumption, and a cousin of her daughter's was insane from epilepsy.

107. A marriage between first-cousins, which produced at least eight children. One of these, a boy, died of consumption, and the eighth-born was an idiot. The father was sound in mind, but delicate and intemperate; his relations were healthy. The mother had twice given birth to twins; she was healthy, but in the seventh month of her pregnancy she was frightened by a cat, and was ill for a week in consequence, while the child became an idiot.

108. A marriage between first-cousins, which produced six children. Of these, four boys and a girl died of consumption, and the youngest, a girl, is an idiot. The father was sound in mind, but died of pulmonary hæmorrhage. One of his sisters died of consumption. The mother had lost a sister from consumption, and was consumptive herself.

109. A marriage between second-cousins. The mother had been married before to a person who was not related to her, by whom she had three children, two girls who were healthy, and a boy who was epileptic. The second marriage produced four children; first, two healthy girls, then a boy, who is an idiot, and then another of weak

---

[1] See Case No. 103.   [2] See Case No. 104.
[3] Cases Nos. 101 to 104, reported by Ancelon, *Comptes Rendus*, vol. lviii. 1864, pp. 166, 167.

mind. The father was healthy and sound in mind; but his father had lost his sight when young, and two of his brothers had died of consumption. The mother was also healthy and sound in mind, but her father's sister was insane. She believed that the idiocy of the first boy was caused by a fright she had had, and that of the second from constant fear that he also might be an idiot.

110. A marriage between first-cousins, which produced six children, three boys and three girls. One of these girls, the eldest of all the children, was an idiot. At the time of her birth the mother suffered greatly from anxiety regarding pecuniary matters, parturition was difficult, the forceps were employed, and the head greatly crushed. The father was delicate when young; his uncle is imbecile, an aunt died of phthisis, and his father was eccentric and intemperate. The mother is very deaf, and her sister died from cancer.

111. A marriage between second-cousins, which produced at least six children. Of these, the eldest, a boy, died of acute hydrocephalus; the sixth, a girl, is an idiot, but all the rest are bright and healthy. The father had fits when a child, and is weak and ailing, and feeble in mind. The mother is not very strong, and lost a brother from consumption. She says she saw a girl precisely like her idiot daughter, both mentally and physically, at a time when she was seven months advanced in pregnancy.

112. A marriage between third-cousins, which produced eleven children, nine boys and two girls. The seventh birth was twins, a boy and a girl, of which the former is an idiot. All the rest of the children are healthy and intelligent. The father is healthy and sound in mind, but very deaf. The mother is also healthy, but all her relations are consumptive.

113. A marriage between first-cousins, which produced six children, three boys and three girls. The fourth-born, a boy, is an idiot; all the remaining children are particularly intelligent. The father died of Bright's disease, after five years' illness. The mother is healthy, but nervous; had bad health during the pregnancy, and was much distressed at hearing her eldest child had croup at Paris, and could not procure a doctor. She had besides a very bad labour, owing to the large size of the child.

114. A marriage between second-cousins, which produced six children, four boys and two girls. One boy died of scarlatina; one is rather delicate; the fourth-born, a boy, is an idiot; the remaining three are healthy and intelligent. The father is sane and sound; but has lost five brothers and sisters by consumption. The mother

has had a fistula, but is sound in mind. She was very low-spirited during her fourth pregnancy at the prospect of having another child, and but small means to support her family.

115. A marriage between first-cousins, which produced eleven children, five boys and six girls. The first and eleventh births, both boys, are idiots; one boy died of inflammation of the lungs; another of convulsions when a fortnight old; two sisters who were twins died, the one at birth, and the other of convulsions when ten days old; another girl died of fever; and another besides is dead. The remaining three children are all healthy and intelligent. The father was healthy, but below the average in mental power. The mother has always been delicate, is very nervous, had an aunt insane, and all her brothers and sisters died young.

116. A marriage between first-cousins, which produced twelve children. The eleventh-born is an idiot; all the remaining children were endowed with average mental and physical power. The father enjoys good health, but is very irritable and desponding; his mother died of consumption, and he lost one sister from congestion of the brain. The mother also is healthy, but she lost a sister from consumption, and has an uncle who is imbecile. When in the fourth month of her eleventh pregnancy, she suffered severely from sea-sickness; the infant's umbilical cord had to be divided before birth, animation was suspended, and had to be resuscitated artificially.

117. A marriage between first-cousins, which produced nine children, two boys and seven girls. The fourth-born, a boy, is an idiot; one of the others had no parietal bone, and lived but two days; four died at birth, but seemed perfectly constituted; and the first three children were healthy and lived. There was one miscarriage of a child with one leg. The father was healthy, sober, and steady. The mother suffers from chronic bronchitis, but is sound in mind, and comes of a healthy family. The ages of the father and mother at the birth of the idiot were forty-one and twenty-nine; the labour was lingering, ergot of rye was given twice, and the infant's head was mis-shapen.

118. A marriage between first-cousins, which produced six children. Of these, one boy was subject to fits, and died of consumption; another was an idiot; a girl died of epilepsy at seven years of age; and all the rest are bright and healthy. The father is in good health, but of low intelligence. The mother is delicate, and lost a brother and sister from consumption.

119. A marriage between second-cousins, which produced seven children. Of these, one died of hydrocephalus; one of scarlatina; one

of phthisis, and four are idiots. The father has average health, but is a drunkard, and six of his immediate relatives stammer. The mother is very poor in health, suffers from uterine disease, and two of her sisters died young from consumption. She was ill during the whole of her second pregnancy; fell down stairs when in the seventh month; was nervous, and suffered from hallucinations.

120. A marriage between first-cousins, which produced four children. The last-born, a boy, is an idiot. The father is healthy, but his father was eight months in a mad-house, though he ultimately recovered; his mother was very eccentric; and his sister had a spinal disease, and died of consumption. The mother was herself healthy, but all her immediate relatives were consumptive. She had much trouble and business difficulties during her last pregnancy.

121. A marriage between first-cousins, which produced six children, all boys. Of these, the eldest is an idiot; one died in a fit during whooping-cough; one of bronchitis; one is hemiplegic; another, three and a half years old, cannot talk, is only just able to walk, and has a large head. The father is a very weak and nervous man, faints frequently, and his whole bearing is like one suffering from mercurial tremor. The mother is in good health, but she is very nervous, and her mother also is very nervous and bordering on insanity; one uncle of hers died insane, and a cousin is imbecile. In the sixth month of her first pregnancy she tumbled down stairs, and about the same time she was much frightened by her husband falling down in a fainting fit, and this brought on uterine contractions, and a considerable amount of flooding.

122. A marriage between first-cousins, which produced two children. The eldest was born with animation suspended, and he is an idiot. The second, a girl, was born prematurely, and died three days after birth. The father is healthy but irritable; two of his sisters died of consumption, and one of these was also subject to hallucinations. The mother died of consumption; an uncle was insane, brought on as it was supposed by alcoholic abuse. She was frightened two months before her first confinement by stepping on an adder; the labour was lingering, and the child's head much distorted.

123. A marriage between first-cousins, which produced five children, two boys and three girls. The second-born, a boy, was an idiot, as was also one of his sisters. Another sister was quite helpless, and died at the age of fourteen months. The father is healthy and of sound mind. The mother's mother was insane, her aunt died in a mad-house; and a cousin suffers after every confinement from puerperal mania.

124. A marriage between first-cousins, which produced five children, four boys and one girl. Of these, the eldest, a boy, is an idiot; the second and third boys were twins, and are very healthy and sound in mind; the fourth boy died of whooping-cough; the girl is healthy in every respect. The father died of contraction of the bowel, and was quite healthy, but his mother used to stammer. The mother was very nervous, and was frightened during her first pregnancy by an idiotic man. The labour lasted twenty-eight hours, and she was delivered by instruments.[1]

125. A marriage between uncle and niece, proved barren.

126. A marriage between uncle and niece, produced three children with goîtres. Of these, two were also crétins, and one was intelligent.

127. A marriage between uncle and niece, produced one son who was epileptic and one crétin. The father was epileptic, and the mother was nervous and hysteric.

128. A marriage between first-cousins, which proved barren.

129. A marriage between first-cousins, which also proved barren.

130. A marriage between first-cousins, which produced healthy children.

131. A marriage between first-cousins, which also produced healthy children.

132. A marriage between first-cousins, which resulted in four children, who died young, and two cases of phthisis.

133. A marriage between first-cousins, which produced five children. Of these, one was epileptic and died of phthisis; one died at four years of age of typhus; one, aged six, looks tuberculous; one suffers from hydrocephalus; and one is an idiot and tuberculous.

134. A marriage between first-cousins, which produced nine children and two abortions. Of these nine children, two suffer from hydrocephalus; one died of croup; and the remaining six are healthy.

135. A marriage between first-cousins, which produced twelve children. Of these, seven died very young, and the remaining five are healthy.

136. A marriage between first-cousins, thirteen of the children born from which died young, and five were slightly scrofulous.

137. A marriage between first-cousins, which proved barren.

138. A marriage between first-cousins, which produced two deaf-mutes.

---

[1] Cases Nos. 105 to 124, reported by Down, *London Hospital Reports*, vol. iii. 1866, pp. 226-231. They are taken hap-hazard from a number of investigations made on the parentage of idiots.

139. A marriage between first-cousins, which produced four deaf-mutes.

140. A marriage between third-cousins, which produced two deaf-mutes.

141. A marriage between uncle and niece, which produced healthy offspring.

142. A marriage between uncle and niece, which produced three children, two of whom died young, and one is healthy and lives.[1]

143. A marriage between uncle and niece, which produced healthy, but dwarfish children.

144. A marriage between first-cousins, which also produced healthy, though dwarfish children. In this case the mother was scrofulous.

145. A marriage between first-cousins, which produced one healthy child, a girl.

146. A marriage between first-cousins, which produced healthy children.

147. A marriage between first-cousins, which produced six tuberculous children, five of whom are already dead. The parents in this case were tuberculous.[2]

148. A marriage between uncle and niece, which proved barren.

149. A marriage between first-cousins, which produced healthy offspring.

150. Another marriage between first-cousins, which also produced healthy offspring.

151. A marriage between first-cousins, which produced scrofulous children, and one was also rachitic. The parents in this case were scrofulous.

152. A marriage between first-cousins, which produced scrofulous children. The parents were scrofulous.

153. A marriage between first-cousins, which produced two children; one still-born, and the other healthy, and still alive. The one parent was club-footed and the other scrofulous.

154. A marriage between second-cousins. Of the children born, one child, a girl, is scrofulous.[3]

155. A marriage between first-cousins, which resulted in two abor-

---

[1] Cases Nos. 125 to 142, reported by Lombroso, see Mantegazza, *Studj sui Mat. Cons.*, etc., pp. 20-26. In these cases and the following cited by M. Mantegazza, it is a pity that the total number of children is rarely given, and still more rarely the state of health of the parents and relations.

[2] Cases Nos. 143 to 147, reported by Sargenti, see Mantegazza, *ut sup.*, p. 20.

[3] Cases Nos. 148 to 154, reported by Moretti, see Mantegazza, *ut sup.*, p. 22.

tions in the fourth months, four premature still-births, and two children who died before they reached their fifth year.

156. A marriage between first-cousins, which produced ten children. The first birth was twins, premature and deformed; the second birth was again twins, who died before they were five months old; after these the remaining six children were born, who are all healthy.[1]

157. A marriage between first-cousins, which produced ten children. Of these, three were either imbecile, paralytic, or died young; and the remaining seven are perfectly healthy.

158. A marriage between first-cousins, which produced two epileptic children.[2]

159. A marriage between first-cousins, which produced scrofulous children.

160. A marriage between first-cousins, which produced scrofulous, tuberculous, puny children.

161. A marriage between first-cousins, which produced two children, who were deformed and died young.

162. A marriage between first-cousins, which produced four children, who all died of phthisis.

163. A marriage between first-cousins, which produced scrofulous and dwarfish children.

164. A marriage between first-cousins, which resulted in scrofulous and rachitic children.

165. A marriage between first-cousins, which produced one scrofulous child, one albino, and one daughter, who was a crétin.

166. A marriage between first-cousins, which produced healthy offspring.

167. A marriage between first-cousins, which produced only one child, a boy, very puny, and who died at the age of six months.

168. A marriage between first-cousins, which proved barren.

169. A marriage between second-cousins, which proved barren.

170. A marriage between uncle and niece, which produced twins, who died a few days after birth, and one deaf-mute.[3]

171. A marriage between uncle and niece, which produced children of average health.

---

[1] Cases Nos. 155 to 156, reported by Longhi, see Mantegazza, *Studj sui Mat. Cons.*, etc., p. 22.

[2] Cases Nos. 157 to 158, reported by De Orchi, see Mantegazza, *ut sup.*, pp. 22, 24.

[3] Cases Nos. 159 to 170, reported by Demeva, see Mantegazza, *ut sup.*, pp. 22, 24.

172. A marriage between uncle and niece, which produced many, but scrofulous children.

173. A marriage between uncle and niece, which produced healthy offspring.

174. A marriage between first-cousins, which produced many, and healthy children.

175. ⎫
176. ⎬ Marriages between first-cousins, which produced healthy children.
177. ⎭

178. A marriage between first-cousins, which produced only few children, and these died young.

179. A marriage between a man and his half-sister, which produced six children. Of these, two who died young were tuberculous; and three girls are married. The first of these marriages proved barren. The second produced three children who died young; but the mother in this case was tuberculous. The third marriage produced, together with healthy children, some who were tuberculous, and one consumptive; but the mother was tuberculous in this instance also. The mother of the six children was tuberculous.[1]

180. A marriage between first-cousins, doubly related, which produced six children. Of these, two died almost immediately after birth, and four are alive and healthy.

181. A marriage between first-cousins, which produced healthy offspring.

182. A marriage between first-cousins, which produced three children. Of these, one was healthy, and two were tuberculous.

183. ⎫
184. ⎪
185. ⎬ Marriages between second-cousins, all resulting in healthy
186. ⎪   offspring.
187. ⎪
188. ⎭

189. A marriage between third-cousins, and the wife was also derived from a consanguineous marriage. There were five abortions, and one girl stillborn. The mother was slightly rachitic.

190. ⎫ Marriages between third-cousins, which resulted in healthy
191. ⎭   offspring.

192. A marriage between fourth-cousins, which produced healthy offspring.[2]

---

[1] Cases Nos. 171 to 179, reported by Roboletti, see Mantegazza, *Studj sui Mat. Cons.*, etc., p. 24.

[2] Cases Nos. 180 to 192, reported by Liberali, see Mantegazza, *ut sup.*, p. 24.

193. A marriage between first-cousins, which produced three children. Of these, one was a deaf-mute and the two others were healthy. Deaf-mutism was previously unknown in the family.[1]

194. A marriage between first-cousins, which proved barren. Both were healthy.

195. A marriage between first-cousins, which also proved barren. Both were healthy. The husband when sixty years of age married again, and has a child every year.

196. A marriage between first-cousins, which produced six children. Of these, four who were suckled by their mother died in their first year, scrofulous and rachitic; the two others were suckled by a healthy nurse, and, though ailing, they are alive.

197. A marriage between first-cousins, which produced many children, who are all either scrofulous or rachitic.[2]

198. A marriage between first-cousins, which produced scrofulous children.

199. A marriage between first-cousins, which produced five children, two boys and three girls. Of these, all three girls are in bad health, and one died at the age of ten years, scrofulous, and very small for her age. The two boys are healthy.

200. A marriage between first-cousins, which produced four children. Of these, one is imbecile and rachitic; the remaining three children are healthy. The father in this case was tuberculous.

201. A marriage between first-cousins, which produced one scrofulous girl and two abortions.

202. A marriage between first-cousins, which proved barren.

203. A marriage between first-cousins, which produced four children and one abortion. Of these four children, two died young, one girl is scrofulous and ailing, and one boy is healthy. The mother is healthy, but the father was slightly scrofulous.

204. A marriage between first-cousins, which proved barren.

205. A marriage between first-cousins, which produced three children. Of these, two were healthy and one was epileptic. No case of epilepsy was previously known in the family.

206. A marriage between first-cousins, which produced healthy offspring.

207. A marriage between first-cousins, which produced healthy offspring.

---

[1] Case No. 193, reported by Bizzozero, see Mantegazza, *Studj sui Mat. Cons.*, p. 26.

[2] Cases Nos. 194 to 197, reported by Zaniboni, see Mantegazza, *ut sup.*, p. 26.

208. A marriage between first-cousins, which produced two children, who were crétins.

209. A marriage between first-cousins, in which there were many abortions and four robust children born.

210. A marriage between first-cousins, which produced many deaf-mutes and imbeciles.

211. A marriage between first-cousins, which produced two children still-born, and one epileptic boy.

212. A marriage between uncle and niece, in which there were four children born and six abortions. Of these four children, two died in early youth.

213. A marriage between uncle and niece, which produced six children and five abortions. Of these six children, one was paralytic.

214. A marriage between uncle and niece, which produced three children and four abortions. Of these three children, one female infant was deaf.

215. A marriage between uncle and niece, which has produced five children, still young, and all healthy. The parents are healthy.[1]

216. A marriage between nephew and aunt, which produced healthy children.

217. An "incestuous" marriage, which produced healthy children.[2]

218. A marriage between first-cousins. The hands of the parents were deformed, but those of their children were normal, and the children themselves perfectly healthy.[3]

219. A marriage between first-cousins, which produced two children, who both died in their babyhood.

220. A marriage between first-cousins, which produced seven children. Of these, three died in their infancy, and not one of the remaining four was "bright."

221. A marriage between first-cousins, which produced three children. Of these, two boys died young, and the remaining girl has bad eyes.

222. A marriage between first-cousins, which produced a boy with club-foot, and another who died and was an idiot.

223. A marriage between first-cousins, which produced two children, who both died young.

224. A marriage between first-cousins, which produced children who were blear-eyed and feeble.

---

[1] Cases Nos. 198 to 215, reported by Mantegazza, *Studj sui Mat. Cons.*, pp. 22-26.

[2] Cases Nos. 216 to 217, reported by Turck, see Mantegazza, *ut sup.*, p. 20.

[3] Case No. 218, reported by Dobell, see Mantegazza, *ut sup.*, p. 20.

225. A marriage between first-cousins, which produced two children, who were idiots.

226. A marriage between first-cousins, which produced two children, who were deaf-mutes.

227. A marriage between first-cousins, which produced two children, who were blind.

228. A marriage between first-cousins, which produced three children, who were hermaphrodites.

229. A marriage between first-cousins, which produced one child, who was club-footed, one an idiot, one near-sighted, one ailing and irritable, one blind, and the rest feeble.

230. A marriage between first-cousins, which produced ten children, eight sons and two daughters. Of these, two sons and one daughter cannot walk, and the youngest child is a deaf-mute.

231. A marriage between first-cousins, which produced eight children, one son and seven daughters. Of these, three daughters are mentally deranged, and the rest are nervous.

232. A marriage between first-cousins, which produced eight children, three sons and five daughters. Of these, one daughter is an idiot, and two others are feeble in mind.

233. A marriage between first-cousins, which produced five children. The first is healthy and bright, the rest are idiots.[1]

234. Two brothers, tall, strong, and vigorous, married two sisters, their first-cousins. The eldest still lives in the place of his birth, and has several children, of which only the eldest is a deaf-mute, aged twenty.

235. The other has been employed on the railway for six years at Besançon, and has at present six children. Of these, the first, a girl aged twelve, is weak, small, and very timid, but otherwise healthy. The second, a girl aged ten, is slim, tall, and vigorous, but a deaf-mute. The third died young, and appeared to have been healthy. The fourth, a boy aged seven, is strong and big, but a deaf-mute. The fifth a girl, aged four and a half, is very small, talks badly, but hears well. The sixth, only three months old, seems but slightly sensible to noise.[2]

236. A marriage between first-cousins, which produced four children. Of these, one was a deaf-mute, one scrofulous, one an idiot, and one

---

[1] Cases Nos. 219 to 233, reported by the Rev. C. Brooks, see Allen, *The Intermarriage of Relations*, pp. 13, 14.

[2] Cases Nos. 234 to 235, reported by Perron, see Boudin, *Comptes Rendus*, vol. lv. 1862, pp. 659, 660.

epileptic. There was a strong idiosyncrasy in the family of the parents, but no definite taint.[1]

237. A marriage between first-cousins, which produced four children. Of these, two were twins, albinoes, and only lived forty-eight hours; one also an albino, only lived a year. The fourth is well formed, healthy, and not an albino.[2]

238. A marriage between first-cousins, which produced one congenital deaf-mute, besides another child, who became dumb and idiotic at the age of six years without assignable cause.

239. A marriage between first-cousins, which produced two congenital deaf-mutes. Seven of the cousins of these children were also deaf-mutes. The mother was herself the produce of a marriage between first-cousins.

240. A marriage between first-cousins, the parents of the wife being also first-cousins, which produced five congenital deaf-mutes. Two first-cousins of these children were also deaf-mutes. One of the five deaf-mutes married, but none of his children are deaf-mutes.

241. A marriage between third-cousins, which produced eight children. Of these, two became dumb and idiots at the ages respectively of four and five years, without assignable cause.

242. A marriage between first-cousins, which produced one child who became dumb when three years old, without assignable cause; another a girl was born dumb, and became paralyzed shortly after birth.

243. A marriage between second-cousins, which produced five children. Of these, two were congenital deaf-mutes. The father married again a person not related to him, and none of the six children resulting from this second marriage were deaf-mutes.

244. A marriage between second-cousins, which produced eight children. Of these, three were mutes, one of the three being also deaf, another only dumb, and the condition of the third could not be ascertained.

245. A marriage between first-cousins, a child born from which became dumb at the age of two and a half years from fright in his sleep.

246. A marriage between persons not related, but the parents and grand-parents of the husband were first-cousins. Two children born from it were congenital deaf-mutes.

[1] Case No. 236, reported by Carpenter, *Human Physiology*, p. 863, note.
[2] Case No. 237, reported by Goux, see Boudin, *Mém. de la Soc. d'Anthrop. de Paris*, vol. i. 1863, p. 514.

247. A marriage between persons not related, but the wife's parents were first-cousins. They had one child, who was born dumb and an idiot.

248. A marriage between third-cousins; the parents of the husband were besides first-cousins. A child born to this marriage was congenitally dumb; but the mother had been frightened during her pregnancy.[1]

249. A marriage between first-cousins, which produced five children. The eldest of these was sound in mind and body; the second was imbecile; the third died, but at what age and the cause of death is not known; the fourth was imbecile; and the fifth became insane.

The eldest son was married twice, and each time to a person not related to him. By his first marriage he had four children, of which two were healthy, one became insane in adult life, and one died in early infancy. By his second, he had five children, of which one was healthy, one eccentric, one imbecile, and two insane.

250. A marriage between first-cousins, in which apparently the husband was derived from the family described in Case No. 249. Nine children were born, of whom the first was sound in mind, but scrofulous and weak-sighted; the second was sane, but short, had an impediment in his speech, and a cleft palate; the third was also sane, but suffered from spinal curvature; the fourth was an idiot; the fifth was sane, but suffered from bad sight; the sixth was healthy in mind and body; the seventh was imbecile, deformed, and obliged to use crutches; the eighth was sound in mind and body, but short. The ninth was an idiot and a dwarf.

The second married a person not related to him, and has several children, whose condition, however, is not known.

The third was unmarried.

The fifth married a lady not related to him, and the marriage proved barren.

The sixth also married a lady not related to him, and in this case also the marriage proved barren.

The eighth married a person not related to her, and has three children, whose condition is not known.

251. The eldest child of the last case (No. 250) married his first-cousin, and the marriage proved barren.

252. A marriage between first-cousins, which produced five children. Of these the first was eccentric, and remained unmarried; the second was sound in mind and a good business man, but rachitic; the third

---

[1] Cases Nos. 238 to 248, reported in the *Irish Census Returns for* 1871, *Status of Disease*, p. 21.

was healthy; the fourth was imbecile, dwarfish, and impossible to educate; the fifth was an idiot and a dwarf.

The third married a gentleman not related to her, and had two children, of which one was an idiot, but the condition of the other was not ascertained.

253. A marriage between first-cousins, which produced several children. Of these, one boy was an epileptic idiot, and all the rest of the children are sane and sound.

One of the daughters married a man not related to her, and had two boys, one an epileptic idiot, and one healthy.

This last one married a woman not related to him, and had one girl, an epileptic idiot, but all the rest of his children were sane and sound.[1]

254. A marriage between first-cousins, whose fathers were brothers, both salt-workers by profession, and in good health. They had six children, two boys and four girls, all of whom were healthy and intelligent.

255. A brother of the father in the former case (No. 254) also married his first-cousin; both were salt-workers in the marshes, and they had four children, one boy and three girls. The children are quite healthy, and are all very intelligent.

256. A marriage between first-cousins, the children of two sisters. They were fishers by profession, healthy, and had four children, one boy and three girls. Of these, one girl died of a chest disease after she was married, and another girl became slightly deaf after she was seventeen years of age, in consequence of a sore throat, which, in its turn, was caused by exposure to the mists of the marshes while she was engaged in her work at night. The other children are very healthy, and all four are very intelligent. The father was rather dull, but the mother was intelligent.

257. A marriage between first-cousins, the children of a brother and of his sister. Both were intelligent, handsome, and healthy, and salt manufacturers by profession. By the time they had been married six years, they had three children, one boy and two girls, all healthy and very intelligent.

258. A marriage between first-cousins, the children of a brother and of his sister. They were workers in the salt marshes, both handsome and intelligent, and had six children, three boys and three girls. Of these, one boy died at nine years of age from an abscess in the thigh, followed by necrosis of the femur; the others are all healthy

[1] Cases Nos. 249 to 253, reported by Mitchell, *Mem. read before the Anthrop. Soc. of London*, vol. ii. 1866, pp. 405, 406, 408.

and intelligent. The father had had inflammation of the lungs, and was since subject to catarrh; the mother had never had an illness.

259. A marriage between second-cousins. The husband was a salt-worker, and the wife a worker in the marshes. Three children were born to them, two boys and one girl. Of these, one boy died during service with the army; a second boy is married, and has a healthy child; and the girl is in perfectly good health.

260. A marriage between second-cousins, both dealers by profession, and the husband subject to arthritic rheumatism. Five children were born to them, of which one boy died at the age of fourteen months from a spinal disease; a daughter died at the age of seventeen years from typhoid fever; and the other children are all healthy. One daughter, aged twenty, is very big and tall, and a child only twelve years old is very big and strong. The father was very tall.

261. A marriage between second-cousins: in which the husband was a muleteer, and the wife a worker in the salt marshes. When the parents were aged respectively thirty-two and twenty-three years, they had one child, very intelligent, and in good health.

262. A marriage between second-cousins: the husband a salt factor, and the wife a worker in the marshes. The health of both was good. They had already six children by the time the husband was aged thirty-four, and the wife thirty-six, two boys and four girls. There was besides one abortion. All the children are healthy.

263. A marriage between second-cousins. The husband was a worker in the salt marshes, and healthy; the wife was a salt factor, older than her husband by one year, and very nervous. They had no children.

264. A marriage between second-cousins. The husband was a worker in the salt marshes, rather a drunkard, and of poor intelligence. The wife was a dealer, fifteen years younger than her husband, intelligent, and worked for both. They had six children, two boys and four girls, all healthy and intelligent.

265. A marriage between cousins of the seventh degree of relationship. The husband was a lighterman, and died, when only thirty-three years old, of brain fever. The wife was a salt-marsh worker, healthy, and intelligent. They had two children, one boy and one girl, who are both healthy, intelligent, and tall.

266. A marriage between persons related in the seventh degree. The husband was a shoemaker, and healthy; the wife died of a liver disease. They had eight children, two boys and six girls, and there was also an abortion. One boy died at the age of two years.

267. A marriage between second-cousins. The husband was a fund-holder, in good health. They had three children, one boy and two girls. These three children are healthy; one girl is married, and has four healthy children, while the other is a tall, handsome woman.

268. A marriage between second-cousins. The husband was a shoemaker. They had three children, all in good health, and besides, one abortion.

269. A marriage between second-cousins. The husband was a salt factor, in good health; the wife, a worker in the salt marshes, and four years younger than her husband. They had three children, of which one child died the fifth day after birth, and another of croup; the third was, however, quite healthy. The mother's mother is mad.

270. A marriage between second-cousins. The husband, a dealer, and Mayor of Batz, was healthy, but not very intelligent. The wife was the same age as her husband, and only married when aged thirty-eight. She was rather stout, and rather deaf. They had five children, of which two died at birth, two of convulsions at the respective ages of eight and ten days, and the remaining child, a girl, is now aged twenty, and is perfectly healthy.

271. A marriage between second-cousins. The husband was a salt factor, in good health; but his wife was poorly, and frequently bled by means of leeches, and five years younger than her husband. They married young, and had one child, a boy, who is healthy. The father of the wife was rather a drunkard

272. A marriage between second-cousins. The husband was a salt factor, healthy, and four years his wife's senior. The wife was married when aged twenty-seven. They have four children, two boys and two girls, all healthy, and very intelligent.

273. A marriage between relatives of the fifth degree, cousins. The husband was a salt factor, and three years his wife's senior. The wife, like her husband, was perfectly healthy, and married at the age of twenty-eight. They had four children, two boys and two girls, all healthy and intelligent.

274. A marriage between second-cousins. The husband was a worker in the salt marshes, rather a drunkard, and four years his wife's senior. The wife followed the same profession as her husband. They had four children, one boy and three girls, of which one girl, aged nineteen, is scrofulous and consumptive. Nothing is known concerning the health of the father's parents.

275. A marriage between second-cousins. The husband was a salt factor, healthy, and two years his wife's senior. The wife was also healthy, and worked in the salt marshes. They had five children, two boys and three girls, all healthy.

276. A marriage between second-cousins. The husband was a worker in the salt marshes, very healthy, but had been rather addicted to drunkenness ever since his twentieth year. The wife followed the same profession as her husband, and was five years his junior. They had nineteen children, of which eight are living, the rest having all died at ages ranging from two years to thirty. Not one, however, was malformed or infirm; they were all carried off by acute diseases.

277. A marriage between second-cousins. The husband was a worker in the salt marshes; two years his wife's senior, healthy, but fond of drinking. The wife was of the same profession as her husband. They had two children by the time the wife was thirty years old, both boys, and healthy.

278. A marriage between second-cousins, both workers in the salt marshes, and both healthy. They married when the wife was aged twenty-three, and have eleven children, eight boys and three girls. There was also an abortion. All the children were healthy and intelligent.

279. A marriage between second-cousins, both workers in the salt marshes, and both healthy. The wife was three years her husband's junior. They had seven children, five boys and two girls. All of them healthy and intelligent.

280. A marriage between second-cousins, which proved barren. Husband and wife were workers in the salt marshes, healthy and intelligent, and the wife was two years her husband's junior. They have been married twenty years.

281. A marriage between second-cousins. The husband was a salt factor, who ultimately died of asthma. His wife was a worker in the marshes and healthy. They had two children, both boys, and are healthy and intelligent.

282. A marriage between second-cousins. The husband was a salt factor. The wife was of the same profession and healthy. They had four children, all boys, of which two died of croup, one is an advocate, and one a salt factor.

283. A marriage between second-cousins. The husband was a salt factor, healthy, and three years his wife's senior The wife was a worker in the salt marshes, married at the age of twenty-four, and was

healthy. By the time they had been married five years, they had two children, both boys, healthy, and big for their age.

284. A marriage between second-cousins. The husband was healthy and a salt factor by profession. The wife was a worker in the marshes. They had five children, of which three died at the ages respectively of three, four, and seven years, all of acute diseases. One of the two survivors is married, and has four healthy children. The father of the husband was rather a drunkard.

285. A marriage between second-cousins. The husband was a sailor, in good health. His wife was a worker in the salt marshes, and the same age as her husband. They had but one child, who died of croup, aged four years.

286. A marriage between second-cousins. The husband was a drunkard, and two years his wife's senior. The wife was healthy, intelligent, and worked for both. They had three children, one boy and two girls, all healthy and intelligent.

287. A marriage between second-cousins. Husband and wife were workers in the salt marshes, both were healthy, and the husband was four years his wife's senior. They had one child, a girl, healthy, and now aged four years.

288. A marriage between second-cousins. The husband was a salt factor, in good health, and two years his wife's senior. She was a worker in the salt marshes, and also healthy. They have been married five years, and have as yet one child, a boy, very healthy and intelligent.

289. A marriage between second-cousins. The husband was a salt factor, and one year his wife's senior. They had one child, a girl, who died, aged eight days, from an illness called *céran*, characterized by fever, and purple spots on the skin.

290. A marriage between third-cousins. The husband was a carpenter, in good health; the wife was also in good health, and four years the husband's junior. They had four children, three boys and one girl, of which one child died from an accident, and the others are very healthy.

291. A marriage between third-cousins. Husband and wife were workers in the salt marshes, and in good health; the wife was eight years her husband's junior. They had four children, all girls, and very intelligent.

292. A marriage between third-cousins, both of them workers in the salt marshes, and both in the enjoyment of good health. The husband used occasionally to get drunk, and was six years his wife's senior. They had two children by the time the wife was aged thirty, both boys, very strong and intelligent.

293. A marriage between third-cousins, both workers in the salt marshes. The husband was not particularly healthy, and was twenty-two years his wife's senior. They had six children, and one abortion. These children, two boys and four girls, are very healthy and intelligent.

294. A marriage between third-cousins, both workers in the salt marshes. The husband on some occasions got drunk, and was twelve years his wife's senior. They had one child, a boy, healthy and intelligent.

295. A marriage between third-cousins, both workers in the salt marshes, in good health, and of the same age. By the time they were thirty-six years old, three children had been born to them, two boys and one girl. Of these, one child died of croup, but the others are all healthy.

296. A marriage between third-cousins, both workers in the salt marshes, and in good health. The husband was eleven years his wife's senior. By the time he was aged forty they had one child, a girl, very healthy and intelligent.

297. A marriage between cousins related in the seventh degree. The husband was a landowner, at least fifteen years his wife's senior, in good health, but was already dead when his wife was aged thirty-seven. The wife was also healthy. They had six children, one of whom died of croup when two years old, and the rest are very healthy.

298. A marriage between third-cousins. The husband was eight years his wife's senior, a customs' officer by profession, and in the enjoyment of good health. The wife was a worker in the salt marshes, healthy, and intelligent. When she was aged twenty-five, one child had been born to them, healthy and intelligent.

299. A marriage between third-cousins. The husband was a day labourer, in good health, addicted to occasional drunkenness, and six years his wife's senior, who was a worker in the salt marshes, and also in good health. When the wife was twenty-three, one child had been born to them; it is aged fifteen months, is very strong, intelligent, and already speaks.

The other marriages which have taken place in the Commune of Batz were all between relatives, but in more distant degrees, generally between fifth-cousins.[1]

---

[1] Cases Nos. 254 to 299, reported by Voisin, who examined *all* the inhabitants of the Commune of Batz, see the *Mém. de la Soc. d'Anthrop. de Paris*, vol. ii. 1865, pp. 448-459.

APPENDIX. xxxiii

ANALYSIS OF THE ABOVE CASES.

Idiocy: Cases Nos. 1, 2, 4, 19, 22, 59, 67, 105-124, 133, 157, 200, 210, 222, 225, 229, 232, 233, 236, 238, 241, 247, 249, 250, 252, 253.
Insanity: Cases Nos. 22, 68, 231, 249.
Hydrocephalus: Cases Nos. 10, 23, 49, 69, 111, 119, 121, 133, 134.
Meningitis: Cases Nos. 12, 22, 88.
Epilepsy: Cases Nos. 27, 30, 51, 52, 59, 102, 118, 127, 133, 158, 205, 211, 236, 253.
Chorea: Case No. 3.
Convulsions: Cases Nos. 21, 23, 29, 34, 50, 53, 67, 69, 106, 115, 118, 121, 270, 279.
Croup: Cases Nos. 29, 92, 94, 134, 269, 282, 285, 295, 297.
Whooping-cough: Cases Nos. 121, 124.
Paralysis: Cases Nos. 121, 157, 213, 242.
Deaf-mutism: Cases Nos. 2, 5, 7, 8, 20, 22, 30-33, 35-47, 57, 58, 64, 68, 69, 138, 139, 140, 170, 193, 210, 226, 230, 234-236, 238-240, 243, 244, 246.
Mutism: Cases Nos. 121, 238, 241, 242, 244, 245, 247, 248.
Defective-hearing: Cases Nos. 22, 34, 69, 88, 106, 214, 256.
Defective-speech: Cases Nos. 7, 25, 49, 67, 68, 235, 250.
Blindness: Cases Nos. 66, 227, 229.
Injuries to the eye: Cases Nos. 36, 221, 224, 229, 250.
Goître: Case No. 126.
Crétinism: Cases Nos. 126, 127, 165, 208.
Polydactylism: Cases Nos. 14, 56.
Anencephalous: Case No. 13.
Want of cranial-arch: Cases Nos. 15, 117.
Hermaphroditism: Case No. 228.
Hare-lip: Cases Nos. 13, 61-63, 65.
Cleft-palate: Cases Nos. 61, 65, 250.
Club-foot: Cases Nos. 13, 222, 229.
Lameness: Cases Nos. 22, 44, 230, 250.
Malformations in general: Cases Nos. 2, 3, 8, 13, 25, 26, 117, 156, 161, 250.
Rickets: Cases Nos. 3, 28, 151, 164, 196, 197, 200, 252.
Spinal disease: Cases Nos. 24, 55, 66, 250, 260.
Dwarfing: Cases Nos. 30, 143, 144, 163, 199, 235, 250, 252.
Scrofula: Cases Nos. 25, 28, 30, 52, 67-69, 136, 151, 152, 154, 159, 160, 163-165, 172, 196-199, 201, 203, 236, 250, 274.
Tuberculosis: Cases Nos. 21, 35, 83, 107, 108, 118, 119, 132, 133, 147, 160, 162, 179, 182, 274.
Albinoïsm: Cases Nos. 1, 9, 16-18, 165, 237.
Hæmorrhage: Case No. 29.
Psoriasis: Case No. 23.
Early death in the family: Cases Nos. 3, 8-12, 17, 18, 21-25, 29, 32, 35,

44, 50, 53, 54, 56, 61-65, 67, 69, 79, 82, 83, 91, 92, 94, 111, 115, 117-119, 121-124, 132-136, 142, 147, 155-157, 161, 167, 170, 178-180, 196, 199, 203, 212, 219-223, 235, 237, 258, 260, 266, 269, 270, 276, 282, 284, 285, 289, 295, 297.

Healthy children only in the family: Cases Nos. 6, 48, 70-73, 75, 76, 78, 80, 84, 85, 89, 90, 93, 95-98, 100, 103, 104, 130, 131, 141, 145, 146, 149, 150, 166, 171, 173-177, 181, 183-188, 190-192, 206, 207, 209, 215-218, 254, 255, 257, 261, 262, 264, 265, 267, 268, 271-273, 275-279, 281, 283, 286-288, 290-294, 296, 298, 299.

In which there is a double marriage of the same person, one consanguineous, and one not: Cases Nos. 18, 55, 56, 60, 98, 109, 195, 243.

In which children born from consanguineous marriages marry persons not related to them: Cases Nos. 2, 7, 8, 25, 48, 49, 68, 69, 93, 102, 179, 240, 246, 247, 249, 250, 252, 253, 259, 267, 284.

In which deaf-mutes marry: Cases Nos. 7, 8, 68, 69, 240.

Still-born: Cases Nos. 3, 13, 69, 74, 117, 122, 153, 155, 189, 211, 270.

Abortion: Cases Nos. 99, 117, 134, 155, 189, 201, 203, 209, 212-214, 262, 266, 268, 278, 293.

The number of children born to each marriage:—[1]

No. of Case.	No. of Births.	No. of Case.	No. of Births.	No. of Case.	No. of Births.	No. of Case.	No. of Births.	No. of Case.	No. of Births.	No. of Case.	No. of Births.
1	2∞	24	10	47	4	70	2	93	3	116	12
2	4	25	7	48	2	71	1	94	7	117	10
3	1∞	26	2∞	49	1	72	2	95	2∞	118	6
4	2	27	1	50	3	73	4	96	3∞	119	7
5	8∞	28	6	51	1	74	1	97	8	120	4
6	3	29	1	52	2	75	2∞	98	1∞	121	6
7	5	30	4	53	4	76	1∞	99	3	122	2
8	4	31	2∞	54	1	77	1	100	5	123	5
9	1∞	32	2	55	2	78	1∞	101	0	124	5
10	11	33	3	56	2	79	2	102	5	125	0
11	8	34·	1	57	2	80	1	103	2∞	126	3∞
12	5	35	7	58	3∞	81	0	104	3∞	127	2∞
13	1∞	36	4	59	4	82	4	105	4	128	0
14	1∞	37	2	60	0	83	2	106	2∞	129	0
15	1∞	38	2	61	2	84	2	107	8∞	130	1∞
16	1∞	39	8	62	7	85	2∞	108	6	131	1∞
17	2	40	2	63	3	86	1	109	4	132	6∞
18	5	41	3	64	5	87	2	110	6	133	5
19	2	42	2	65	4	88	2	111	6∞	134	11
20	2∞	43	2	66	3	89	3	112	11	135	12
21	14	44	2	67	5	90	7	113	6	136	13∞
22	8	45	4	68	5	91	7	114	6	137	0
23	8	46	5	69	8	92	3∞	115	11	138	2∞

[1] Each sign of fertility, such as an abortion, is put down; and the sign ∞ shows that probably more children were born than are reported in that case.

APPENDIX. XXXV

No of Case.	No. of Births.	No. of Case.	No. of Births.	No. of Case.	No. of Births.	No. of Case.	No. of Births.	No. of Case.	No. of Births.	No. of Case.	No. of Births.
139	4∞	166	1∞	193	3	220	7	247	1∞	274	4
140	2∞	167	1	194	0	221	3	248	1∞	275	5
141	1∞	168	0	195	0	222	2	249	5	276	19
142	3	169	0	196	6	223	2∞	250	9	277	2∞
143	1∞	170	3∞	197	1∞	224	1∞	251	0	278	12
144	1∞	171	1∞	198	1∞	225	2∞	252	5	279	7
145	1	172	1∞	199	5	226	2∞	253	2∞	280	0
146	1∞	173	1∞	200	4	227	2∞	254	6	281	2
147	6∞	174	1∞	201	3	228	3∞	255	4	282	4
148	0	175	1∞	202	0	229	6∞	256	4	283	2∞
149	1∞	176	1∞	203	5	230	10	257	3∞	284	5
150	1∞	177	1∞	204	0	231	8	258	6	285	1
151	1∞	178	1∞	205	3	232	8	259	3	286	3
152	1∞	179	6	206	1∞	233	5	260	5	287	1
153	2	180	6	207	1∞	234	2∞	261	1∞	288	1∞
154	1∞	181	1∞	208	2	235	6∞	262	7∞	289	1
155	8	182	3	209	4∞	236	4	263	0	290	4
156	10	183	1∞	210	1∞	237	4	264	6	291	4
157	10	184	1∞	211	3∞	238	2∞	265	2	292	2∞
158	2∞	185	1∞	212	10	239	2∞	266	9	293	7
159	1∞	186	1∞	213	11	240	5∞	267	3	294	1
160	1∞	187	1∞	214	7	241	8	268	4	295	3∞
161	2∞	188	1∞	215	5∞	242	2∞	269	3	296	1∞
162	4∞	189	6∞	216	1∞	243	5	270	5	297	6
163	1∞	190	1∞	217	1∞	244	8	271	1	298	1∞
164	1∞	191	1∞	218	1∞	245	1∞	272	4	299	1∞
165	3∞	192	1∞	219	2∞	246	2∞	273	4		

Total, 1,155 children, when every ∞ counts as one, of which 37 were abortions, and 17 out of the 299 marriages were barren.

Cases No.			Reported by	Cases No.			Reported by
1	to	4,	Balley.	57,			Duteval.
5,			Brochard.	58	to	59,	Trousseau.
6	,,	8,	De Ranse.	60,			Lisle.
9,			Aubé.	61,			Potier-Duplessy.
10	,,	21,	Devay.	62	,,	65,	Rizet.
22,			Forestier.	66,			Robillard.
23,			Potton.	67	,,	69,	Ponsin.
24	,,	25,	Viennois.	70	,,	92,	Bourgeois.
26	,,	27,	Bondet.	93	,,	96,	Devic.
28,			Teissier.	97,			Livy.
29,			Doyon.	98	,,	100,	Dally.
30	,,	48,	Chazarain.	101	,,	104,	Ancelon.
49	,,	56,	Chipault.	105	,,	124,	Down.

Cases No.	Reported by	Cases No.	Reported by
125 to 142,	Lombroso.	216 to 217,	Turck.
143 ,, 147,	Sargenti.	218,	Dobell.
148 ,, 154,	Moretti.	219 ,, 233,	Brooks.
155 ,, 156,	Longhi.	234 ,, 235,	Perron.
157 ,, 158,	De Orchi.	236,	Carpenter.
159 ,, 170,	Demeva.	237,	Goux.
171 ,, 179,	Roboletti.	238 ,, 248,	Reported in the Irish Census for 1871.
180 ,, 192,	Liberali.		
193,	Bizzozero.		
194 ,, 197,	Zaniboni.	249 ,, 253,	Mitchell.
198 ,, 215,	Mantegazza.	254 ,, 299,	Voisin.

Dr. Bemiss gives an account of 833 marriages between relations, which need not here be given in full as the account is not difficult to get at,[1] and it would take up a great deal of space. The results, however, are given in the following table:—[2]

Classes of Relationship.	No. of Observations.	Average No. of Births.	Per Cent.							
			Defective.	Deaf-mutes.	Blind.	Idiots.	Insane.	Epileptic.	Scrofulous.	Deformed.	Died Young.
A. Incest with parent, or between brother and sister	10	3·1	93·5	...	...	61·2	...	3·2	16·1	35·4	...
B. Marriage with niece, or aunt	12	4·42	75·4	1·9	5·6	5·6	1·0	1·9	20·7	26·4	43·3
C. Marriages between blood-relations, the issue of blood-relations... ...	56	4·18	53·8	4·2	5·1	12·8	1·2	1·7	18·8	3·8	26·9
D. Marriages between double first-cousins	27	5·7	27·2	1·2	1·2	2·5	3·8	1·2	6·3	1·2	35·0
E. Marriages between first-cousins	580	4·8	24·9	4·2	2·2	8·3	...	1·6	6·2	1·9	22·5
F. Marriages between second-cousins	112	4·58	13·0	1·7	...	3·3	...	1·1	2·0	1·7	16·5
G. Marriages between third-cousins	12	4·92	27·0	5·0	...	1·7	1·7	3·4	16·0	..	13·5
H. Marriages between first-cousins, irregularly reported	24	5·0	17·5	2·5	...	2·5	1·6	...	12·5	...	10·0
Total	833	4·6	28·7	3·6	2·1	7·0	2·04	1·5	7·6	2·4	22·4
Marriages between persons in no way related ...	125	6·7	2·1	0·35	0·1	0·71	0·1	0·35	0·1	...	16·0

This "table of aggregates," as Dr. Bemiss calls it, does not include all

[1] Bemiss, *Report on Influence of Marriages of Consanguinity upon Offspring*, *Trans. of the Amer. Med. Assoc.*, 1858, vol. xi. pp. 324-419.
[2] *Ibid.*, pp. 420-423; but I have condensed it slightly.

the cases he gives; for these are 873, producing 4,124 children, while the table only gives 833 producing 3,942 children. I have besides noticed that the averages of births are not quite correct, especially in class A, but have amended this in my copy of the table.¹ Of course these results are open to the objections already given: Dr. Bemiss himself says, "It is natural for contributors to overlook many of the more fortunate results of family intermarriage, and furnish those followed by defective offspring or sterility."² M. Boudin quotes the above statistics of Dr. Bemiss, but without giving any authority, and because, as we shall see, he did not copy them first hand, he has given them wrongly.³ He must, indeed, have obtained them from the *Annales d'Hygiène*,⁴ but he has contrived to make the further blunder of attributing their collection to Mr. Morris, whereas Professor O. W. Morris appears only to have been president of the American Medical Association, who appointed a committee to inquire into the subject,⁵ and M. Dally seems even to have doubted his existence.⁶ If M. Boudin had inquired a little further, he would have found that the *Annales d'Hygiène* obtained its information from Ranking's *Abstracts*,⁷ which oddly enough took its information from the *Dublin Hospital Gazette*, which got its information from the *Art Médicale*,⁸ which reprinted it from the *Nouvelliste de Rouen*.⁹ But M. Boudin did not look any further, and disdaining alike the doubts of Dally, and the notice of his authority, reprints his statement intact in the *Mém. de la Soc. d'Anthropologie*.¹⁰

In another article, Dr. Bemiss says of 34 marriages between relations which he examined, 7 or 20·5 per cent. proved barren. The remaining

¹ M. Devay is incorrect as usual in referring to these statistics. He says 873 marriages between relations produced 3,900 children; whereas they produced 4,124. These 873 marriages, he says, were contracted in the State of Ohio; whereas they were spread over twenty-five different States. As we have already seen, he falsely asserts that marriages between first-cousins are prohibited (see p. 59 of this work), and as we shall see directly is guilty of further carelessness (Devay, *Du Danger*, etc., pp. 141-142).

² Bemiss, *Report on Influence of Marriages of Consanguinity upon Offspring*, Trans. of the Amer. Med. Assoc., 1858, vol. xi. p. 323.

³ Boudin, Ann. d'Hygiène, vol. xviii. 1862, p. 13, where he says *Morris obtained 883 returns of marriages between relations, which produced 4,013 children*!

⁴ Annales d'Hygiène, vol. xvii. 1862, p. 227.

⁵ Ibid.; and Bemiss, ut sup., p. 322; and Ranking's *Abstracts*, vol. xxxiii. 1861, p. 12.

⁶ Dally, Gaz. Heb., vol. ix. 1862, p. 515, where he says that neither M. Perier, M. Broca, nor M. Giraldés, could after the most diligent inquiry find a trace of his existence.

⁷ Ranking's *Half-Yearly Abstract of the Med. Sciences*, vol. xxix. 1859, p. 10.

⁸ December 1st, 1858.

⁹ Ranking, ut sup., vol. xxix. 1859, p. 10.

¹⁰ Mém. de la Soc. d'Anthrop. de Paris, 1863, vol. i. p. 511.

27 marriages produced 192 children,[1] or 5·6 children per marriage, barren and fertile. Of these 192 children, 58 died young, from causes which were given in 24 cases, to wit: 15 from consumption, 8 from "spasmodic affections," and 1 from hydrocephalus. Of the remaining 134, 46 were healthy, 32 are said to be " deteriorated, but without absolute indications of disease," of 9 nothing was known, while of the remaining 47, 23 were scrofulous, 4 were epileptic, 2 were insane, 2 were deformed, 4 were idiots, 2 were dumb, 2 were blind, 6 had defective sight, 5 were albinoes, and 1 suffered from chorea. These make together 51 cases, but Dr. Bemiss says they were only 47; hence several must have been affected with one or more of these diseases.[2]

Dr. Howe obtained from Massachusetts statistical tables, 17 cases of consanguineous marriage. "Most of the parents were intemperate or scrofulous; some were both the one and the other." These marriages produced 95 children, or about 5·5 per marriage; 44 were idiots, or nearly 50 per cent.; 12 were scrofulous or puny, 1 was deaf, and 1 was a dwarf.[3] In some cases all the children were very scrofulous and puny, and in one case, in a family of eight children, 5 were idiots.[4]

Dr. Mitchell collected 45 cases of consanguineous marriage, of which 8 were either virtually or actually barren. The 37 fertile marriages produced 146 children, or an average of about 4 per fertile marriage, and about 3 for all the marriages, barren and fertile. Of these children,

8, or 5·5 per cent.	were idiots.			3, or 2·0 per cent.	were deformed (spinal, etc.)		
5 ,, 3·4	,,	,,	imbecile.				
11 ,, 7·5	,,	,,	insane.	6 ,, 4·1	,,	,,	lame.
2 ,, 1·4	,,	,,	epileptic.	1 ,, 0·7	,,		was rachitic.
4 ,, 3·0	,,	,,	paralytic.	22 ,, 15·0	,,		were consumptive, scrofulous, or manifestly of weak constitution.
2 ,, 1·4	,,	,,	deaf-mutes.				
3 ,, 2·0	,,	,,	blind (congenital?)				
2 ,, 1·4 per cent.	,,		"defective" in vision.				

In 8 cases, none of the above diseases or defects appeared among the

---

[1] In the *Journal of Psych., Med., and Mental Path.* for April, 1857 (vol. x.), p. 369, it makes this number 191, and on the following page 192. I do not know whether this mistake occurs also in the *North Amer. Med. Chir. Review*, for January, 1857, from which this article is taken; but it is copied with great exactitude by M. Devay (*Du Danger*, etc., pp. 140-141).

[2] Bemiss, *Journal of Psych., Med., and Mental Path.*, vol. x. 1857, pp. 369, 370.

[3] M. Devay says: "44 sont idiots et 14 scrofuleux," ignoring the classification of Dr. Howe (*Du Danger*, etc,, p. 142).

[4] Howe, *On the Causes of Idiocy, in the Journal of Psych., Med., and Mental Path.*, July, 1858, pp. 393-394.

children; and therefore 29, or 64·5 per cent. of all the marriages produced children who were affected in some way injuriously. Dr. Mitchell, however, adds, that though his notes show a total of only 146 children born, there were probably many more, since some of the marriages were very prolific.[1]

M. Cadiot collected 54 cases of marriage between persons related in the third or fourth degree with this result:—

	No.	Per Cent.
Barren	14	25·9
Marriages producing children who were scrofulous, rachitic, idiots, or deaf-mutes	18	33·4
Marriages which produced children who all died before reaching adult age	7	13·0
Marriages which produced healthy children	15	28·0[2]

M. Ancelon, however, as a complement to the above statement, gives the results of an inquiry into the whole population of Dieuze, a town of some 3,700 persons, composed of 800 families. The *non-consanguineous* marriages gave the following results:—

	Per Cent.
Barren	7·50
Marriages producing children who were scrofulous, rachitic, consumptive, idiots, or deaf-mutes	47·33
Marriages producing children who all died before reaching adult age	0·69
Marriages which produced healthy children	44·93

There were only four consanguineous marriages, which are given above. Cases Nos. 101 to 104.[3]

Of 13 cases collected by M. Devay 8 were barren, 4 produced scrofulous children who all died before they attained their fourteenth year, and 1 produced a child afflicted with ichthyosis. Of another set of 26 cases he collected, 11 turned out unfortunately, 1 produced an epileptic child, 3 others produced children who died of hydrocephalus or convulsions, 2 were barren, and 5 "produced *two* children whose state of health left much to be desired;" 4 only produced healthy offspring. To these observations M. Devay adds 82 more, which "offer much analogy to the preceding ones." Of these 14, he says, were barren, "which makes with the *eight* mentioned above, in all 22." Of these 22 cases of sterility, he continues, 16 were absolutely sterile, and in 6 there were miscarriages, but the marriages remained barren, and on the whole 121 cases of consanguineous marriage, he found 17 cases in which there had been mis-

---

[1] Mitchell, *Mem. read before the Anthrop. Soc. of London*, vol. ii. 1866, p. 403.

[2] Cadiot, *Comptes Rendus*, vol. lvii. 1863, p. 978.

[3] Ancelon, *Comptes Rendus*, vol. lviii. 1864, pp. 166, 167.

carriages, of which 11 were preceded or succeeded by births. In the 82 cases, there was another example of ichthyosis.[1]

Of 19 cases recorded of marriages between relatives by M. Courtans, from Upper Savoy, 6 idiots were born, 5 imbeciles, and 6 rachitic children.[2]

Dr. Viennois found that among 6 marriages of persons related to each other, 1 was barren, and the 5 fertile unions produced 14 children, or an average of 2·8 children per marriage. Of these, 1 was an idiot, 2 were blind, 3 were lame, and 1 had hydrocephalus, or rather, there was a disproportion between the head and body.[3]

M. Dally reports a case of continual intermarriage between two families. These marriages never took place in a more distant degree than first-cousins, except in two instances where the daughter of first-cousins has married her second-cousin. This has continued for five generations, and the mean number of children per marriage was 3 or 4. The total number of branches, direct and collateral, is 120 to 140. This number of branches he considers surprising, since a great many of the family have devoted themselves to celibacy. There has not been a single case of idiocy or deaf-mutism; but there were two deaths from consumption, caused in the one case by a cold, and the other from no appreciable cause; and 1 case of senile insanity in a female aged 68, who died three years after. There was no predisposition to disease manifested in the family, except perhaps to rheumatism, and that was limited to a few individuals.[4]

M. Bourgeois gives a genealogical table from his own family, in which, in the course of five generations there were more than eight consanguineous marriages. Only one of these marriages proved barren, and that was not a marriage of consanguinity: in this case the fault probably lay with the wife, who was no relation; the husband was the offspring of cousins three generations higher up. Of all the eight consanguineous marriages here given, only the grand-children and great-grand-children of Flavien and Emilie G. are at all scrofulous; all the others are healthy. The last generation is as yet incomplete[5]

M. Seguin gives the results of ten consanguineous marriages which have occurred in his own family. These give an average of 6·1 children per marriage, of which the average life was 30 years. One of these marriages was barren, which makes the fertile marriages average 6·78 children per marriage. There was not a single case of deaf-mutism, hydrocephalus, impediment of speech, or supernumerary fingers or toes, among all these children; and this notwithstanding that the relationship of the parents

---

[1] Devay, *Du Danger*, etc., pp. 89, 90, 93, 104.
[2] Mantegazza, *Studj sui Matrim. Cons.*, p. 20.
[3] Devay, *Un Mot sur le Danger des Mar. Cons.*, p. 30.
[4] Dally, *Anthrop. Review*, May, 1864, pp. 95, 96.
[5] Bourgeois, *Comptes Rendus*, vol. lvi. 1863, pp. 180, 181; and *Quelle est l'Influence*, etc., Thèse No. 91, Paris, 1859, pp. 27, 28.

```
                M. G.
             1690. d. 1745.
              │
    ┌─────────┴──────────┐
    │                    │
= St. Martin G.     G. the elder   =   Charlotte D.
  1738. d. 1814.      brother.             Her husband's first-
                    1725. d. 1805.         cousin once removed.
                                           She had 15 children.
```

Delphin G. 1779, d. of cholera, 1832	9 others who died unmarried. Of these, 5 lived to old age, especially 2 priests.
Flavien G., 1764, d. 1834.	
= Emilie G., 1775, d. 1812. She was doubly first-cousin to her husband.	

- Ernest G.
- Paul G. ⎫
- Sylvain G. ⎬ Each several children
- Narcisse G. ⎭
- Joseph G. — 1 child
- Théophile G. — 5 children
- Ernestine G. — 2 children — 5 children
- Antyme G., 1808 = 6 children four times related
- Flavie G., 1809 — 7 daughters — 6 children from two of the daughters
- Ernestine G., 1802 ⎧ Adolphe G., 1807. d. 1812, of marsh fever
- Alphonse G., 18 d. 1830, unmarried.
- Jules G., 1799 ⎧ Jules G. I., 1825. d. 1829, of meningitis.
                ⎨ Jules G. II., 1843. d. 1857, of typhoid fever.
                ⎩ Adolphe G., 1827. Unmarried.
- Flore G., 1830.

's Genealogical Table: "COMPTES RENDUS," vol. lvi., 1863, pp. 180, 181.

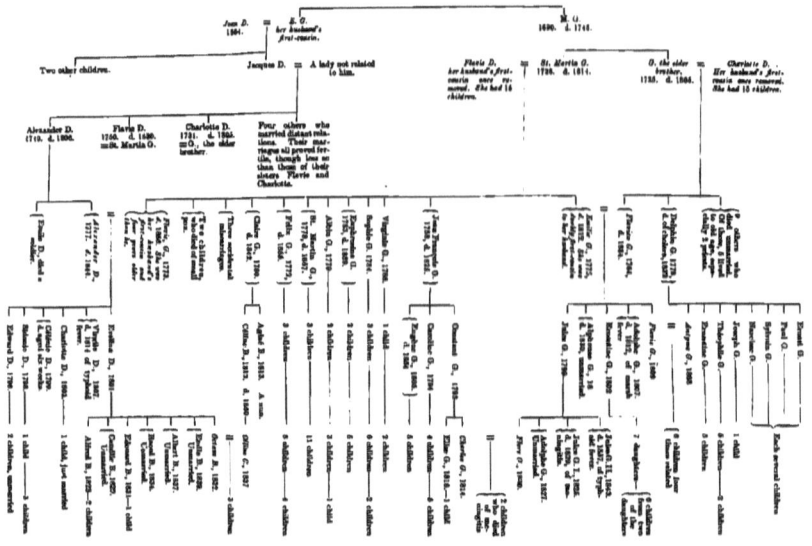

APPENDIX.

was very close and complicated, as may be seen from the following table:—

	Degree of the Relationship of the Parents.	Date of the Parent's Marriage.	Total No. of Children.	No. of Children living 1863.	Sum of Years of the Children's Ages up to 1863.
I. Jean Baptiste de Montgolfier, married Méranie de Montgolfier	First-cousins	1805	10	7	520
II. Elie de Montgolfier, married Pauline Duret, daughter of Jeanne de Montgolfier. .	Ditto	1812	9	8	320
III. Raymond, son of Elie de Montgolfier, married Julie Seguin, daughter of Augustine, who was the daughter of Jeanne de Montgolfier	Ditto	1840	5	3	55
IV. Laurent, son of Elie de Montgolfier, married Hélène Seguin, daughter of Augustine, daughter of Jeanne de Montgolfier	Ditto	1840	3	2	40
V. Eugène de Montgolfier, married Jenny de Montgolfier.	Ditto	1845	5	5	55
VI. Marc Seguin, married Augustine, daughter of Jeanne de Montgolfier	Ditto	1813	13	5	450
VII. The same, married a second time to Augustine, daughter of Elie de Montgolfier	Uncle and niece	1838	6	6	70
VIII. Camille Seguin, married Célie, daughter of Jeanne Seguin	First-cousins	1814	8	8	320
IX. Paul Seguin, married Thérèse, daughter of Camille Seguin	Uncle and niece	1840	2	2	15
X. Joseph Seguin, married Marie, daughter of Lydie de Montgolfier	First-cousins	1858	0	0	0[1]

Mr. Power says that from five marriages between first-cousins among his own immediate relations, 33 children were born, or 6·6 per marriage. Of these, 8 died young, or 24·3 per cent; 1 from teething, 2 from croup,

---

[1] Seguin, *Comptes Rendus*, vol. lvii. 1863, pp. 253, 254.

and 1 from whooping-cough (all injudiciously fed); 1 from accident, 1 from cyanosis, and 2 from well-marked scrofulous disease. The last two deaths occurred in the same family, and were the only offspring of an extremely obese father, and a highly scrofulous mother. The surviving children are of an unusually fine and healthy growth.[1]

M. Perier says that for some years he has noted down every case of consanguineous marriage which he has heard of. These cases now, he says, amount to 26, for the most part between first-cousins and second-cousins, and generally among the well-to-do class and where the parents are in good health; nor was there among the whole 26 a single case of misadventure fairly attributable to consanguinity, and isolated from every other cause.[2]

[1] Carpenter's *Human Physiology*, 1869, p. 863, note.
[2] Perier, *Mém. de la Soc. d'Anthrop. de Paris*, vol. i. 1863, p. 236.

WORKS CITED.

Adam, Wm.: "Consanguinity in Marriage." In the Fortnightly Review, Nos. 12 and 13, for November 1st and 15th, 1865. London.
Adams, Wm.: Club Foot; its Causes, Pathology, and Treatment. London, 1866.
Allen, Nathan: The Intermarriage of Relations. New York, 1869. (A reprint from the Quarterly Journal of Psychological Medicine and Medical Jurisprudence, April, 1869.)
Amyraut, Moyse: Considérations sur les Droits par lesquels la Nature a reiglé les Mariages. Saumur, 1648.
Ancelon: In the Comptes Rendus Hebdomadaires des Séances de l'Académie des Sciences, vol. lviii., 1864, pp. 166, 167.
Annales d'Hygiène, 2nd série, vol. xvii., January, 1862, pp. 222-229, No. 33: "Des Mariages Consanguins—examen des travaux recent sur ce sujet."
Aubé, C.: "Note sur les inconvénients qui peuvent résulter du défaut de croisement dans la propagation des espèces animales." Read before the Société d'Acclimatation, February 6th, 1857. (Quoted by M. Devay, in his "Du Danger des Mariages Consanguins," Paris, 1862, pp. 50-64.)
Augustine: Citie of God; with the learned comments of Jo. Lod Vives; Englished by J. Healy. London, 1610.
Balley: "Note sur les inconvénients des alliances consanguines," in the Comptes Rendus Hebdomadaires des Séances de l'Académie des Sciences, vol. lvi., 1863, pp. 135, 136. Paris.

"Sur la surdi-mutité et ses rapports avec les alliances consanguines; recherches faites à l'asile des sourds-muets de Rome." *Ibid.*, vol. lvii., 1863, pp. 870, 871. See also, the Gazette Médicale de Paris, vol. xviii., p. 804, 3rd Series, December 5th, 1863.
Beaudouin: In the Comptes Rendus Hebdomadaires des Séances de l'Académie des Sciences, vol. lv., 1862, pp. 236-238. Paris.
Beddoe, John: On the Stature and Bulk of Man in the British Isles. London, 1870. (A reprint from vol. iii. of the Memoirs of the Anthropological Society of London.)

Beigel, Hermann: Beitrag zur Geschichte und Pathologie des Albinismus partialis und der Vitiligo. Dresden, 1864.
Bemiss, S. M.: "Report on Influence of Marriages of Consanguinity upon Offspring," in the Transactions of the American Medical Association, vol. xi., 1858, pp. 319-425. Philadelphia.
"On Marriages of Consanguinity," in the Journal of Psychological Medicine and Mental Pathology, New Series, No. vi., April, 1857; vol. x., pp. 368-379.
Biographie Universelle, Ancienne et Moderne. Paris, 1811.
Blunt, J. H.: Dictionary of Sects, Heresies, Ecclesiastical Parties, and Schools of Religious Thought. London, Oxford, and Cambridge, 1874.
Boccaccio: Decamerone. London, 1825.
Boudin: "Dangers des Unions Consanguines et Nécessité des Croisements dans l'espèce humaine et parmi les animaux," in the Annales d'Hygiène publique et de Médecine légale, vol. xviii., No. 35, 2nd series, July, 1862, pp. 5-82. Paris. Also in the Mémoires de la Société d'Anthropologie de Paris, vol. i., 1863, pp. 505-557.
A Reply to a Letter from the Grand Rabbi Isidore, in the Annales d'Hygiène publique et de Médecine légale, vol. xviii., 2nd series, No. 36, October, 1862, pp. 460-463. Paris.
"Études statistiques sur les dangers des unions consanguines dans l'espèce humaine et parmi les animaux," in the Journale de la Société de Statistique de Paris, Nos. 3 and 4, for March and April, pp. 69-84, and 103-120. 1862.
Bounty: The Eventful History of the Mutiny and Piratical Seizure of H.M.S. *Bounty;* its Cause and Consequences. London, 1831.
Bourgeois, Alfred: "Quelle est l'influence des mariages consanguins sur les générations?" in the collection of Thèses of the Ecole de Médecine, Paris, vol. ii. Thèse No. 91, 1859. Also in the Comptes Rendus Hebdomadaires des Séances de l'Académie des Sciences, vol. lvi., pp. 177-181, January 26th. Paris, 1863.
Brand, J.: Popular Antiquities. London, 1849.
Broca, P.: On the Phenomena of Hybridity in the Genus Homo. Edited by Blake. London, 1864.
Brochard: In the Comptes Rendus Hebdomadaires des Séances de l'Académie des Sciences, vol. lv., 1862, pp. 43, 44.
Brodhurst, B. E.: The Deformities of the Human Body. London, 1871.
Browne, J. C.: Indian Infanticide; its Origin, Progress, and Suppression. London, 1857.
Buckle, H. T.: History of Civilization in England, vol. ii. London, 1861.
Miscellaneous and Posthumous Works. Edited by Helen Taylor. London, 1872.

Burton, Robert: Anatomy of Melancholy. Oxford, 1621.
Cadiot: "Effets des alliances consanguines," in the Comptes Rendus Hebdomadaires des Séances de l'Académie des Sciences, vol. lvii., 1863, p. 978. Paris.
Calmet: Dictionary of the Bible, with Biblical Fragments by Charles Taylor. London, 1847.
Campbell, Hugh: Deafness; its various causes, and their successful removal by electrolysis, etc. London, 1872.
Campbell, James: "Polygamy; its Influence on Sex and Population," in the Journal of Anthropology, for October. London, 1870.
Carpenter, W. B.: Principles of Human Physiology. Edited by Henry Power. London, 1869.
Cecil: *Pseudonym*. See Tongue, Cornelius.
Chalmers, Alex.: Biographical Dictionary. 1812-17.
Chateauneuf, B. de: "Mémoire sur la durée des familles nobles de France," in the Annales d'Hygiène, for January, 1846, vol. xxxv., pp. 27-56. Paris.
Chazarain, L. T.: "Du mariage entre consanguins considéré comme cause de dégénérescence organique, et plus particulièrement de surdi-mutité congéniale," in the collection of Thèses of Montpellier. Thèse No. 63. 1859.
Child, G. W.: Essays on Physiological Subjects. London, 1869.
Chipault, Antony: "Etude sur les mariages consanguins et sur le croisement dans les règnes animal et végétal," in the collection of Thèses of the School of Medicine of Paris. Thèse No. 150. Also in the Comptes Rendus Hebdomadaires des Séances de l'Académie des Sciences de Paris, vol. lvi., 1863, p. 1,001.
Codice Civile del Regno d'Italia. Firenze, 1866.
Colebrooke, H. S.: A Digest of Hindu Law. London, 1801.
Crossman, Edward: "On Intermarriage of Relations as a Cause of Degeneracy of Offspring," in the British Medical Journal, for April 13th, 1861, vol. i., pp. 401, 402.
Culloch, J. R. Mc: A Dictionary—Geographical, Statistical, and Historical, of the Various Countries, Places, and Principal Natural Objects of the World. London, 1866.
Dally, E.: A review, "Sur les dangers attribués aux mariages consanguins," in the Gazette Hebdomadaire de Médecine et de Chirurgie, vol. ix., 1862. Paris. Nos. 32, 33, 34, of the 8th, 15th, and 22nd of August, pp. 499-502, 513-516, 531-534.
"An Inquiry into Consanguineous Marriages and Pure Races." Translated by H. J. C. Beavan, from a Paper read before the Anthropological Society of Paris, November 5th, 1863, in the Anthropological Review for May. London, 1864, pp. 65-108.
Darwin, Charles: On the Various Contrivances by which British and Foreign Orchids are Fertilized by Insects, and on the Good Effects of Intercrossing. London, 1862.

Darwin, Charles: On the Origin of Species by Means of Natural Selection. London, 1859.
 The Variation of Animals and Plants under Domestication. London, 1868.
 The Descent of Man, and Selection in relation to Sex. London, 1871.
Devay, Fr.: Du Danger des Mariages Consanguins sous le Rapport Sanitaire. Paris, 1862.
 Un Mot sur le Danger des Mariages Consanguins; réponse à une attaque; état de la question. Paris, 1863.
Devic, O. F.: "Note sur les Unions Consanguins," in the Gazette Médicale de Paris, 3rd series, No. 10, vol. xviii., pp. 158, 159, March 7th, 1863.
Disraeli, J.: Curiosities of Literature. London, 1834.
Dobson, J. R.: The Ox, his Diseases and their Treatment. London, 1864.
Down, J. L. H.: "Marriages of Consanguinity in Relation to Degeneration of Race," in the Clinical Lectures and Reports by the Medical and Surgical Staff of the London Hospital, vol. iii., 1866, pp. 224-236.
Dugard, Saml.: The Marriages of Cousin-Germans vindicated from the Censures of Unlawfulness and Inexpediency." Oxford, 1673. (Published anonymously.)
Duncan, J. M.: Fecundity, Fertility, Sterility, and Allied Topics. Edinburgh, 1871.
Elam, Charles: A Physician's Problems. London, 1869.
Eliot, J.: "Observations on the Inhabitants of the Garrow Hills," in the Asiatic Researches, vol. iii. pp. 17-37. Calcutta, 1792.
Emmerton: Mr. Emmerton's marriage with Mrs. Bridget Hyde considered. Wherein is discoursed the Rights and Nature of Marriage. What authority the *Curia Christianitatis* hath in Matrimonial Causes at this day. The Levitical Degrees, the Bounds of a Legal Marriage, and the Reasons thereof, etc. In a Letter from a Gentleman in the Country to one of the Commissioners Delegates in that Cause, desiring his opinion therein. London, printed for the Authour, and published by Richard Baldwin, 1682.
Encyclopædia of Anatomy and Physiology. *See* "Todd."
Encyclopädie, Algemeine, der Wissenchaften, etc. *See* "Ersch."
Encyclopédie Methodique. Paris et Liege, 1784. Articles: "Empêchements du Mariage, Affinité," and "Incest;" Part. "Jurisprudence."
Ersch: Algemeine Encyclopädie der Wissenschaften und Künste. Leipsic, 1838.
Esquirol: Mental Maladies, a Treatise on Insanity. Translated with additions by E. K. Hunt. Philadelphia, 1845.
Faber, J. H.: Vicissitudines Juris Romani de Incestis Nuptiis. Leipsic, 1763.

Flourens: in the Comptes Rendus Hebdomadaires des Séances de l'Académie des Sciences de Paris, vol. lv., 1862, pp. 238, 239.
Friensch: edition of Q. Curtius, II. Snakenburg, 1724.
Fry, John: The Case of Marriages between Near Kindred, particularly Considered with Respect to the Doctrine of Scripture, the Law of Nature, and the Laws of England, etc. London, 1773.
Galton, Francis: Hereditary Genius: an Inquiry into its Laws and Consequences. London, 1869.
Gardner, James: "On Intermarriage of Relations as the Cause of Degeneracy of Offspring," in the British Medical Journal, March 16th, 1861, vol. i. pp. 290, 291.
George, H. B.: Genealogical Tables illustrative of Modern History. Oxford, 1874.
Gerland, Georg: Ueber das Aussterben der Naturvölker. Leipsic, 1848.
Gibbon, Edward: The Decline and Fall of the Roman Empire. London, 1868.
Glennie, J. S. S.: "Mr. Buckle in the East," in Frazer's Magazine, August, 1863, pp. 171-189.
Gobineau, M. A. de: Essai sur l'Inégalité des Races Humaines. Paris, 1853.
Godron, D. A.: De l'espèce et des races dans les êtres organisés, et spécialment de l'unité de l'espèce humaine. Paris, 1872.
Gourdon, J.: "Consanguinité chez les animaux domestiques," in the Annales d'Hygiène publique et de Médecine légale, vol. xviii., 2nd series, No. 36; October, 1862, pp. 463, 464. Also in the Comptes Rendus Hebdomadaires des Séances de l'Académie des Sciences de Paris, vol. lv., 1862, pp. 269-273. (It was originally read before this Academy on August 11th, 1862; and is quoted nearly in full by Chipault: "Etude sur les Mariages Consanguins," pp. 97-100; and by Boudin, "Du Croisement des Familles," etc., in the Mém. de la Soc. d'Anthrop., pp. 547, 548. It is also to be found in the Gazette Hebdomadaire de Médecine et de Chirurgie, vol. ix., No. 34; August 22nd, 1862, p. 538.)
Grote, George: A History of Greece, from the Earliest Period to the Close of the Generation Contemporary with Alexander the Great. London, 1862.
Halhed, N. B.: A Code of Gentoo Laws, or Ordinations of the Pundits, from a Persian Translation made from the Original, written in the Sanscrit language. London, 1777.
Hallam, Henry: View of the State of Europe during the Middle Ages. London, 1846.
Hecker, J. F. C.: Die Grossen Volkskrankheiten des Mittelalters. Berlin, 1865.
Hewitt, Graily: The Diseases of Women. London, 1868.
Heywoode, Thomas: History of Women. London, 1624.

H. Londiniensis:—A Broadside on the Prohibited Degrees of Marriage, bearing the date 167⅞, Jan. xxii.

Howe, S. G. : " On the Causes of Idiocy," in the Journal of Psychological Medicine and Mental Pathology, New Series, No. xi., July, 1858, pp. 365-395.

Huzard: "Note sur les accouplements entre consanguins dans les familles ou races des principaux animaux domestiques," in the Annales de l'Agriculture Française, vol. ix., June 15th, 1857, 5th series, No. 11, pp. 497-512. Paris.

Johnston, L. F. C. : Institutes of the Civil Law of Spain, by Doctors D. Ignatius Jordan De Asso Ye Del Rio, and D. Miguel De Manuel Y Rodriguez. Translated from the Spanish. London, 1825.

Isidore, Grand Rabbi: In the Comptes Rendus Hebdomadaires des Séances de l'Académie des Sciences de Paris, vol. lv., 1862, pp. 128-129. Also quoted with M. Boudin's answer, in the Mémoirs de la Société d'Anthropologie de Paris, vol. i., 1863, p. 527. And in the Annales d'Hygiène publique et de médecine légale, vol. xviii., 2nd series, No. 36, Paris, October, 1862, p. 460. Also in other places.

Jurieu, Pierre: A Critical History of the Doctrines and Worships (both good and evil) of the Church, from Adam to our Saviour Jesus Christ. Done into English. London, 1705.

Laboulaye, Edouard : Recherches sur la condition civile et politique des femmes, depuis les Romains jusqu'à nos jours. Paris, 1843.

Lager, J. St. : Etudes sur les causes du crétinisme et du goître endémique. Paris, 1867.

Lane, E. W. : An Account of the Manners and Customs of the Modern Egyptians. Edited by E. S. Poole. London, 1871.

Laverack, Edward: The Setter; with Notices of the most Eminent Breeds now Extant; Instructions how to Breed, etc. etc. London, 1872.

Lawrence: Vindication of Marriage. London, 1680.

Lecky, W. E. H. : History of European Morals from Augustus to Charlemagne. London, 1869.

Legoyt, A.: "Du mouvement de l'aliénation mentale en Europe et dans l'Amérique du Nord," in the Journal de la Société de Statistique de Paris, Nos. iii. and iv., March and April, 1863, pp. 54-80, and 87-101.

Legrain, J. B. : " Recherches critiques et experimentales relatives aux mariages consanguins," in the Bulletines de l'Académie Royal de Médecine de Belgique, vol. ix., 2nd series, No. 3, pp. 180-326, Brussels, 1866.

Lennan, J. F. Mc. : Primitive Marriage; an Inquiry into the Origin of the form of Capture in Marriage Ceremonies. Edinburgh, 1865.

Liebreich, R.: "Abkunft aus Ehen unter Blutsverwandten als Grund von Retinitis Pigmentosa," in the Deutsche Klinik, vol. xiii.,

No. 6, February 9th, 1861, pp. 52-55. Also given in full from the Archives Générales de Médicine, for February, 1862, by Chipault, " Etude sur les Mariages Cons.," pp. 50-56.
Lingard, John: The History and Antiquities of the Anglo-Saxon Church. London, 1845.
Livingstone, D. and C. : Narrative of an Expedition to Zambesi and its Tributaries, and of the Discovery of the Lakes Shirwa and Nyassa, 1858-1864. London, 1865.
Lubbock, Sir J. : The Origin of Civilization and the Primitive Condition of Man. London, 1870.
Lugol, J. G. A.: Recherches et Observations sur les Causes des Maladies Scrofulouses. Paris, 1844.
Macdonald, D. G. F.: Cattle, Sheep, and Deer. London, 1872.
Magnússon, E., and Morriss: The Story of the Volsungs and Niblungs, from the Icelandic. London, 1870.
Mantegazza, Paolo: Studj sui Matrimonj Consanguinei. Milan, 1868.
Martineau, Harriet: Eastern Life, Present and Past. London, 1848.
Mayhew, Henry: London Labour and London Poor. London. Vols. i. and ii., 1851; vol. iii., 1862.
Menière: " Du Mariage entre Parents considéré comme Cause de la surdi-mutité congéniale " (read before the Académie de Médecine, April 29th. 1856), in the Gazette Medicale de Paris, vol. xi., 3rd series, May 17th, 1856, pp. 303-306.
Michel, Fr. : Histoire des Races maudites de la France et de l'Espagne, Paris, 1847.
Mirabeau: Errotika Biblion. Rome, 1783.
Mitchell, Arthur: " Blood-Relationship in Marriage, considered in its Influence upon the Offspring." in the Memoirs read before the Anthropological Society of London. Article No. xxx., vol. ii., 1865-66, pp. 402-456. Also a discussion on this in the Edinburgh Medical Journal, No. lxxxi., March, 1862, pp. 872-878. See also the Edinburgh Medical Journal for June, 1863, and January, 1866; and the Medical Times and Gazette for November 15th, 1862.
Monson, Hon. E. : " Report on the Trade and Commerce of the Azores for the year 1870," in the Commercial Reports received at the Foreign Office from Her Majesty's Consuls in 1871, No. 4.
Montesquieu: Esprit des Lois.
Morris, O. W.: See Ranking's Half-Yearly Abstract of the Medical Sciences, vol. xxix., 1859, p. 10, and vol. xxxiii., pp. 12-13, and pp. xxxvi., xxxvii., Appendix to Huth's " Marriage of Near Kin."
Mosheim, J. L.: Ecclesiastical History, translated by A. Maclaine. Edinburgh and Glasgow, 1839.
Neufville, W. C. de: Lebensdauer und Todesursachen zwei und zwanzig versciedener Stände und Gewerbe, nebst vergleichender Statistik der christlichen und israelitischen Bevölkerung Frankfurts. Frankfurt am Main. 1855.

Niebuhr, M.: Vorträge über alte Geschichte. Berlin, vols. i. and ii., 1848; vol. iii., 1851.
Oesterlen, Fr.: Handbuch der Medicinischen Statistik. Tübingen, 1865.
Pennefather, J. P.: Deafness, its Early Cause, with Practical Directions for its Treatment. London, 1871.
Percy, Bishop: Folio Manuscript, English Ballads, edited by F. J. Furnivall. London, 1868.
Perier, J. A. N.: "Essai sur les Croisements Ethniques," in the Mémoirs de la Société d'Anthropologie de Paris, vol. i., 1863, pp. 69-92, and 187-236; vol. ii., 1865, pp. 261-374; vol. iii., 1870, pp. 211-296.
Perron, Anquetil du: Zend-Avesta, Ouvrage de Zoroastre. Traduit en Française. Paris, 1761.
Randall, H. S.: Fine Wool Sheep Husbandry. New York, 1863.
Ranking, W. H.: Half-Yearly Abstract of the Medical Sciences. London.
Ranse, Q. de: In the Comptes Rendus Hebdomadaires des Séances de l'Académie des Sciences de Paris, vol. lv., 1862, pp. 405, 406.
Reich, Eduard: Geschichte, Natur-und Gesundheitslehre des ehelichen Lebens." Cassel, 1864.
Reynolds, J. R.: A System of Medicine. London, 1870, etc.
Report of the Commissioners appointed to inquire into the State and Operation of the Law of Marriage as relating to the Prohibited Degrees of Affinity, and to Marriages Solemnized Abroad or in the British Colonies. London, 1848. Reprinted, 1856.
Revillout: In the Journal de Médecine et de Chirurgie pratiques, vol. xxxvi., 2nd series, February, 1865. Article No. 6818, p. 53 note. (This is an extract from the Gazette des Hôpitaux Civ. et Milit.)
Ritchie, C. G: Contributions to Assist the Study of Ovarian Physiology and Pathology. London, 1865.
Sale, George: The Koran; translated from the original Arabic. London (undated).
Salt, T. P.: A Treatise on the Deformities of the Lower Extremities. London, 1866.
Sandars, Th. C.: The Institutes of Justinian; with English Introduction, Translation, and Notes. London, 1869.
Sanson, A.: "Questions de Zootechnie, à propos des Mariages Consanguins," in the Gazette Hebdomadaire de Médecine et de Chirurgie, vol. ix., September 12th, 1862, No. 37, pp. 584, 585. Also in the Comptes Rendus Hebdomadaires des Séances de l'Académie des Sciences de Paris, vol. lv., 1862, pp. 121-125.
Seguin, l'aîné: In the Comptes Rendus Hebdomadaires des Séances de l'Académie des Sciences de Paris, vol. lvii., 1863, pp. 253, 254.
Sharpe, Samuel: The History of Egypt from the Earliest Times till the Conquest by the Arabs. London, 1870.

Shirley, Evn. P.: On Deer and Deer Parks, or some Account of English Parks; with Notes on the Management of Deer. London, 1867.
Simpson, Sir J. Y.: On Hermaphroditism, in Todd's Cyclopædia of Anatomy and Physiology, vol. ii. London, 1836-1839.
Smith, Wm.: Dictionary of Greek and Roman Biography and Mythology. London, 1850.
   Dictionary of Greek and Roman Antiquities. London, 1851.
   Dictionary of the Bible. London, 1861 and 1863.
Spencer, Herbert: The Principles of Biology. London, 1865.
Stark, J.: "Contribution to the Vital Statistics of Scotland," in the Journal of the Statistical Society of London, vol. xiv., 1851, p. 61.
Steele, Arthur: The Law and Custom of Hindoo Castes, within the Dekhun Provinces subject to the Presidency of Bombay, chiefly affecting civil suits. London, 1868.
Stephens, H.: The Book of the Farm. London and Edinburgh, 1871.
Sterne, Laurence, the Works of. London, 1819.
Stonehenge: *Pseudonyme*. See Walshe, J. H.
Story, W. W.: Roba di Roma. London, 1864.
Taylor, Jeremy: Ductor Dubitantium; or, the rule of conscience in all her general measures, serving as a great instrument for the determination of cases of conscience. London, 1676.
Taylor, John: Elements of Civil Law. London, 1828.
Taylor, W. C.: The History of Mohammedanism, and its sects; derived chiefly from Oriental sources. London, 1834.
Thibault: "Mariages Consanguins dans la Race noire," in the Archives de Médecine navale, vol. i., 1864, p. 310.
Thompsen, J.: Ueber Krankheiten und Krankheitsverhältnisse auf Island und den Färoër-Inseln. Schleswig, 1855.
Thorpe, B.: Ancient Laws and Institutes of England. London, 1840.
Todd: Encyclopædia of Anatomy and Physiology. London, 1836-1839.
Tongue, Cornelius: The Stud Farm; or, Hints on Breeding for the Turf, the Chase, and the Road. London, 1865.
Turner, Sharon: History of the Anglo-Saxons. Paris, 1840.
Tylor, E. B.: Researches into the Early History of Mankind, and the Development of Civilization. London, 1870.
Voisin, A.: "Contribution à l'histoire des mariages entre consanguins," in the Mémoirs de la Société d'Anthropologie de Paris, vol. ii., 1865, pp. 433-459. A reprint of the same, Paris, 1866. Also in Comptes Rendus Hebdomadaires des Séances de l'Académie des Sciences de Paris, vol. lvx., 1865, pp. 105-108. Also in the Journal de Médecine et de Chirurgie pratiques, vol. xxxvi. February, 1865, pp. 52, 53, Article No. 6,818. Also in the Union Médicale for October 3rd, 1868.
Volney, C. F.: Voyage en Egypt et en Syrie pendant les années 1735, etc. Paris, 1737.

Waitz, Th.: Anthropologie der Naturvölker. Leipzig, 1859, etc., and continued by Dr. Gerland from 1870.
Walker, Alexander: Intermarriage; or, the Natural Laws by which Beauty, Health, and Intellect result from certain Unions, and Deformity, Disease, and Insanity from others, etc. London, 1841.
Walsh, J. H.: The Greyhound in 1864. London, 1864.
    The Horse in the Stable and in the Field: his Varieties, Management in Health and Disease, Anatomy, Physiology, etc. London, 1862.
Watson, Sir Thomas: Lectures on the Principles and Practice of Physic. London, 1871.
Wells, T. S.: Diseases of the Ovaries; their Diagnosis and Treatment. London, 1872.
Wilkinson, Sir G.: Manners and Customs of the Ancient Egyptians, etc. First Series, London, 1837. Second Series, London, 1841.
Williams, C. J. B., and C. Th.: Pulmonary Consumption; its Nature, Varieties, and Treatment. London, 1871.
Youatt, Wm.: On the Horse. Edited by Cecil. London, 1855.
    The Pig. Edited by S. Sidney. London, 1860.
Zend-Avesta: Ouvrage de Zoroastre. Traduit en Française par M. Anquetil du Perron. Paris, 1761.

INDEX.

ABELONITES, doctrines of the, 65
Abipones, marriage among the, 124
Abstinentes, tenets of the, 66
Abyssinian-Arab half-breeds, 328
Adamites, doctrines of the, 65, 68, 71
Æthiopia, marriage in, 18
Affghans, marriage among the, 91; physical state of the, *ib.*
Africa, marriage in, 17, 18, 126-131
Agapetæ, doctrines of the, 67
Akombwi, marriage among the, 128, 129
Albanenses, doctrines of the, 69
Albigenses, *ib.*
Albinoïsm, attributed to consanguineous marriage, 259, 271, 297, 302; some causes of, 259, 298, 299; experiments of M. Legrain for the production of, 298-302
Alcoholism, a cause of idiocy, 219, 220; of epilepsy, 220, 221; of paraplegia, 220; of phthisis, *ib.*; of scrofula, 222
Aleuts, marriage among the, 117
Algonquin, marriage among the, 117
Amalricians, doctrines of the, 69
Amboyna, half-breeds of, 312
Ambrose, doctrines of St., 43, 74

Amenites, doctrines of the, 72
America, marriage prohibitions in the United States of, 59; among the savages of North, 117-119; among the savages of Central, 119, 120, 125, 126; among the savages of South, 120-125
Anabaptists, doctrines of the, 70
Andorra, Republic of, an isolated community, 181, 182
Angelic Brothers, doctrines of the, 72
Anglo-Saxons, marriage among the, 49, 50
Angoumois, paper-makers of the, 186, 187
Animals, the habits of, an illustration of the relationship of the sexes in the most barbarous times, 132, 133; have no horror of incest, 144-146; polygamous, 145, 146; observations on the breeding of, applicable to man, 264-271; and together with observations on the breeding of plants, the only direct means of determining whether in-and-in breeding is harmful of itself, and not through inheritance only, 261, 269-271; unwillingness of to cross, 278, 281, 287, 295, 296; diseases of, 263 note 2, 264 note 1; small races of, not created by in-and-in breeding, 277 note 4

Apachalites, marriage among the, 125
Apaches, marriage among the, 120
Apostolicals, doctrines of the, 69, 70
Apotactics, doctrines of the, 66
Aquapim, marriage in, 127, 128
Arab-Abyssinian half-breeds, 328; Arab-Negro, 327; Arab-Persian, 328
Arabia, marriage in, 18, 86-89
Archontics, doctrines of the, 65
Ardæus founds a reforming sect, 67
Argentine Republic, marriage among the savages of the, 125
Ariana, marriage in, 17
Armenians, little value of female infants among the, 90
Arreoi, the society of the, 106, 107, 107 note 2, 108, 109
Arromanches, isolated community at, 181
Arrowak, marriage among the, 121
Ascension Island, marriage in, 105, 106
Asceticism, the cause of modern prohibitions on marriage, 60; a means of moral reform, 61; origin of, 62
Assubo-Galla, marriage among the, 129
Assyrians, marriage of the ancient, 17
Asylums, untrustworthiness of returns from, 202, 203
Atmah, marriage among the, 117
Audouère, the divorce of, 82
Auses, marriage among the, 18
Australia, marriage in, 113-115; female infanticide in, 115; punishment of incest, *ib.*; state of health in, *ib.*; half-breeds in, 314, 315.
Auvergne, paper makers of the, 186, 187; Marans of, 187
Ayowas, marriage among the, 118
Azanæa, marriage in, 18
Azores, consanguineous marriages in the, 190; state of health, *ib.*

BABYLONIA, marriage in, 17
Bagnolonses, doctrines of the, 69
Bali, marriage in, 104, 105
Balnabruiach, isolated community at, 166, 167
Bambarras, marriage among the, 127
Barbarians, marriage among nations known to the ancients as, 17-19
Bardesanes, doctrines of, 65
Barrenness, see *Sterility*
Bas-Bretons, consanguineous marriages of, and state of health, 181
Basilides, doctrines of, 65, 73
Basques, consanguineous marriages and state of health of the, 181
Bass's Straits, isolated community on islands in, 191, 192
Bastaards, history of the, 316, 317
Batta, marriage among the, 104
Batz, isolated community at, 179, 180
Bedouins, marriage among the, 88, 89; female infanticide among, 89
Boghards, doctrines of the, 69
Beguines, doctrines of the, 69, 70; their spread, and various names, 69
Beloochistan, marriage in, 92
Benkulen, marriage in, 104
Bentham, isolated communities in the neighbourhood of, 162
Berneray, isolated community at, 168
Bicorni, doctrines of the, 69
Bigamy, original meaning of the word, 75, note 2; instances of excommunication for, 79

INDEX. lv

Biguttes, doctrines of the, 69
Birth, a first or last more likely to be an idiot, 218, 218 note 3
Black Death, accusation against the Jews of causing the, 186 note 2
Bodo, marriage among the, 97
Bogo, marriage among the, 129; superior value of boys among, 129
Bogomiles, doctrines of the, 68
Bonny, twins killed in, 251
Borneo, marriage in, 105; punishment of incest, ib.
Boulmer, isolated community at, 163
"Bounty," the mutineers of the, 158-160
Boyndie, isolated community at, 166
Brahoos, the, 92
Brazil, marriage in, 123, 124
Bretagne, health of the population of, 180
Brethren of the Free Spirit, doctrines of the, 69
Bretons, consanguineous marriages, and state of health of the, 180, 181
Brighton, the fishermen of, 162
Buchanites, doctrines of the, 72
Buckhaven, isolated community at, 166
Burins, isolated community of the, 188
Burnmouth, isolated community at, 163-165

CADZOW CASTLE, the wild cattle of, 279
Cafusos, the, 326
Cagots, history of the, 183-185; consanguineous marriages of the, 184; state of health, 185, 186
Cainites, doctrines of the, 65

Canaan, marriage in, 18
Canadian Indians, marriage among the, 117
Canaries, marriage in the, 126
Capots, see *Cagots*
Caraïbs, marriage among the, 125, 126
Caroline Islands, marriage in the, 105, 106
Carpocrates, doctrines of, 65
Caste, the spirit of, leads to endogamy, 137
Cathari, doctrines of the, 69, 70; various names of the, 70
Cattle, the results of in-and-in breeding of, 279-284; the wild herds of Cadzow Castle, Chillingham, and Chartly, 279, 280; of South America, 279-281
Cava drinking, the results of, 111, 112
Celebes, marriage in the, 105
Census, the use of a, 6, 261, 354-356
Cerdo, doctrines of, 64
Ceylon, marriage in, 97, 98; statistics on the number of insane, deaf, and blind in, 98
Chaouia Berbers, lobeless ears of the, 186, note 1
Charruas, marriage among the, 125
Chartly, wild cattle at, 279
Cherokee, marriage among the, 118
Chichemeks, marriage among the, 119
Chiefs, why stronger than commoners, 112, 113
Chillingham, wild cattle at, 279
China, marriage in, 99, 100; female infanticide in, 100, 101, 135 note; punishment of incest in, 99
Chinese half-breeds, 329
——— Turkestan, marriage in, 101
Choctaw, marriage among the, 118

Chorea, some causes of, 227
Chrestiaa, see *Cagots*
Christianity, effect of, on marriage, 42, 43; prohibitions of marriage under the Christian emperors and Roman Pontiffs, 43-47
Chrysostom, opinions of St., on marriage, 74
Chuetas, isolated community of the, 189, 190
Chukmas, marriage among the, 97
Circassians, marriage among the, 89, 90
Cleft-palate, said to be caused by consanguineous marriage, 250; some causes of, 250-252
Clemens of Alexandria, doctrines of, 73
Club-foot, said to be caused by consanguineous marriage, 250; some causes of, 250-252
Cœlibacy not recommended in the Gospels, 62; recommended by St. Paul, *ib.*; early feeling in its favour for the priesthood, 73; when imposed on the priesthood, 76, 77; injurious consequences, 77, 78
Comanche, marriage among the, 120
Comnenus, persecution of the Paulicians by the Emperor Alexius, 68
Concorezences, doctrines of the, 70
Concordences, *ib.*
Concorenses, *ib.*
Concoretii, *ib.*
Convulsions, a cause of insanity, 221, 256; of epilepsy, 222; some causes of, 255, 256
Cornwall, isolated communities of, 162
Coroado, marriage among the, 124
Cousins, marriage with first, encouraged by the ancient Greeks, 23, 24; ancient Jews, 33; ancient Romans, 39; Mohammedans, 88; Kykaree, 93; Yerkala, 97; Caraïbs, 125, 126; more fertile than other marriages, 244
Creeks, marriage among the, 118
Crétinism, said to be caused by consanguineous marriages, 212; has a common cause with goître, 213; causes of, 213, 214; impossibility that it is caused by consanguineous marriage, 214, 215; difference of, and idiocy, 215-217
Crosses, in animals frequently harmful, 268, 277, 278, 294, 300; dislike of animals to, 278, 281, 287, 295, 296; tendency to reversion given by, 285, 305, 308 note 1, 349; if beneficial, the greater the cross the greater the benefit, 307; between different races generally sterile, 308, 310, 311, 312, 313, 314, 315, 319, 350; reasons for, of hermaphrodites, 346-348; effect of, on the reproductive organs the cause of the greater development of hybrids, 349, 350, 351
Crusades, said to have given origin to the Beguins, 69
Customs, the identity of curious, not necessarily a proof of the common origin of tribes practising them, 138-140
Cyprian, opinions of on marriage, 74
Cyrenaica, marriage in, 18
Cyrill of Jerusalem, opinions of, on marriage, 74

DAMASUS condemns the Venustians, 67
Danes, marriage among the ancient, 49; prohibitions on marriage among the modern, *ib.*

Darien, marriage among the Indians of, 120
David Georgians, doctrines of the, 71
Deaf-mutism said to be caused by marriages between near kin, 228; some causes of congenital, 228, 229; manner in which statistics on, are gathered, 229, 230; worthlessness of statistics on, 230-239
Deer, the results of in-and-in breeding of, 296, 297; rickback in, 296
Delaware Indians, marriage among the, 117
Dermoid cysts, distinction between, and ovarian cysts of the same nature, 341, 342
Dhumal, marriage among the, 97
Dilectæ, doctrines of the, 67
Dogs, the results of in-and-in breeding of, 290-296; natural division of, into tribes, 295, 296
Doinguaks, marriage among the, 97
Dolcinists, doctrines of the, 70
Donkeys, the results of in-and-in breeding of, 290
Druses, marriage among the, 89
Ducks, the results of in-and-in breeding of, 304
Dutch-Hottentot half-breeds, 315-317
Dwarfing, said to result from breeding in-and-in, 112 note 2, 178, 259, 275, 276, 277 note 4, 290, 303
Dyaks, marriage among the, 105

EARS, lobeless, common to the Chaouia Berbers, 186 note 1; and to the Cagots, 184
Ecuadore, marriage in, 121
Egypt, sterility of foreigners in, 328
Egyptians, ancient, influence of on modern marriage laws, 9; marriage among the, 10-13, 18; decline and fall of the, due to their disregard of the marriage tie, 13; asceticism among, 61
Elands, results of the in-and-in breeding of, 284
Elchasaites, doctrines of the, 65
Encratites, doctrines of the, 64, 66
Endogamy, not the result of any observed evil effect of crosses, 85, 86, 142; caused by pride of race, 86, 137, 138, 140-142; not of traditional origin, 138-142; how it may change to exogamy, 137
England, history of the degrees within which marriage was prohibited in, 49-56
Epilepsy a predisposing cause to insanity, 220; is hereditary, 221, 222; some causes of, 222; not caused by marriage between near kin, 223
Epiphanes, doctrines of, 65
Epiphanius, opinions of, on marriage, 74
Eskimo, marriage among the, 117; imaginary relationship among persons brought up together, ib.; half-breeds, 330
Essenes, doctrines of the, 62
Euchites, doctrines of the, 66, 68
Eurasians, physical and moral state of the, 308-311
Europeans, bad influence of, on savages, 111, 113
Eustathius, opinions of, on marriage, 74
Exogamy, not the result of any observed evil effect of marriages between near kin, 85, 86, 136 note, 138, 141, 142; caused by female infanticide, 86, 134-137; male births do not sufficiently exceed the female to account for

it, 134 note; not of traditional origin, 138-141; rarely changes to endogamy, 137; does not account for the horror of incest in the nearest degrees, 143, 144

Eye, diseases of the, attributed to marriage between near kin, 239; some causes of, 239, 240; worthlessness of the statistics on, 240, 241

Eyemouth, isolated community at, 162

FAMILISTS, doctrines of the, 70, 71, 73

Fantis, marriage among the, 127

Faroë Islands, state of health in the, 177

Female infanticide a cause of exogamy, 86, 134-137; among the Arabs, 89; Hindoos, 94, 95, 134 note; Chinese, 100, 101; Kalmucks, 103; Hawaiians, 108; Tahitians, *ib.*; the people of Vate, 110; the New Zealanders, *ib.*, Papuans, 116; Panches, 120; Guanias, 124, 125; causes of, 134-136

Fertility, comparative, of different pure races together, 319

Fetu, one of twins killed in, 251 note 2

Fiji, marriage in, 110; licence accorded to chiefs in, *ib.*; size of the chiefs compared with the common people, 112, 113

Fish, said to become albinoes and barren if bred in-and-in, 302

Flinder's Island, the Tasmanians in, 191

Florinians, doctrines of the, 65

Forèatines, isolated community of the, 180

Formosa, marriage in, 105

Fowls, the results of in-and-in breeding of, 303, 304

France, the marriage prohibitions of, 57, 58

Fraticelli, doctrines of the, 69

Free Lovers, the sect of the, 73

Fright, as a cause of idiocy, 217, 218; of epilepsy, 222

Fuertaventura, marriage in, 126

GAHETS, see *Cagots*

Galatians, marriage among the, 18

Galla, marriage among the, 129; inferior value of women, 129, 130; levirate law, *ib.*

Garamantes, marriage among the, 18

Garrows, marriage among the, 97

Gauchos, physical and moral state of the, 324

Gaust, isolated community at, 181

Gedrosia, marriage in, 17

Genesis, Mr. Herbert Spencer's theory on, 334, 337; Mr. Darwin's theory on, 335, 336; little difference of the sperm and germ cells, 337, 338; hermaphroditism a possible phase in the development to dual sex, 338-340; parthenogenesis an argument against the radical difference of the sexes, 341; ovarian dermoid cysts examples of parthenogenesis, 341-343; the advantages of dual sex, 344-346

Georgians, inferior value of female infants among the, 90

Germany, marriage prohibitions of, 56, 57

Ghagar, marriage among the, 130

Ghawazee, marriage among the, 130; direct descendants of the ancient Egyptians, *ib.*

Gichtel, George, doctrines of, 72

Gnostics, doctrines of the, 63; Asiatic, 63-65; Egyptian, *ib.*

Goats, the results of in-and-in

INDEX. lix

breeding of, 290; weakness of the Angora, 290
God-parentage made a bar to marriage, 82-84
Goitre, connection of, and crétinism, 213; causes of, 213-215; connection of with idiocy and deaf-mutism, 215, 216; occurs in animals, 265 note.
Gomera, marriage in, 126
Gosawee, marriage among the, 95
Granville, isolated community at, 181
Greeks, influence of the ancient, on modern marriage laws, 9; marriage among the ancient, 19-27; mythological incest, 26-27; asceticism among the, 61; marriage prohibitions of the modern, 47, 48
Greenland Eskimo, marriage among the, 117; imaginary relationship of persons brought up together, 117, 142; half-breeds, 330
Gregorius of Nyssa, opinions of, on marriage, 74
Griquas, history of the, 316, 317
Guanaco, results of in-and-in breeding of, 284
Guanias, female infanticide among the, 124, 125
Guiana, marriage in, 121
Gypsies, marriage among the Luri, 92; Ghagar, Nuri, Helebi, and Ghawazee, 130

H AL ben Ali, marriage among the, 130, 131
Half-breeds, generally worse than pure races, 308-332; often barren or nearly so, 308, 310, 311, 312, 313, 322, 328, 329, 332; European-Hindoo, 308-310; Dutch-Singhalese, 310; Dutch-Malays, 311; Spanish-Malays, 312; of the Ladrone Islands, 312; European-Polynesians, 312, 313; European-Maoris, 313, 314; European-Tasmanians, 314; European-Australians, 314-315; European-Chinese, 315; European-Hottentots, 315, 316; the Griquas, 316, 317; Mulattoes, 318-320, 321; Mestisos, 320-324; Paulistas, 323-324; Gauchos, 324; Zamboes, 324-327; Cafusos, 326; Negro-Arab and Negro-Hottentot, 327; Maroons, 327, 328; Arab-Persian, 328; Arab-Abyssinian, ib.; Kuruglis, ib.; Chinese-Cambojia, and Chinese-Tagal, 329; European-Eskimo, 330; the general sterility of, does not preclude the theory that mankind was derived from one original race, 330, 331
Hautponnais, isolated community of the, 187, 188
Hawaii, female infanticide in, 108; marriage in, 109; fertility in, 112
Helebi, marriage among the, 130
Helvidius, opinions of on marriage, 67, 74
Hermaphrodites excluded from caste in India, 251, note 2; ancient belief of the existence of, 338 note; not very rare, 339; why do they cross? 346-348
Hermaphroditism, a probable phase in the development of dual sex, 338-340
Herods, marriages of the, 34
Hieracites, doctrines of the, 67
Hieronymus, opinions of, on marriage, 74
Hindoos, marriage prohibitions of the, 92-95
Ho, marriage among the, 96, 97
Holland, prohibited degrees in, 57
Horses, the results of in-and-in breeding of, 287-290; origin of our race-horses, 287-289

Hottentots, marriage among the, 128; female infanticide practised by, *ib.*; one of female twins killed, 251 note 2; half-breeds, 315-317, 327
Hovas, marriage among the, 131
Hybrids, see *Crosses*
Hydrocephalus, a cause of insanity, 221; connected with scrofula, 254, 255; some causes of, 257, 258
Hysteria, a predisposing cause to insanity, 220

IBU, the, accustomed to kill one of twins, 251 note 2
Iceland, marriage in ancient times in, 49; state of the population of, 171-176
Idiocy, difference between, and crétinism, 216-217; causes of, 217-227; impossibility of showing that it is due to marriage between near kin, 222; worthlessness of the statistics on, 226
Ignatius, opinions of, on marriage, 74
Illegitimacy, a cause of idiocy, 218; of malformations and dwarfing, 218 note 4
Illuminati, doctrines of the, 71
In-and-in breeding of sheep, 271-279; of cattle, 279-284; of the eland, 284; of guanaco, *ib.*; of pigs, 284-286; of horses, 287-290; of donkeys, 290; of goats, *ib.*; of dogs, 290-296; of deer, 296, 297; of rabbits, 297-303; of fowls, 303, 304; of ducks, 304; of pigeons, 304-305; selection in, increases the value of the experiment, 266-271; if harmful, it is only so through inheritance, 305
Incest, practised in consequence of the vow of cœlibacy, 77, 78; of Pope John XXIII., 77 note 8; of Pope Alexander VI., *ib.*; how punished in China, 99; in Borneo, 105; in Australia, 115; is the horror of, innate, or only the result of custom? 132-149; if innate, this horror should be universal, 143, 146-149; no horror of, in animals, 144-146; imperfect definition of, 287
India, marriage in, 17, 18, 92-97; relative value of the sexes in, 94, 95; female infanticide in, 134 note; horror of hermaphrodites in, 251 note 2; half-castes in, 308-310
Infanticide of females, see *Female Infanticide*
Ingeburga, the divorce of, 81
Insanity, hereditary, 220; some causes of, 220, 221; worthlessness of the statistics on, 226, 227
Irish, marriage among the ancient, 18; fabulous story of a colony in Sligo and Mayo, 177, 178
Iroquese, marriage among the, 117
Italy, marriage prohibitions in, 58; marriages between first-cousins in, said to be rare, *ib.* note 2
Itchinferry, isolated community at, 162
Izeaux, isolated community at, 182, 253, 254

JAKUTS, marriage among the, 101
Jamparika, marriage among the, 120
Jarejah, female infanticide practised by the, 134 note
Java, marriage in, 104; isolated community in, 161
Jerome, opinions of St., on marriage, 74
Jews, influence of, on modern marriage laws, 9; marriage among

the, 27-34; the levirate law, 30, 33; Tamar and Amnon, 31; the curse of sterility, 32; the Herods, 34; asceticism of the, 61; accused of causing the *Black Death*, 186 note 2; of the introduction of syphilis, 187; cosmopolitanism of the, 194, 195; difference of their statistical laws from those of other races inhabiting Europe, 195; fertility and viability of the, 195-197; greater liability to mental disease, and the cause, 197-200; accusation against; of furnishing an undue proportion of deaf-mutes, 201-203; of the Ghetto at Rome, 204, 205

Jolofs, marriage among the, 127

Jovinian, the doctrines of, 67; condemned by St. Ambrose, 74; by St. Jerome, *ib.*

Justin Martyr, doctrines of, 64, 73

KADJAK Eskimo, marriage among the, 117
Kaffirs, marriage among the African, 128
Kalang, marriage among the, 104
Kalmucks, marriage among the, 103; are dying out, 103, 171; female infanticide practised by, 103
Kamburani, the, 92
Kenai, marriage among the, 117, 118
Khlisti, doctrines of the, 71
Khonds, marriage among the, 95, 96; female infanticide practised by, 96
Kilda, state of the population of the Island of St., 169-171
Kirghiz, marriage among the, 103
Koch, marriage among the, 97
Kolosh, marriage among the, 117
Koombees, infanticide practised by the, and its cause, 133, 134

Koupooe, marriage among the, 97
Kroomen, marriage among the, 127
Kuruglis, physical state of the, 328

LABADISTS, doctrines of the, 72
Lactantius, opinions of, on marriage, 73
Ladrone Islands, marriage in the, 106; half-breeds of the, 312
Lampong, marriage among the, 104
Lanzarota, marriage in, 126
Lapps, marriage among the, 104
Lausanne, doctrines of Henry of, 68
Leutardus, the vision and doctrines of, *ib.*
Libertines, the sect of the, 71
Limousin, paper-makers of, 186, 187
Lipplapen, sterility of, among themselves, 311
Lules, accustomed to kill one of twins, 251 note 2
Luri, marriage among the, 92
Lyzelards, isolated community of the, 187, 188

MACAO, Portuguese-Chinese half-breeds at, 315
Macedonia, marriage in, 25
Machlyses, marriage among the, 18
Macusi, marriage among the, 121
Madagascar, marriage in, 131; inferior value of female infants in, *ib.*; one of twins killed in, 251 note 2
Magar, marriage among the, 97
Malays, marriage among the, 104; punishment for incest, 105
Malformations said to be caused by marriages between near kin, 250; some causes of, 218 note 4, 250, 252; statistics on, 200 note 3, 253-255
Mandrucu, marriage among the, 123
Mangarova, polydactylism, in, 111
Manichæans, doctrines of the, 66

their division into the *Perfect*, and *Hearers*, *ib.*; their spread, *ib.*; of the fourth century, *ib.*; their further history, 67-69
Mantchu Tartars, marriage among the, 101
Mapodzo, marriage among the, 128, 129
Marans, isolated community of the, 187
Marcion, doctrines of, 64, 68
Marcus, doctrines of, 64
Marianne Islands, marriage in the, 106; half-breeds of the, 312
Markesas, marriage in the, 109; state of health in, 111
Marmarica, marriage in, 18
Marriage prohibitions among various nations, *see* under separate headings; danger of throwing difficulties in the way of, 2, 3, 81, 357; opinions of early Christians on, 62-75; second, 74-76; restrictions on married persons, 75 note 3; what was the first stage, 132, 133; early, the most productive, 244; premature, produces puny offspring, 252; what should be prohibited and what not, 357, 358

————, consanguineous, why prohibited by modern laws, 3, 4; accused of causing crétinism, 213; idiocy, epilepsy, chorea, and insanity, 223-229; deaf-mutism, 228; diseases of the eye, 239, 240; sterility, 242; low viability, 244; malformations, 250; polydactylism, 253; rickets, 254; scrofula, 258; ichthyosis and leprosy, 259; hydatis, 6, 259; albinoïsm, 259; St. Augustine on, 43, 150, 151; St. Athanasius on, 43; Ovid on, 145; Socrates on, 149; Plato on, *ib.*; Novatian on, *ib.*; Aristotle on, *ib.*; Chrysippus on, 149, 150; Zeno on, *ib.*; Philo on, 150; Agathias on, *ib.*; Statius on, *ib.*; Plutarch on, *ib.*; Chrysostom on, *ib.*; Pope Gregory I. on, 151, 155; Aquinas on, 151-152; Luther on, 152; Beza on, *ib.*; Burton on, 152, 153; Jeremy Taylor on, 153-155; Amyraut on, 155; Dugard on, 155, 156; Lawrence on, 156; Montesquieu, *ib.*; made difficult for the sake of lucre, 78, 99; and to the end that noble families should be dependent on the Pope, 79-81; cases of, in the royal Spanish line, 80; sad results of wide prohibitions on, 81; said to be a cause of mortality by the Kenai Indians, 117, 118, 141 note; of degeneracy by the East Indians, 141 note; of Tamehameha's small family by the Hawaiians, *ib.*; reasons why they should be prohibited, 149, 156; statistics on the proportion of, to non-consanguineous marriages, 206-212; collection of cases of, useless except negatively, 260, 261, 286, 353, 354; inability of any Government to prevent, 356
Marshall Islands, marriage in the, 106, 107
Media, marriage in, 17, 18
Memlouks, sterility of the, in Egypt, 328
Mennonites, doctrines of the, 71; the *Hook* and the *Button*, 72
Men of Understanding, the sect of, 70
Merino sheep, origin of the, 273, 274
Mesopotamia, marriage in, 17
Messalians, doctrines of the, 66
Mestisos, the mental and physical

INDEX.

lxiii

condition of, 320-324; their viability and power of reproduction, 322
Mexico, marriage among the ancient people of, 119; among the modern, 120
Micronese Islands, marriage in the, 106, 107
Mingrelese, marriage among the, 91; their beauty, *ib.*
Misteks, marriage among the, 119
Mohammed, marriage prohibitions of, 86, 87; the wives of, 87; favour shown by, to marriages between cousins, 87, 88; the levirate law ordained by, 88
Mohammedans, inferior value of female infants among the, 89
Mongols, marriage among the, 101; the Turks not descended from the, 328
Montanus, doctrines of, 64, 67
Moondah, marriage among the, 97
Moors, marriage among the, 130, 131; half-breeds, 328
Mormons, doctrines of the, 73
Mousehole, isolated community at, 162
Mow, marriage among the, 97
Moxos, accustomed to kill one of twins, 251 note
Mulattoes, the mental and physical state of, 318, 319, 321; viability and power of reproduction, 319, 320
Munipuree, marriage among the, 97
Muram, marriage among the, 97
Murring, marriage among the, 97

NATCHES, marriage among the, 118
Nauni, marriage among the, 120
Naura, marriage in, 18
Nawer, *see* Nuri

Negro-Arab half-breeds, 327
——— Hottentot half-breeds, 327
Neuralgia, a predisposing cause to insanity, 220
New-born, the sect of the, 72
New England, marriage among the savages of, 118, 119
New Granada, marriage in, 120; female infanticide in, *ib.*
New Guinea, marriage in, 116; female infanticide in, *ib.*
New Leicester sheep, origin of the, 276
Newlyn, isolated community at, 162
New Zealand, marriage in, 110; female infanticide in, *ib.*; state of health in, 110, 111; fertility in, 112; half-breeds in, 313, 314
Nicaragua, marriage in, 120
Nicolaitanes, doctrines of the, 63, 65
Nicolas, Henry, doctrines of, 71
Nitendi Islands, marriage in the, 110
Niva, strength of the chiefs in, 112
Niyuna, marriage among the, 120
Noble families said to degenerate and die out on account of their consanguineous marriages, 245-248
Nogais, marriage among the, 103
Novatians, doctrines of the, 66, 67
Nukuhiva, marriage in, 109
Nuri, marriage customs of the, 130
Nutka, trade carried on by the, in wives, 118

OBESITY, a cause of sterility, 249; a cause of malformations, 250; incompatible with wool in sheep, 277
Omaha, marriage among the, 117
Omish Church, doctrines of the, 72
Ophites, doctrines of the, 65
Oraon, marriage among the, 97

Origen, opinions of, on marriage, 73
Orinoko, marriage among the Indians of the, 120; Indians of, accustomed to kill one of twins, 252 note.
Ortlibenses, doctrines of the, 69
Ossetes, marriage among the, 90
Ostyaks, marriage among the, 101, 102
Ouled Sidi Sheikh, marriage among the, 131
Ovarian dermoid cysts an example of parthenogenesis in mankind, 341-343

PALEMBANG, marriage among the, 104
Palma, marriage in the Island of, 126
Palmas, marriage at Cape, 127
Panches, marriage among the, 120; female infanticide practised by the, *ib.*
Pangenesis, Mr. Darwin's theory of, 335, 336
Papels, marriage among the, 127
Papua, marriage in, 116; female infanticide practised in, *ib.*
Paraguay, marriage in, 124, 125
Parsees, marriage among the, 91; their beauty, *ib.*; relative value in which the sexes are held by, *ib.*
Parthenogenesis, an argument against the radical difference of the sexes, 341; ovarian dermoid cysts an example of in mankind, 341-343
Parthia, marriage in, 17
Patagonia, marriage in, 125
Paterini, doctrines of the, 69
Paterniani, doctrines of the, 67
Pauillac, isolated community at, 181
Paul, St., praise of cœlibacy by, 62

Paulicians, doctrines of the, 66-68
Paulistas, origin of the, 323; character, 323, 324
Perfectionists, doctrines of the, 73
Persians, marriage among the ancient, 13-17, 18, 19; the law of Zoroastre, 14; Arab half-breeds, 328
Peruvians, marriage among the ancient, 121, 122; progress of, in the sciences, 122; marriage among the modern, 123; ancient, accustomed to kill one of twins, 252 note
Petits-Blancs, *see* Petits Créoles
———Créoles, isolated community of, at Réunion, 190, 191
Petrobrusians, doctrines of the, 69
Phœnicia, marriage in, 18, 19
Phrygia, marriage in, 18
Phthisis, some causes of, 220, 256, 257; connected with scrofula, 254, 255
Picards, doctrines of the, 69
Pigeons, the results of in-and-in breeding of, 304, 305
Pigs, the results of in-and-in breeding of, 284-286
Polydactylism, said to be caused by marriages between near kin, 182, 253, 254; in Mangareva, 111
Polygamous animals, 145, 146
Polygamy, the place of, in the history of marriage, 133
Polynesia, marriage in, 107-109; mythological incest, 108; health in, 110-113; fertility in, 112; superior size of the chiefs in, 112, 113; half-breeds in, 312, 313
Ponape, marriage in, 105, 106
Portel, isolated community at, 181
Portland Island, isolated community in, 162

Portmaholmack, isolated community at, 166
Portugal, marriage prohibitions in, 59
Priscillian, doctrines of, 67
Prodicians, doctrines of the, 65
Prohibited degree, the only natural, 86, 156, 157, 357; the reasons which have been given for a, 149-156
Proverbs, the identity of, not necessarily a proof of their common derivation, 140
Ptolemies, married their sisters, 10, 11; not sterile, or short-lived, 11 note.

RABBITS, the results of in-and-in breeding of, 297-303
Radack, marriage in, 106, 107
Rajpoots, marriage among the, 93; female infanticide practised by the, 134 note
Ralick, marriage in, 106, 107
Ranters, doctrines of the, 71
Rathen, isolated community in, 166
Relationship; no, between a mother and her child, 26; by civil and by canon law, 44 note; by God-parentage, 82-84; not *felt* when unknown, 146-148.
Religion, influence of, on the frequency of consanguineous marriage, 58 note 3, 200 note 3.
Retinitis pigmentosa said to be caused by marriages between near kin, 240; the accusation not proved, 240, 241; evidence that it is not so caused, 241
Réunion, isolated community at, 190, 191
Rickets, some causes of, 174, 255; said to be caused by marriages between near kin, 254
Roman Pontiffs, marriage under the, 43-47, 73-84

Romans, influence of the ancient, on modern marriage law, 9; relationship among the, how influenced by marriage, 34-37; marriage prohibitions of the, 37-42; adoption, 41, 42; marriage with a sister-in-law, 42
Rosenfelders, doctrines of the, 72
Russia, marriage prohibitions in, 48; of the Muscovites and Livonians, 48, 50

SAKKOPHORI, doctrines of the, 66
Salivas, accustomed to kill one of twins, 251 note 2
Salomon Islands, marriage in the, 110
Samaritans, isolated community of the, 192, 193
Samoa, marriage in, 109; fertility in, 112
Samoyeds, marriage among the, 102, 103
Sandwich Island, marriage in, 109, 110; infanticide in, 110
Saturninus, the doctrines of, 63
Scilly Islands, state of health on the, 178
Scotch, marriage among the ancient, 18; consanguineous marriages in small communities of, 162-171
Scrofula, connected with rickets, phthisis, tabes mesenterica, meningitis, hydrocephalus, and spina bifida, 254, 255; said to be caused by marriages between near kin, 258
Scythians, marriage among the, 18
Selection of animals does not depreciate the value as an experiment on breeding in-and-in, 266-271
Seniavin, marriage in, 105, 106

Sermoyers, isolated community of the, 188, 189
Sethites, doctrines of the, 65
Severianus, doctrines of, 64
Sexes, male births do not sufficiently exceed the female to account for exogamy, 134 note; preponderance of the male among the Jews, 196 note; why are there two? 333, 344-346; Mr. Herbert Spencer's theory for the, 334; Mr. Darwin's, 335, 336; why not three, or more? 348
Shakers, doctrines of the, 72
Sheep, the results of in-and-in breeding of, 271-279; origin of the merino, 273, 274; of the New Leicester, 276; of the Mauchamp, 272, 279; of the Ancon, 279
Siam, marriage in, 98, 99
Singhalese, marriage among the, 97, 98
Sister-in-law, marriage with a, among the Jews, 30, 33; Romans, 42; in Sweden, 48; in Denmark, 49; among the Livonians, 50; English law on, 51-56; the vulgar opinion that it is unlawful has been productive of great wrongs and much misery, 56, 356 note 3; in Germany, 57; in Holland, *ib.*; in France, 58; in Italy, *ib.*; Mohammedan law on, 86; among the Affghans, 91; the Hindoos, 93, 94; the Garrows, 97; the Mongols, 101; the Kalmucks, 103; in the Caroline Islands, 106; among Australian savages, 115; among Australian colonists, 357 note; Papuans, 116; Indians of the Orinoko, 120; Macusi, 121; Warrou, *ib.*; Mandrucu, 123; in Western Equatorial Africa, 127; among the Fantis, *ib.*; the Papels, *ib.*; the people of Cape Palmas, *ib.*; the Bambarras, *ib.*; the Somali, 129; the Galla, 130
Skoptzi, doctrines of the, 71
Sodha, marriage among the, 96
Solitaries, doctrines of the, 66
Somali, marriage among the, 129; levirate law enforced by the, *ib.*
Souza, the family of, 161
Spain, marriage prohibitions in, 58, 59
Spirituals, doctrines of the, 71
Stedingers, doctrines of the, 70
Sterility, said to be caused by marriages between near kin, 11 note 3, 30 note 2, 141 note, 155, 241; average of, for all marriages, 242; average of, for consanguineous marriages, 242, 243; of noble families, 244-248; inheritance of a tendency to, 246; some causes of, 248, 268; in animals said to be caused by breeding in-and-in, 272, 279, 282, 285, 289, 291, 302, 303, 304
Stuarts of Glenfinlas, consanguineous marriages of the, 162
Superfœtation, believed in by the Salivas, Lules, and Moxos, 251 note 2
Sweden, marriage prohibitions in, 48, 49
Syphilis, a predisposing cause to insanity, 220, 221; to malformations, 252, 257

TAHITI, marriage in, 108, 109; female infanticide in, 108; fertility in, 112; custom of killing one of twins, 252 note
Tanna, marriage in, 110
Tarquins, marriage of the, 37, 38
Tartars, marriage among the, 103 note 4; for the various tribes of, *see* under their names

INDEX.

Tasmania, marriage in, 116; half-breeds in, 314
Tatian, doctrines of, 64
Tertullian, opinions of, on marriage, 73
Tetaus, marriage among the, 120
Theodore of Canterbury, opinions of, on marriage of widows, 74, 75
Therapeutæ, the sect of the, 62
Thibet, marriage in, 101
Tinneh, marriage among the, 118
Todas, marriage among the, 97
Tonga, marriage in, 109; fertility in, 112
Topas, mental and physical state of the, 310
Torres Straits, marriage in the Islands of, 110
Troglodytica, marriage in, 18
Tuarik, marriage among the, 131
Tukopia, male infanticide in, 109; fertility in, 112
Tunguz, marriage among the, 101
Tunis, marriage in, 130
Tunkers, the sect of the, 72
Tupi, marriage among the, 124
Tupinamba, adoptive relationship among the, 124
Turlupins, doctrines of the, 69
Turkestan, Chinese, marriage in, 101
Turks, origin of the, 328, 329
Tuscans, marriage among the ancient, 38 note
Twana, marriage among the, 96
Twinning, period at which, is most common, 251; a cause of malformations, ib.; regarded as monstrous by many nations, 251 note 2

UDROPARASTATÆ, doctrines of the, 66
Uganda, marriage in, 129

VALENTINIAN, the doctrines of, 63, 64
Valesians, doctrines of the, 66
Vaquéros, isolated community of the, 59, 189
Vate, marriage in, 109, 110; infanticide in, 110
Venustians, the sect of the, 67
Vigilantius, the doctrines of, 67, 74

WAJO, marriage in, 105
Waldenses, the doctrines of the, 69
Warali, marriage among the, 97
Warrou, marriage among the, 121
Westmannoë, state of health in, 176, 177
White Quakers, the sect of the, 73
Widows prohibited a second marriage, 75, 76

YAMANOS, marriage among the, 121
Yerkala, marriage among the, 97
Yorkshire, isolated communities in, 162
Yucatan, marriage in, 120
Yuracares, marriage among the, 125

ZAMBOES, mental and physical characteristics of the, 324-327
Zoroastre, the law of, concerning the prohibited degrees, 14
Zulu, marriage among the, 128